ADVANCED LINEAR ALGEBRA FOR ENGINEERS WITH MATLAB®

ADVANCED LINEAR ALGEBRA FOR ENGINEERS WITH MATLAB®

Sohail A. Dianat

Rochester Institute of Technology,
New York, U.S.A.

Eli S. Saber

Rochester Institute of Technology,
New York, U.S.A.

CRC Press
Taylor & Francis Group
Boca Raton London New York

CRC Press is an imprint of the
Taylor & Francis Group, an **informa** business

MATLAB® and Simulink® are trademarks of The MathWorks, Inc. and are used with permission. The MathWorks does not warrant the accuracy of the text of exercises in this book. This book's use or discussion of MATLAB® and Simulink® software or related products does not constitute endorsement or sponsorship by The MathWorks of a particular pedagogical approach or particular use of the MATLAB® and Simulink® software.

CRC Press
Taylor & Francis Group
6000 Broken Sound Parkway NW, Suite 300
Boca Raton, FL 33487-2742

© 2009 by Taylor & Francis Group, LLC
CRC Press is an imprint of Taylor & Francis Group, an Informa business

No claim to original U.S. Government works
Printed in the United States of America on acid-free paper
10 9 8 7 6 5 4 3 2 1

International Standard Book Number-13: 978-1-4200-9523-4 (Hardcover)

This book contains information obtained from authentic and highly regarded sources. Reasonable efforts have been made to publish reliable data and information, but the author and publisher cannot assume responsibility for the validity of all materials or the consequences of their use. The authors and publishers have attempted to trace the copyright holders of all material reproduced in this publication and apologize to copyright holders if permission to publish in this form has not been obtained. If any copyright material has not been acknowledged please write and let us know so we may rectify in any future reprint.

Except as permitted under U.S. Copyright Law, no part of this book may be reprinted, reproduced, transmitted, or utilized in any form by any electronic, mechanical, or other means, now known or hereafter invented, including photocopying, microfilming, and recording, or in any information storage or retrieval system, without written permission from the publishers.

For permission to photocopy or use material electronically from this work, please access www.copyright.com (http://www.copyright.com/) or contact the Copyright Clearance Center, Inc. (CCC), 222 Rosewood Drive, Danvers, MA 01923, 978-750-8400. CCC is a not-for-profit organization that provides licenses and registration for a variety of users. For organizations that have been granted a photocopy license by the CCC, a separate system of payment has been arranged.

Trademark Notice: Product or corporate names may be trademarks or registered trademarks, and are used only for identification and explanation without intent to infringe.

Library of Congress Cataloging-in-Publication Data

Dianat, Sohail A.
   Advanced linear algebra for engineers with MATLAB / Sohail A. Dianat, Eli Saber.
       p. cm.
   Includes bibliographical references and index.
   ISBN 978-1-4200-9524-1 (alk. paper)
   1. Algebras, Linear. 2. Matrices. 3. MATLAB. 4. Engineering mathematics. I. Saber, Eli. II. Title.

QA184.2.D53 2008
620.001'51--dc22   2009000664

Visit the Taylor & Francis Web site at
http://www.taylorandfrancis.com

and the CRC Press Web site at
http://www.crcpress.com

To my wife Mitra Nikaein and my son Arash Dianat

– Sohail Dianat

To my wife Debra Saber and my sons Daniel, Paul, and Joseph Saber

– Eli Saber

Contents

Preface xiii
Authors xvii

1 Matrices, Matrix Algebra, and Elementary Matrix Operations 1
1.1 Introduction 1
1.2 Basic Concepts and Notation 1
    1.2.1 Matrix and Vector Notation 1
    1.2.2 Matrix Definition 1
    1.2.3 Elementary Matrices 3
    1.2.4 Elementary Matrix Operations 5
1.3 Matrix Algebra 6
    1.3.1 Matrix Addition and Subtraction 7
        1.3.1.1 Properties of Matrix Addition 7
    1.3.2 Matrix Multiplication 7
        1.3.2.1 Properties of Matrix Multiplication 8
    1.3.3 Applications of Matrix Multiplication in Signal and Image Processing 8
        1.3.3.1 Application in Linear Discrete One Dimensional Convolution 9
        1.3.3.2 Application in Linear Discrete Two Dimensional Convolution 14
        1.3.3.3 Matrix Representation of Discrete Fourier Transform 18
1.4 Elementary Row Operations 22
    1.4.1 Row Echelon Form 23
    1.4.2 Elementary Transformation Matrices 24
        1.4.2.1 Type 1: Scaling Transformation Matrix (E_1) 24
        1.4.2.2 Type 2: Interchange Transformation Matrix (E_2) 25
        1.4.2.3 Type 3: Combination Transformation Matrices (E_3) 26
1.5 Solution of System of Linear Equations 27
    1.5.1 Gaussian Elimination 27
    1.5.2 Over Determined Systems 31
    1.5.3 Under Determined Systems 32
1.6 Matrix Partitions 32
    1.6.1 Column Partitions 33
    1.6.2 Row Partitions 34
1.7 Block Multiplication 35

	1.8	Inner, Outer, and Kronecker Products 38
		1.8.1 Inner Product 38
		1.8.2 Outer Product 39
		1.8.3 Kronecker Products 40
	Problems 40
2	**Determinants, Matrix Inversion and Solutions to Systems of Linear Equations** 49
	2.1	Introduction 49
	2.2	Determinant of a Matrix 49
		2.2.1 Properties of Determinant 52
		2.2.2 Row Operations and Determinants 53
		2.2.2.1 Interchange of Two Rows 53
		2.2.2.2 Multiplying a Row of A by a Nonzero Constant 54
		2.2.2.3 Adding a Multiple of One Row to Another Row 55
		2.2.3 Singular Matrices 55
	2.3	Matrix Inversion 58
		2.3.1 Properties of Matrix Inversion 60
		2.3.2 Gauss–Jordan Method for Calculating Inverse of a Matrix 60
		2.3.3 Useful Formulas for Matrix Inversion 63
		2.3.4 Recursive Least Square (RLS) Parameter Estimation 64
	2.4	Solution of Simultaneous Linear Equations 67
		2.4.1 Equivalent Systems 69
		2.4.2 Strict Triangular Form 69
		2.4.3 Cramer's Rule 70
		2.4.4 LU Decomposition 71
	2.5	Applications: Circuit Analysis 75
	2.6	Homogeneous Coordinates System 78
		2.6.1 Applications of Homogeneous Coordinates in Image Processing 79
	2.7	Rank, Null Space and Invertibility of Matrices 85
		2.7.1 Null Space $N(A)$ 85
		2.7.2 Column Space $C(A)$ 87
		2.7.3 Row Space $R(A)$ 87
		2.7.4 Rank of a Matrix 89
	2.8	Special Matrices with Applications 90
		2.8.1 Vandermonde Matrix 90
		2.8.2 Hankel Matrix 91
		2.8.3 Toeplitz Matrices 91
		2.8.4 Permutation Matrix 92
		2.8.5 Markov Matrices 92
		2.8.6 Circulant Matrices 93
		2.8.7 Hadamard Matrices 93
		2.8.8 Nilpotent Matrices 94

- 2.9 Derivatives and Gradients 95
  - 2.9.1 Derivative of Scalar with Respect to a Vector 95
  - 2.9.2 Quadratic Functions 96
  - 2.9.3 Derivative of a Vector Function with Respect to a Vector 98
- Problems 99

3 Linear Vector Spaces 105
- 3.1 Introduction 105
- 3.2 Linear Vector Space 105
  - 3.2.1 Definition of Linear Vector Space 105
  - 3.2.2 Examples of Linear Vector Spaces 106
  - 3.2.3 Additional Properties of Linear Vector Spaces 107
  - 3.2.4 Subspace of a Linear Vector Space 107
- 3.3 Span of a Set of Vectors 108
  - 3.3.1 Spanning Set of a Vector Space 110
  - 3.3.2 Linear Dependence 110
  - 3.3.3 Basis Vectors 113
  - 3.3.4 Change of Basis Vectors 114
- 3.4 Normed Vector Spaces 116
  - 3.4.1 Definition of Normed Vector Space 116
  - 3.4.2 Examples of Normed Vector Spaces 116
  - 3.4.3 Distance Function 117
  - 3.4.4 Equivalence of Norms 118
- 3.5 Inner Product Spaces 120
  - 3.5.1 Definition of Inner Product 120
  - 3.5.2 Examples of Inner Product Spaces 121
  - 3.5.3 Schwarz's Inequality 121
  - 3.5.4 Norm Derived from Inner Product 123
  - 3.5.5 Applications of Schwarz Inequality in Communication Systems 123
    - 3.5.5.1 Detection of a Discrete Signal "Buried" in White Noise 123
    - 3.5.5.2 Detection of Continuous Signal "Buried" in Noise 125
  - 3.5.6 Hilbert Space 129
- 3.6 Orthogonality 131
  - 3.6.1 Orthonormal Set 131
  - 3.6.2 Gram–Schmidt Orthogonalization Process 131
  - 3.6.3 Orthogonal Matrices 134
    - 3.6.3.1 Complete Orthonormal Set 135
  - 3.6.4 Generalized Fourier Series (GFS) 135
  - 3.6.5 Applications of GFS 137
    - 3.6.5.1 Continuous Fourier Series 137
    - 3.6.5.2 Discrete Fourier Transform (DFT) 144
    - 3.6.5.3 Legendre Polynomial 145
    - 3.6.5.4 Sinc Functions 146

    3.7 Matrix Factorization 147
        3.7.1 *QR* Factorization 147
        3.7.2 Solution of Linear Equations Using *QR* Factorization 149
  Problems 151

4 Eigenvalues and Eigenvectors 157
    4.1 Introduction 157
    4.2 Matrices as Linear Transformations 157
        4.2.1 Definition: Linear Transformation 157
        4.2.2 Matrices as Linear Operators 160
        4.2.3 Null Space of a Matrix 160
        4.2.4 Projection Operator 161
        4.2.5 Orthogonal Projection 162
            4.2.5.1 Projection Theorem 163
            4.2.5.2 Matrix Representation of Projection
                    Operator 163
    4.3 Eigenvalues and Eigenvectors 165
        4.3.1 Definition of Eigenvalues and Eigenvectors 165
        4.3.2 Properties of Eigenvalues and Eigenvectors 168
            4.3.2.1 Independent Property 168
            4.3.2.2 Product and Sum of Eigenvalues 170
        4.3.3 Finding the Characteristic Polynomial of a Matrix 171
        4.3.4 Modal Matrix 173
    4.4 Matrix Diagonalization 173
        4.4.1 Distinct Eigenvalues 173
        4.4.2 Jordan Canonical Form 175
    4.5 Special Matrices 180
        4.5.1 Unitary Matrices 180
        4.5.2 Hermitian Matrices 183
        4.5.3 Definite Matrices 185
            4.5.3.1 Positive Definite Matrices 185
            4.5.3.2 Positive Semidefinite Matrices 185
            4.5.3.3 Negative Definite Matrices 185
            4.5.3.4 Negative Semidefinite Matrices 185
            4.5.3.5 Test for Matrix Positiveness 185
    4.6 Singular Value Decomposition (SVD) 188
        4.6.1 Definition of SVD 188
        4.6.2 Matrix Norm 192
        4.6.3 Frobenius Norm 195
        4.6.4 Matrix Condition Number 196
    4.7 Numerical Computation of Eigenvalues and Eigenvectors 199
        4.7.1 Power Method 199
    4.8 Properties of Eigenvalues and Eigenvectors of Different
        Classes of Matrices 205
    4.9 Applications 206
        4.9.1 Image Edge Detection 206

Contents xi

              4.9.1.1   Gradient Based Edge Detection of Gray Scale Images 209
              4.9.1.2   Gradient Based Edge Detection of RGB Images 210
        4.9.2   Vibration Analysis 214
        4.9.3   Signal Subspace Decomposition...217
              4.9.3.1   Frequency Estimation...217
              4.9.3.2   Direction of Arrival Estimation 219
    Problems 222

5   Matrix Polynomials and Functions of Square Matrices............ 229
    5.1   Introduction 229
    5.2   Matrix Polynomials 229
        5.2.1   Infinite Series of Matrices 230
        5.2.2   Convergence of an Infinite Matrix Series..........................231
    5.3   Cayley–Hamilton Theorem 232
        5.3.1   Matrix Polynomial Reduction..234
    5.4   Functions of Matrices 236
        5.4.1   Sylvester's Expansion 236
        5.4.2   Cayley–Hamilton Technique..240
        5.4.3   Modal Matrix Technique 245
        5.4.4   Special Matrix Functions 246
              5.4.4.1   Matrix Exponential Function e^{At} 247
              5.4.4.2   Matrix Function A^k 249
    5.5   The State Space Modeling of Linear Continuous-time Systems 250
        5.5.1   Concept of States..250
        5.5.2   State Equations of Continuous Time Systems 250
        5.5.3   State Space Representation of Continuous LTI Systems..254
        5.5.4   Solution of Continuous-time State Space Equations 256
        5.5.5   Solution of Homogenous State Equations and State Transition Matrix 257
        5.5.6   Properties of State Transition Matrix 258
        5.5.7   Computing State Transition Matrix 258
        5.5.8   Complete Solution of State Equations...............................259
    5.6   State Space Representation of Discrete-time Systems 263
        5.6.1   Definition of States...263
        5.6.2   State Equations 263
        5.6.3   State Space Representation of Discrete-time LTI Systems..264
        5.6.4   Solution of Discrete-time State Equations.........................265
              5.6.4.1   Solution of Homogenous State Equation and State Transition Matrix 266
              5.6.4.2   Properties of State Transition Matrix....................266

                5.6.4.3 Computing the State Transition Matrix 267
                5.6.4.4 Complete Solution of the State Equations 268
        5.7 Controllability of LTI Systems 270
                5.7.1 Definition of Controllability 270
                5.7.2 Controllability Condition 270
        5.8 Observability of LTI Systems 272
                5.8.1 Definition of Observability 272
                5.8.2 Observability Condition 272
        Problems 276

6 **Introduction to Optimization** **283**
        6.1 Introduction 283
        6.2 Stationary Points of Functions of Several Variables 283
                6.2.1 Hessian Matrix 285
        6.3 Least-Square (LS) Technique 287
                6.3.1 LS Computation Using *QR* Factorization 288
                6.3.2 LS Computation Using Singular Value Decomposition
                         (SVD) 289
                6.3.3 Weighted Least Square (WLS) 291
                6.3.4 LS Curve Fitting 293
                6.3.5 Applications of LS Technique 295
                         6.3.5.1 One Dimensional Wiener Filter 295
                         6.3.5.2 Choice of Q Matrix and Scale Factor β 298
                         6.3.5.3 Two Dimensional Wiener Filter 300
        6.4 Total Least-Squares (TLS) 302
        6.5 Eigen Filters 304
        6.6 Stationary Points with Equality Constraints 307
                6.6.1 Lagrange Multipliers 307
                6.6.2 Applications 310
                         6.6.2.1 Maximum Entropy Problem 310
                6.6.3 Design of Digital Finite Impulse Response (FIR) Filters 312
        Problems 316

Appendix A: The Laplace Transform **321**
        A1 Definition of the Laplace Transform 321
        A2 The Inverse Laplace Transform 323
        A3 Partial Fraction Expansion 323

Appendix B: The *z*-Transform **329**
        B1 Definition of the *z*-Transform 329
        B2 The Inverse *z*-Transform 330
                B2.1 Inversion by Partial Fraction Expansion 330

Bibliography **335**

Index **339**

Preface

The language and concepts of linear algebra and matrix methods have seen widespread usage in many areas of science and engineering especially over the last three decades due to the introduction of computers and computer information systems. Matrices have been employed effectively in many applications ranging from signal processing, controls, finite elements, communications, computer vision, electromagnetics, social and health sciences to name a few. This fueled the development of several matrix based analysis type packages such as MATLAB® (MATLAB® is a registered trademark of The Math Works, Inc. For product information, please contact: The Math Works, Inc., 3 Apple Hill Drive, Natick, MA. Tel: +508-647-7000; Fax: +508-647-7001; E-mail: info@mathworks.com; Web: http://www.mathworks.com) to help engineers and scientists simulate and solve large systems using fundamental linear algebra concepts. It is also responsible for the development and addition of matrix methods and/or linear algebra type courses at the undergraduate and graduate levels in many engineering and scientific curriculums across many universities worldwide. This effect has been amplified by the rapid transition from analog to digital type systems that we have seen across the spectrum in many areas of engineering where matrices form the fundamental pillar for system simulation, analysis and understanding.

This underlying textbook is intended to provide a comprehensive and practical approach for the study of advanced linear algebra for engineers and scientists in different disciplines. It is the result of ten years of lectures in a first-year graduate level course in matrix theory for electrical engineers. We have taken great care to provide an appropriate blend between theoretical explanations/illustrations that build a strong foundation for the topic at hand and corresponding real life applications in circuit analysis, communications, signal and image processing, controls, color science, computer vision, mechanical systems, traffic analysis and many more. To this effect, each chapter is designed to convey the theoretical concepts therein followed by carefully selected examples to demonstrate the underlying ideas arrayed in a simple to complex manner. Exercises and MATLAB type problems are also included at the end of each chapter to help in solidifying the concepts at hand. A complete step by step solution manual for each of the exercises is also available for instructors.

The topics discussed in this book are organized into six chapters.

Chapter 1 outlines the basic concepts and definitions behind matrices, matrix algebra, elementary matrix operations and matrix partitions. It also describes their potential use in several applications in signal and image processing such as convolution and discrete Fourier transforms.

Chapter 2 introduces the reader to the concepts of determinants, inverses and their use in solving linear equations that result from electrical and mechanical type systems. The chapter ends by introducing special type matrices followed by a great selection of exercises that build a strong platform for digesting the underlying material as well as preparing for follow-on chapters.

Chapter 3 builds on the foundation laid in the previous chapters by introducing the reader to linear vector spaces and the fundamental principles of orthogonality. This is done in a systematic fashion with an appropriate blend between abstract and concrete type examples. Applications stemming from these concepts in communications and signal processing are discussed throughout the chapter to help in visualizing and comprehending the material. The chapter ends with a discussion on matrix factorization followed by a rich set of exercises and MATLAB assignments.

Chapter 4 introduces the reader to the pertinent concepts of linear operators, eigenvalues, and eigenvectors and explores their use in matrix diagonalization and singular value decomposition. This is followed by a demonstration of their effective use in many applications in shape analysis, image processing, pattern recognition and the like. A special section is also included on the numerical computation of eigenvalues for large matrices. Relevant exercises and MATLAB assignments are provided at the end of the chapter to describe the uses of the above mentioned techniques in a variety of applications.

Chapter 5 extends the concepts developed in Chapter 4 to define matrix polynomials and compute functions of matrices using several well known methods such as Sylvester's expansion and Cayley–Hamilton. This is followed by the introduction of state space analysis and modeling techniques for discrete and continuous linear systems where the above described methods are utilized to provide a complete solution for the state space equation. Applications in control and electromechanical systems are explored throughout to demonstrate the effective use of state space analysis. Exercises and MATLAB type assignments are once again provided to help solidify the concepts described in above four chapters.

Given the above, it can be easily seen that Chapters 1–5 are designed to provide a strong foundation in linear algebra and matrix type methods by striking an appropriate balance between theoretical concepts and practical real life applications. Here, in Chapter 6, we extend the concepts derived in the previous five chapters to solve engineering problems using least square, weighted least square and total least square type techniques. This is clearly demonstrated in many applications in filtering and optimization followed by several exercises and MATLAB assignments designed to illustrate the use of the concepts described in Chapters 1–6 to solving engineering type problems.

As mentioned earlier, the material described in this book is intended for upper undergraduate or first-year graduate students. The book is carefully designed to provide an appropriate balance between mathematical concepts in matrix theory and linear algebra and pertinent engineering type applications. It is intended to illustrate the effective widespread use of matrices in solving many real life problems. The theoretical and practical treatments found within provide the reader with the knowledge and ability to begin to utilize matrix techniques to model and solve many engineering and scientific type problems.

The authors are indebted to the following individuals who provided valuable suggestions through reviews of various chapters of this book: Bhargava Chinni, Prudhvi Gurram, Mustafa Jaber, Siddharth Khullar, Pooja Nanda, Manoj Kumar Reddy, Abdul Haleem Syed, and Shilpa Tyagi.

Sohail A. Dianat
Department of Electrical Engineering
Rochester Institute of Technology
Rochester, NY

Eli Saber
Department of Electrical Engineering
Rochester Institute of Technology
Rochester, NY

Authors

Sohail A. Dianat is a professor in the Electrical Engineering Department and the Imaging Science Center at the Rochester Institute of Technology. He received the BS in Electrical Engineering from the Arya-Mehr University of Technology, Tehran, Iran in 1973, and his MS and DSc in Electrical Engineering from the George Washington University in 1977 and 1981, respectively. In September 1981, he joined the Rochester Institute of Technology, where he is a professor of Electrical Engineering and Imaging Science. Dr. Dianat has taught many graduate and undergraduate courses in the area of digital signal and image processing and digital communication. Dr. Dianat received the "best unclassified paper award" at the 1993 Annual IEEE Military Communication Conference (MILCOM '93). His current research interests include digital signal and image processing and wireless communication and he has published numerous papers an these subjects.

He holds nine patents in the field of control for digital printing. He is a senior member of the Institute of Electrical and Electronic Engineers and a member of the IEEE Signal Processing and Communication Society. He is also a Fellow of SPIE—The International Society for Optical Engineering. He is an associate editor for *IEEE Transaction on Image Processing*. Dr. Dianat was organizer and cochairman of the Fifth IEEE Signal Processing Society's Workshop on Spectrum Estimation and Modeling, Rochester, NY, October 1990. He has served as a technical program committee member for ICIP 2003. Dr. Dianat has also served as chairman for the track on Communications and Networking Technologies and Systems, SPIE Defense and Security Symposium, Orlando, FL, 2000–2007. He is currently serving as technical committee member on Industry DSP Technology. He holds numerous publications and patents in the fields of signal/image processing and digital communications.

Eli Saber is a professor in the Electrical Engineering and the Imaging Science Departments at the Rochester Institute of Technology. Prior to that, he worked for Xerox Corporation from 1988 until 2004 in a variety of positions ending as Product Development Scientist and Manager at the Business Group Operations Platform Unit. During his 16 years at Xerox, he was responsible for delivering color management, image processing innovations, architectures and algorithms, and xerographic subsystems for a variety of color products. He received a BS in Electrical and Computer Engineering from the University of Buffalo in 1988, and an MS and PhD in the same discipline from the University of Rochester in 1992 and 1996, respectively. From 1997 until 2004, he was an adjunct faculty member at the Electrical Engineering Department of the Rochester Institute of Technology and at the Electrical and Computer Engineering Department of the University of Rochester responsible for teaching undergraduate and graduate coursework in signal, image, video processing, pattern recognition and communications and performing research in multimedia applications, pattern recognition, image understanding and color engineering.

Dr. Saber is a senior member of the Institute of Electrical and Electronic Engineers (IEEE), and a member of the IEEE Signal Processing (SP) Society, the Electrical Engineering Honor Society, Eta Kappa Nu, and the Imaging Science and Technology (IS&T) Society. He is an associate editor for the *IEEE Transaction on Image Processing*, the *Journal of Electronic Imaging*, and has served as guest editor for the special issue on color image processing for the *IEEE SP Magazine*. He is currently a member of the IEEE Image and Multi-dimensional Digital Signal Processing (IMDSP) Technical Committee. He is also a member and past chair of the Standing Committee on Industry Digital Signal Processing Applications. Dr. Saber was appointed the finance and tutorial chair for the 2002 and 2007 International Conference on Image Processing (ICIP), respectively. He was also the general chair for the Western New York Imaging Workshop in 1998. He is currently serving as the general chair for ICIP 2012. He has served as the chairman, vice-chairman, treasurer, and secretary for the IEEE Rochester Chapter of the SP society in 1998, 1997, 1996 and 1995, respectively. He has also served as a technical program committee member for several International Conferences on Image Processing (ICIP) as well as International Conferences on Acoustics, Speech and Signal Processing (ICASSP). Dr. Saber holds numerous publications and patents in the field of color image processing; signal, image and video processing; pattern recognition; and computer vision.

1

Matrices, Matrix Algebra, and Elementary Matrix Operations

1.1 Introduction

The beginning of matrices goes back to the 2nd century BC. The real development in this field, however, did not start until the 17th century. Initially, matrices were used to solve simultaneous linear equations. Today, matrices have widespread applications from solving large scale dynamic systems to filtering images, modeling communication systems, solving electric circuits, and analyzing signal processing algorithms. They are widely used in almost every engineering discipline. This chapter aims to explain the basic concepts that form the fundamental basis for linear algebra.

1.2 Basic Concepts and Notation

1.2.1 Matrix and Vector Notation

Throughout this book, the following matrix notations will be used. Matrices will be denoted by capital letters (e.g. matrix A). Row and column vectors will be denoted by lower case letters (e.g. vector x). The (i^{th}, j^{th}) element of matrix A will be denoted by a_{ij} and the ith element of vector x will be denoted by x_i.

1.2.2 Matrix Definition

A rectangular array of numbers of the form:

$$A = \begin{pmatrix} a_{11} & \cdots & a_{1n} \\ \vdots & & \vdots \\ a_{m1} & \cdots & a_{mn} \end{pmatrix} \qquad (1.1)$$

is called an $m \times n$ matrix, where m and n represent the number of rows and columns, respectively. A can be written as $A = [a_{ij}]$, where a_{ij} is the (i^{th}, j^{th}) element of matrix A. The elements of a matrix can be real or complex numbers. If the elements are real, it is called a real matrix. On the other hand, if the elements are complex, it is called a complex matrix. Below are some examples of real and complex matrices:

$$A = \begin{bmatrix} 2 & 9 & -23 \\ 12 & 4 & 7 \\ -2 & 0 & 8 \\ 15 & 6 & -4 \end{bmatrix} \quad (1.2)$$

$$B = \begin{bmatrix} 1-j & -2+j & j & 3 \\ 6-j7 & 4-j4 & 3-j & 1+j \\ 2+j & 8-j & 6 & 1+j \\ -6-j7 & 2+j7 & -j & 1+j4 \end{bmatrix} \quad (1.3)$$

A is a real 4×3 matrix and B is a complex 4×4 matrix.

An example of the application of matrices in image processing is embodied in the description and manipulation of images. Images are defined as matrices and most image processing techniques are based on matrix operations. A monochromatic digital image is a matrix (an array of numbers) of size $N \times M$ denoted by $f(n,m)$, where the value of f at spatial coordinates (n,m) gives the intensity of the image at that location. This is shown in Equation 1.4. Each element of f is a discrete integer quantity having a value between 0 and 255 for an 8-bit image.

$$f(n,m) = \begin{bmatrix} f(0,0) & f(0,1) & \cdots & f(0, M-1) \\ f(1,0) & f(1,1) & \cdots & f(1, M-1) \\ \vdots & \vdots & & \vdots \\ f(N-1,0) & f(N-1,1) & \cdots & f(N-1, M-1) \end{bmatrix} \quad (1.4)$$

On the other hand, a digital RGB color image of size $N \times M$ consists of three channels: red (R), green (G) and blue (B). Each channel is represented by a matrix of size $N \times M$. At a given spatial coordinate (n,m), the image has three components: each of which is between 0 and 255 for a 24-bit image (8 bits per channel). That is:

$$0 \leq R(n,m) \leq 255 \quad (1.5)$$

$$0 \leq G(n,m) \leq 255 \quad (1.6)$$

$$0 \leq B(n,m) \leq 255 \quad (1.7)$$

Example 1.1

The 256×256 digital gray image of LENA and an 8×8 block of that image are shown in Figure 1.1.

Example 1.2

The 256×256 digital color image of LENA is shown in Figure 1.2. An 8×8 block of this RGB image and the corresponding R, G, and B channel matrices are also shown in the same figure.

1.2.3 Elementary Matrices

Matrices with specific structures occur frequently in applications of linear algebra. In this section, we describe some of the most useful elementary matrices.

1. **Square matrix**: If the number of rows of a matrix is equal to the number of columns of the matrix (i.e. $m=n$), it is called a square matrix.

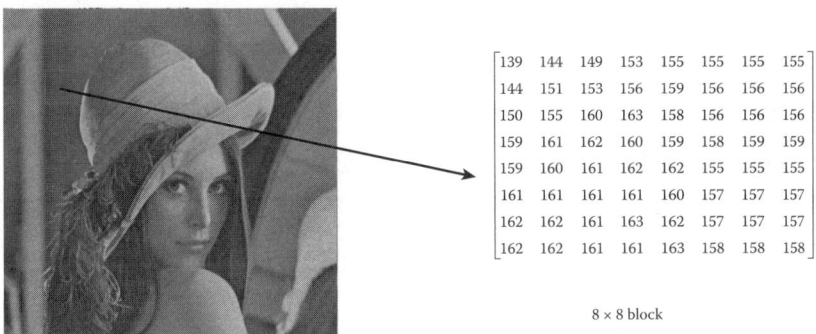

FIGURE 1.1
Gray scale LENA image.

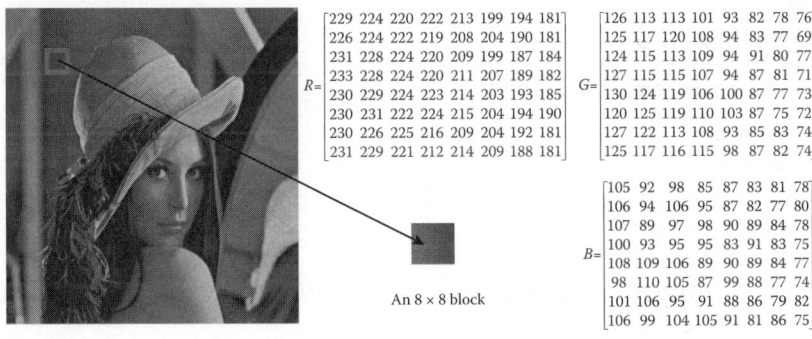

FIGURE 1.2
(See color insert following page 174.) RGB LENA image.

Example 1.3

The matrix A is a 3×3 square matrix

$$A = \begin{bmatrix} 2 & 3 & -9 \\ 3 & 5 & 8 \\ -7 & 4 & -3 \end{bmatrix}$$

2. **Upper triangular matrix**: An $m \times n$ matrix for which $a_{ij}=0$ for $i>j$ is called an upper triangular matrix.

Example 1.4

An upper triangular matrix

$$A = \begin{bmatrix} 2 & 7 & 8 & -3 & 1 \\ 0 & 4 & 6 & 2 & -4 \\ 0 & 0 & 2 & 3 & 5 \end{bmatrix}$$

3. **Lower triangular matrix**: An $m \times n$ matrix for which $a_{ij}=0$ for $j>i$ is called a lower triangular matrix.

Example 1.5

A lower triangular matrix

$$A = \begin{bmatrix} 2 & 0 & 0 & 0 \\ -3 & 4 & 0 & 0 \\ 5 & 1 & 2 & 0 \\ -3 & -4 & 2 & 9 \end{bmatrix}$$

4. **Diagonal matrix**: A square matrix whose nondiagonal elements are equal to zero is called a diagonal matrix.

Example 1.6

A 3×3 diagonal matrix

$$A = \begin{bmatrix} 2 & 0 & 0 \\ 0 & 4 & 0 \\ 0 & 0 & -3 \end{bmatrix}$$

5. **Identity matrix**: A diagonal matrix with all diagonal elements equal to one.

Example 1.7

Identity matrices of different sizes

$$I_2 = \begin{bmatrix} 1 & 0 \\ 0 & 1 \end{bmatrix}; \quad I_3 = \begin{bmatrix} 1 & 0 & 0 \\ 0 & 1 & 0 \\ 0 & 0 & 1 \end{bmatrix}; \quad I_4 = \begin{bmatrix} 1 & 0 & 0 & 0 \\ 0 & 1 & 0 & 0 \\ 0 & 0 & 1 & 0 \\ 0 & 0 & 0 & 1 \end{bmatrix}$$

The MATLAB® command to create a $N \times N$ identity matrix is **eye(N)**.

1.2.4 Elementary Matrix Operations

In this section, we discuss several elementary matrix operations. These are utilized in a number of matrix manipulation techniques.

1. **Matrix transpose and Hermitian**: The transpose of matrix A is obtained by interchanging the rows and columns of A. It is denoted by A^T. If matrix A has dimensions $n \times m$, its transpose will be $m \times n$. Therefore:

$$B = A^T \text{ if } b_{ij} = a_{ji} \tag{1.8}$$

The matrix A^H is obtained by transposing and then conjugating every element of matrix A. That is:

$$B = A^H \text{ if } b_{ij} = \bar{a}_{ji} \tag{1.9}$$

where "bar" stands for complex conjugation. It should be noted that $A^H = A^T$ if A is real. Finally, a real square matrix is symmetric if $A = A^T$. A complex square matrix is said to be Hermitian if $A = A^H$.

The MATLAB® command to transpose matrix A and store it in matrix B is **B=A'**.

Example 1.8

$$\text{If } A = \begin{bmatrix} 2 & 7 & 8 & -3 & 1 \\ 1 & 4 & 6 & 2 & -4 \\ -2 & 3 & 2 & 3 & 5 \end{bmatrix} \text{ then } B = A^T = \begin{bmatrix} 2 & 1 & -2 \\ 7 & 4 & 3 \\ 8 & 6 & 2 \\ -3 & 2 & 3 \\ 1 & -4 & 5 \end{bmatrix}$$

$$\text{If } C = \begin{bmatrix} 2-j & 1+j3 \\ j4 & 4 \end{bmatrix} \text{ then } C^H = \begin{bmatrix} 2+j & -j4 \\ 1-j3 & 4 \end{bmatrix}$$

2. **Properties of matrix transpose (Hermitian)**

   The following properties hold for transpose and Hermitian operations. The proofs are left as an exercise.
   a. $(A^T)^T = A$ or $(A^H)^H = A$
   b. $(AB)^T = B^T A^T$ or $(AB)^H = B^H A^H$

3. **Symmetric matrix:** An $n \times n$ real matrix A is said to be symmetric if $A^T = A$.

4. **Skew-symmetric matrix:** An $n \times n$ real matrix A is said to be skew-symmetric if $A^T = -A$.

5. **Trace of a matrix:** The trace of a square matrix is the sum of the diagonal elements of the matrix, that is:

$$\text{Trace}(A) = \sum_{i=1}^{n} a_{ii} \qquad (1.10)$$

The MATLAB command to compute the trace of matrix A is **trace(A)**.

Example 1.9

$$\text{If } A = \begin{bmatrix} 4 & 4 & -5 \\ 3 & 4 & 1 \\ 6 & 7 & -3 \end{bmatrix} \text{ then:}$$

$$\text{Trace}(A) = \sum_{i=1}^{n} a_{ii} = 4 + 4 - 3 = 5$$

We state the properties of the trace of a matrix without proof. The proofs are left as an exercise. The following properties hold for square matrices:

   a. $\text{Trace}(A \pm B) = \text{Trace}(A) \pm \text{Trace}(B)$
   b. $\text{Trace}(A^T) = \text{Trace}(A)$
   c. $\text{Trace}(AB) = \text{Trace}(B^T A^T)$
   d. $\text{Trace}(\alpha A) = \alpha \, \text{Trace}(A)$
   e. $\text{Trace}(AB) = \text{Trace}(BA)$

Note that property (e) is valid only for matrices of appropriate dimensions that allow for the multiplication to take place. Matrix multiplication is discussed in detail in the following section.

1.3 Matrix Algebra

Matrix algebra deals with algebraic operations such as matrix addition, matrix subtraction, matrix multiplication and inversion. In this section, we

Matrices, Matrix Algebra, and Elementary Matrix Operations

define addition, subtraction and multiplication. Matrix inversion will be discussed in Chapter 2.

1.3.1 Matrix Addition and Subtraction

If $A=[a_{ij}]$ and $B=[b_{ij}]$ are two matrices of the same size, then:

$$C = A \pm B \tag{1.11}$$

where

$$c_{ij} = a_{ij} \pm b_{ij} \tag{1.12}$$

Example 1.10

Find $C=A+B$ and $D=A-B$ if:

$$A = \begin{bmatrix} 2 & 3 & -4 \\ -2 & 1 & 6 \end{bmatrix} \text{ and } B = \begin{bmatrix} -2 & 1 & 3 \\ 4 & 6 & -5 \end{bmatrix}$$

Solution:

$$C = A+B = \begin{bmatrix} 0 & 4 & -1 \\ 2 & 7 & 1 \end{bmatrix} \text{ and } D = A-B = \begin{bmatrix} 4 & 2 & -7 \\ -6 & -5 & 11 \end{bmatrix}$$

The MATLAB® command to add or subtract two matrices is **C=A±B**.

1.3.1.1 Properties of Matrix Addition

The following properties hold for matrix addition. They can be easily proven using the definition of addition.

a. Commutative property: $A+B=B+A$
b. Associative property: $A+(B+C)=(A+B)+C$

1.3.2 Matrix Multiplication

Let A and B represent an $n \times m$ and $m \times k$ matrix, respectively. Then, the product $C=AB$ is a matrix of size $n \times k$ defined as:

$$c_{ij} = \sum_{l=1}^{m} a_{il} b_{lj} \quad i=1, 2, \ldots, n \text{ and } j=1, 2, \ldots, k \tag{1.13}$$

Note that AB can be defined only if the number of columns of A is equal to the number of rows of B. Also observe that AB is, in general, different from BA, that is:

$$AB \neq BA \tag{1.14}$$

Hence, matrix multiplication is not generally commutative.

The MATLAB® command to premultiply B by A and store it in C is **C=A*B**. The command **C=A.*B** is used for point by point array multiplication, where $c_{ij}=a_{ij}b_{ij}$. In this later case, matrix A and B should have the same size.

Example 1.11

Find $D=AB$, $E=BA$, and $F=A\times C$ (array multiplication) if:

$$A = \begin{bmatrix} 2 & 3 & -4 \\ -2 & 1 & 6 \end{bmatrix}, \quad B = \begin{bmatrix} 2 & 4 \\ -2 & 1 \\ 3 & 2 \end{bmatrix} \text{ and } C = \begin{bmatrix} 1 & 2 & 3 \\ 4 & 5 & 6 \end{bmatrix}$$

Solution:

$$D = AB = \begin{bmatrix} -14 & 3 \\ 12 & 5 \end{bmatrix}, \quad E = BA = \begin{bmatrix} -4 & 10 & 16 \\ -6 & -5 & 14 \\ 2 & 11 & 0 \end{bmatrix}$$

$$\text{and } F = A \times C = \begin{bmatrix} 2 & 6 & -12 \\ -8 & 5 & 36 \end{bmatrix}$$

The first element of matrix D is obtained by the multiplication of the first row of matrix A by the first column of matrix B, i.e. $2\times 2+3\times(-2)+(-4)\times 3=4-6-12=-14$ and so on for the remaining elements of D. Since A is a 2×3 matrix and B is a 3×2 matrix, then D will be a 2×2 matrix where the number of rows and columns of D are equal to the number of rows of A and columns of B, respectively. On the other hand, the number of rows (columns) of E is equal to the number of rows of B (columns of A). Note that $AB \neq BA$. Finally, the point by point array multiplication $F=A\times C=C\times A$ is shown above where each element of A is multiplied by its respective element in C. Note the point by point array multiplication is commutative.

1.3.2.1 Properties of Matrix Multiplication

The following matrix multiplication properties are stated without proof:

a. Distributive law: $A(B+C)=AB+AC$
b. Associative property: $A(BC)=(AB)C$
c. $AI=IA=A$, where I is the identity matrix of appropriate size and A is a square matrix.
d. $A^{m+n}=A^m A^n=A^n A^m$ where $A^n = \overbrace{A \times A \times \cdots \times A}^{n \text{ times}}$

1.3.3 Applications of Matrix Multiplication in Signal and Image Processing

Matrix multiplication has numerous applications in signal processing, image processing, control and communications. In this section, we concentrate on two applications in signal and image processing.

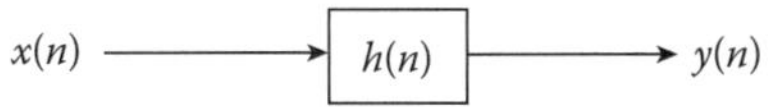

Matrices, Matrix Algebra, and Elementary Matrix Operations

$$x(n) \longrightarrow \boxed{h(n)} \longrightarrow y(n)$$

FIGURE 1.3
Discrete 1-d LTI system.

1.3.3.1 Application in Linear Discrete One Dimensional Convolution

Consider a one dimensional (1-d) discrete linear time invariant (LTI) system with input $x(n)$, output $y(n)$, and impulse response $h(n)$ as shown in Figure 1.3.

The signals $x(n)$, $y(n)$, and $h(n)$ are related through the convolution sum given by:

$$y(n) = x(n) * h(n) = \sum_{k=-\infty}^{+\infty} x(k)h(n-k) = \sum_{k=-\infty}^{+\infty} h(k)x(n-k) \quad (1.15)$$

Now assume that the input signal $x(n)$ is a sequence of length M and the system's impulse response $h(n)$ is of length L. Then, the output signal $y(n)$ is a sequence of length N given by:

$$y(n) = \sum_{k=0}^{M-1} x(k)h(n-k) \quad (1.16)$$

where

$$N = M + L - 1 \quad (1.17)$$

The MATLAB® command to convolve two discrete-time 1-d signals x and h is **y=conv(x,h)**.

The discrete convolution operation can be performed using matrix multiplication as follows: Define the three $N \times 1$ vectors y, h_e and x_e as:

$$y = [y(0) \quad y(1) \quad y(2) \quad \cdots \quad y(N-2) \quad y(N-1)]^T \quad (1.18)$$

$$x_e = [x(0) \quad x(1) \quad \cdots \quad x(M-1) \quad 0 \quad 0 \quad \cdots \quad 0]^T \quad (1.19)$$

$$h_e = [h(0) \quad h(1) \quad \cdots \quad h(L-1) \quad 0 \quad 0 \quad \cdots \quad 0]^T \quad (1.20)$$

Then Equation 1.16 can be written as:

$$y(n) = \sum_{k=0}^{N-1} x_e(k) h_e(n-k) \quad (1.21)$$

if we assume that h_e and x_e are periodic with period N. Since y is of size N, it implies periodic convolution. This is shown graphically in Figure 1.4.

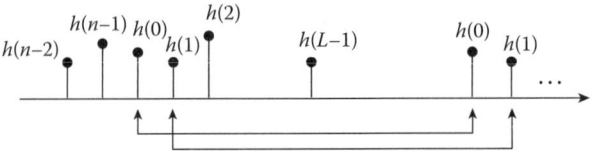

FIGURE 1.4
Periodicity of samples in $h(n)$.

Therefore:

$$h_e(-1) = h_e(N-1); h_e(-2) = h_e(N-2); \cdots; h_e(-N+1) = h_e(1) \qquad (1.22)$$

Expanding the convolution sum of Equation 1.21 results in:

$$y(0) = x_e(0)h_e(0) + x_e(1)h_e(-1) + x_e(2)h_e(-2) + \ldots + x_e(N-1)h_e(-N+1) \qquad (1.23)$$

$$y(1) = x_e(0)h_e(1) + x_e(1)h_e(0) + x_e(2)h_e(-1) + \ldots + x_e(N-1)h_e(1-(N-1)) \qquad (1.24)$$

$$y(2) = x_e(0)h_e(2) + x_e(1)h_e(1) + x_e(2)h_e(0) + \ldots + x_e(N-1)h_e(2-(N-1)) \qquad (1.25)$$

$$\vdots$$

$$y(N-1) = x_e(0)h_e(N-1) + x_e(1)h_e(N-2) + x_e(2)h_e(N-3) + \ldots + x_e(N-1)h_e(0) \qquad (1.26)$$

The above equations can be expressed in matrix form as:

$$y = Hx_e \qquad (1.27)$$

or

$$\begin{bmatrix} y(0) \\ y(1) \\ y(2) \\ \vdots \\ y(N-1) \end{bmatrix}_{N \times 1} = \underbrace{\begin{bmatrix} h_e(0) & h_e(N-1) & h_e(N-2) & \cdots & h_e(2) & h_e(1) \\ h_e(1) & h_e(0) & h_e(N-1) & h_e(N-2) & \vdots & h_e(2) \\ h_e(2) & h_e(1) & h_e(0) & h_e(N-1) & \vdots & h_e(3) \\ h_e(3) & h_e(2) & h_e(1) & h_e(0) & & h_e(4) \\ \vdots & \vdots & \vdots & \vdots & \ddots & \\ h_e(N-1) & h_e(N-2) & h_e(N-3) & h_e(N-4) & & h_e(0) \end{bmatrix}}_{N \times N} \times \begin{bmatrix} x_e(0) \\ x_e(1) \\ x_e(2) \\ \vdots \\ x_e(N-1) \end{bmatrix}_{N \times 1} \qquad (1.28)$$

A closer look at matrix H indicates that the second row is obtained from the first row by shifting it by one element to the right and so on. Note that

Matrices, Matrix Algebra, and Elementary Matrix Operations 11

TABLE 1.1

MATLAB® Code for 1-d Convolution in Matrix Form

```
% f=input signal (1xa)
% h=impulse response (1xb)
% g=output signal (1x(a+b-1))
function [g] =convm1d(f,h)
a=length(h);
b=length(f);
n=a+b-1;
he=[h zeros(1,n-a)];
fe=[f zeros(1,n-b)]';
h1=[he(1) he(n:-1:2)];
H=h1;
for i=1:n-1
h1=[h1(end) h1(1:end-1)];
H=[H;h1];
end
g=H*fe;
g=g';
```

periodic convolution is equivalent to linear convolution if the signal is zero padded to avoid aliasing. The MATLAB implementation of 1-d convolution using matrices is given in the script shown in Table 1.1.

Example 1.12

If $x(n) = [1\ 2\ 4]$ and $h(n) = [2\ -2\ 1]$, compute $y(n) = x(n)*h(n)$ using matrix formulation.

Solution: Since $N = M + L - 1 = 3 + 3 - 1 = 5$, we have:

$$x_e = \begin{bmatrix} 1 \\ 2 \\ 4 \\ 0 \\ 0 \end{bmatrix}, \quad h_e = \begin{bmatrix} 2 \\ -2 \\ 1 \\ 0 \\ 0 \end{bmatrix} \text{ and } H = \begin{bmatrix} 2 & 0 & 0 & 0 & 0 \\ -2 & 2 & 0 & 0 & 0 \\ 1 & -2 & 2 & 0 & 0 \\ 0 & 1 & -2 & 2 & 0 \\ 0 & 0 & 1 & -2 & 2 \end{bmatrix}$$

Therefore:

$$y_e = H x_e = \begin{bmatrix} 2 & 0 & 0 & 0 & 0 \\ -2 & 2 & 0 & 0 & 0 \\ 1 & -2 & 2 & 0 & 0 \\ 0 & 1 & -2 & 2 & 0 \\ 0 & 0 & 1 & -2 & 2 \end{bmatrix} \begin{bmatrix} 1 \\ 2 \\ 4 \\ 0 \\ 0 \end{bmatrix} = \begin{bmatrix} 2 \\ 2 \\ 5 \\ -6 \\ 4 \end{bmatrix}$$

This is equivalent to the result obtained if we use direct convolution using the following steps.

a. Plot x(n) and h(n). See Figure 1.5.
b. Plot x(k) and h(n−k). Note that h(n−k) is the mirror image of h(k) shifted by n samples. See Figure 1.6.
c. Multiply x(k) by h(n−k) for a fixed value of n and add the resulting samples to obtain y(n). Repeat this process for all values of n. The results are illustrated in Figure 1.7.

Then:

$$y(0) = \sum_k x(k)h(-k) = 1 \times 2 = 2 \tag{1.29}$$

$$y(1) = \sum_k x(k)h(1-k) = -1 \times 2 + 2 \times 2 = 2 \tag{1.30}$$

$$y(2) = \sum_k x(k)h(2-k) = 1 \times 1 - 2 \times 2 + 4 \times 2 = 5 \tag{1.31}$$

$$y(3) = \sum_k x(k)h(3-k) = 2 \times 1 - 4 \times 2 = -6 \tag{1.32}$$

$$y(4) = \sum_k x(k)h(3-k) = 4 \times 1 = 4 \tag{1.33}$$

The output signal y(n) is shown in Figure 1.8.

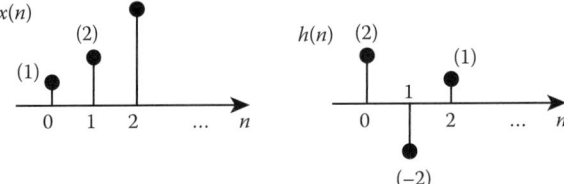

FIGURE 1.5
Plots of x(n) and h(n).

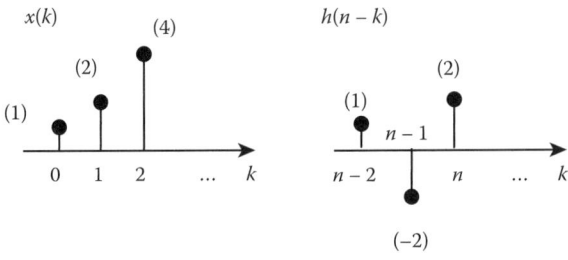

FIGURE 1.6
Plots of x(k) and h(n−k).

Matrices, Matrix Algebra, and Elementary Matrix Operations 13

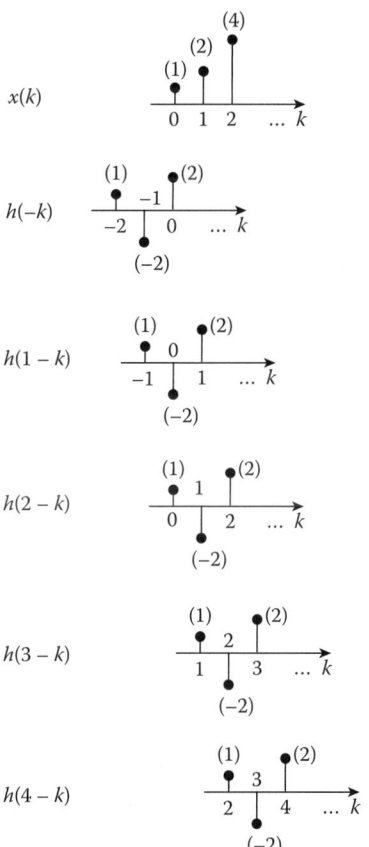

FIGURE 1.7
$x(k)$ and $h(n-k)$ for different values of k.

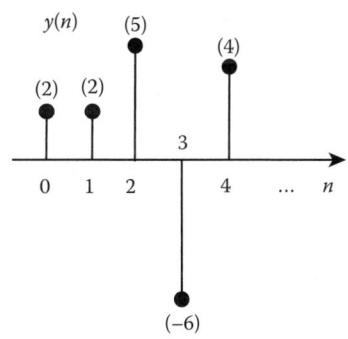

FIGURE 1.8
The output signal $y(n)$.

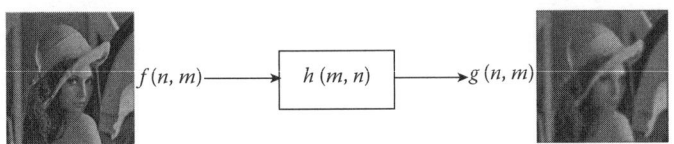

FIGURE 1.9
2-d LSI system.

1.3.3.2 Application in Linear Discrete Two Dimensional Convolution

Another application of matrix multiplication is two dimensional (2-d) convolution in image processing. Consider a 2-d linear shift invariant (LSI) system with input $f(m,n)$, impulse response $h(m,n)$ and output $g(m,n)$ as shown in Figure 1.9.

The input $f(m,n)$ and output $g(m,n)$ are related by the 2-d discrete convolution sum that is represented by the symbol "**" and is given by:

$$g(m,n) = f(m,n) ** h(m,n) = \sum_{k_1=-\infty}^{+\infty} \sum_{k_2=-\infty}^{+\infty} f(k_1,k_2) h(m-k_1, n-k_2) \quad (1.34)$$

If $f(m,n)$ and $h(m,n)$ are $M_1 \times M_2$ and $L_1 \times L_2$, then the output signal $g(m,n)$ is $N_1 \times N_2$, where:

$$\begin{aligned} N_1 &= M_1 + L_1 - 1 \\ N_2 &= M_2 + L_2 - 1 \end{aligned} \quad (1.35)$$

Since $f(m,n)$ has finite support, we have:

$$g(m,n) = \sum_{k_1=0}^{M_1-1} \sum_{k_2=0}^{M_2-1} f(k_1,k_2) h(m-k_1, n-k_2) \quad (1.36)$$

The MATLAB® command for 2-d convolution of two 2-d signals is **g=conv2 (h, f)**.

Similar to the 1-d case, define f_e and h_e as follows (zero pad to reach $N_1 \times N_2$):

$$f_e = \begin{bmatrix} f & | & 0 \\ \hline 0 & | & 0 \end{bmatrix}_{N_1 \times N_2} \quad (1.37)$$

$$h_e = \begin{bmatrix} h & | & 0 \\ \hline 0 & | & 0 \end{bmatrix}_{N_1 \times N_2} \quad (1.38)$$

Then:

$$g(m,n) = \sum_{k_1=0}^{N_1-1} \sum_{k_2=0}^{N_2-1} f_e(k_1,k_2) h_e(m-k_1, n-k_2) \quad (1.39)$$

Matrices, Matrix Algebra, and Elementary Matrix Operations

Let f_L and g_L represent the lexicographic ordering of f_e and g_e, respectively. These can be written as:

$$f_L = \begin{bmatrix} f_e(1,1) \\ f_e(1,2) \\ \vdots \\ \downarrow \\ \vdots \\ \hline 0 \\ 0 \\ \vdots \\ \downarrow \\ \hline 0 \\ \hline 0 \\ 0 \\ \vdots \\ \downarrow \\ \hline 0 \\ \hline 0 \\ 0 \\ \vdots \\ \downarrow \\ 0 \end{bmatrix} ; \text{ and } g_L = \begin{bmatrix} g_e(1,1) \\ g_e(1,2) \\ \vdots \\ \downarrow \\ \vdots \\ \hline 0 \\ 0 \\ \vdots \\ \downarrow \\ \hline 0 \\ \hline 0 \\ 0 \\ \vdots \\ \downarrow \\ \hline 0 \\ \hline 0 \\ 0 \\ \vdots \\ \downarrow \\ 0 \end{bmatrix}$$

Equation 1.39 can then be written in matrix form as:

$$g_L = H f_L \tag{1.40}$$

where g_L and f_L are vectors of size $N_1 N_2 \times 1$.

Matrix H is a block circulant matrix of size $N_1 N_2 \times N_1 N_2$. It consists of $(N_1)^2$ partitions of size $N_2 \times N_2$. Each partition H_j is constructed from the jth row of the extended function h_e as follows:

$$H_j = \begin{bmatrix} h_e(j,0) & h_e(j,N_2-1) & h_e(j,N_2-2) & \cdots & h_e(j,1) \\ h_e(j,1) & h_e(j,0) & h_e(j,N_2-1) & \cdots & h_e(j,2) \\ \vdots & \vdots & h_e(j,0) & & \\ & & & \ddots & \\ h_e(j,N_2-1) & h_e(j,N_2-2) & & & h_e(j,0) \end{bmatrix}_{N_2 \times N_2} \tag{1.41}$$

$$H = \begin{bmatrix} H_0 & H_{N_1-1} & H_{N_1-2} & \cdots & H_1 \\ H_1 & H_0 & H_{N_1-1} & \cdots & H_2 \\ H_2 & H_1 & H_0 & \cdots & H_3 \\ \vdots & \vdots & \vdots & \ddots & \vdots \\ H_{N_1-1} & H_{N_1-2} & H_{N_1-3} & \cdots & H_0 \end{bmatrix}_{(N_1N_2 \times N_1N_2)} \quad (1.42)$$

where each element of H is a matrix. If $N_1 = N_2 = N$, then $g_L = Hf_L$ where:

$$H = \begin{bmatrix} H_0 & H_{N-1} & H_{N-2} & \cdots & H_1 \\ H_1 & H_0 & H_{N-1} & \cdots & H_2 \\ H_2 & H_1 & H_0 & \cdots & H_3 \\ \vdots & \vdots & \vdots & \ddots & \vdots \\ H_{N-1} & H_{N-2} & H_{N-3} & \cdots & H_0 \end{bmatrix}_{(N^2 \times N^2)} \quad (1.43)$$

and

$$H_k = \begin{bmatrix} h_e(k,0) & h_e(k,N-1) & h_e(k,N-1) & \cdots & h_e(k,1) \\ h_e(k,1) & h_e(k,0) & h_e(k,N-1) & \cdots & h_e(k,2) \\ h_e(k,2) & h_e(k,1) & h_e(k,0) & & \\ \vdots & \vdots & \vdots & \ddots & \vdots \\ h_e(k,N-1) & h_e(k,N-2) & & & h_e(k,0) \end{bmatrix}_{N \times N} \quad (1.44)$$

Example 1.13

Let $f(n,m) = \begin{bmatrix} 22 & 33 & 32 \\ 23 & 20 & 18 \\ 16 & 17 & 21 \end{bmatrix}$ and $h(n,m) = \begin{bmatrix} 1 & 2 \\ 2 & -1 \end{bmatrix}$ compute $g(n,m)$.

Solution:

$$g(n,m) = f(n,m) ** g(n,m)$$

$N = M + L - 1 = 3 + 2 - 1 = 4$. Let us begin by zero padding f and h yielding the following 4×4 matrices:

$$f_e = \begin{bmatrix} 22 & 33 & 32 & 0 \\ 23 & 20 & 18 & 0 \\ 16 & 17 & 21 & 0 \\ 0 & 0 & 0 & 0 \end{bmatrix}; \quad h_e = \begin{bmatrix} 1 & 2 & 0 & 0 \\ 2 & -1 & 0 & 0 \\ 0 & 0 & 0 & 0 \\ 0 & 0 & 0 & 0 \end{bmatrix}$$

Matrices, Matrix Algebra, and Elementary Matrix Operations

Now, form f, H_0, H_1, H_2 and H_3 as follows:

$$f = [22 \ 33 \ 32 \ 0 \ 23 \ 20 \ 18 \ 0 \ 16 \ 17 \ 21 \ 0 \ 0 \ 0 \ 0 \ 0]^T$$

$$H_0 = \begin{bmatrix} 1 & 0 & 0 & 2 \\ 2 & 1 & 0 & 0 \\ 0 & 2 & 1 & 0 \\ 0 & 0 & 2 & 1 \end{bmatrix}; \quad H_1 = \begin{bmatrix} 2 & 0 & 0 & -1 \\ -1 & 2 & 0 & 0 \\ 0 & -1 & 2 & 0 \\ 0 & 0 & -1 & 2 \end{bmatrix}$$

$$H_2 = H_3 = \begin{bmatrix} 0 & 0 & 0 & 0 \\ 0 & 0 & 0 & 0 \\ 0 & 0 & 0 & 0 \\ 0 & 0 & 0 & 0 \end{bmatrix}$$

Then:

$$H = \begin{bmatrix} H_0 & H_3 & H_2 & H_1 \\ H_1 & H_0 & H_3 & H_2 \\ H_2 & H_1 & H_0 & H_3 \\ H_3 & H_2 & H_1 & H_0 \end{bmatrix}$$

$$H = \begin{bmatrix}
1 & 0 & 0 & 2 & 0 & 0 & 0 & 0 & 0 & 0 & 0 & 0 & 2 & 0 & 0 & -1 \\
2 & 1 & 0 & 0 & 0 & 0 & 0 & 0 & 0 & 0 & 0 & 0 & -1 & 2 & 0 & 0 \\
0 & 2 & 1 & 0 & 0 & 0 & 0 & 0 & 0 & 0 & 0 & 0 & 0 & -1 & 2 & 0 \\
0 & 0 & 2 & 0 & 0 & 0 & 0 & 0 & 0 & 0 & 0 & 0 & 0 & 0 & -1 & 2 \\
2 & 0 & 0 & -1 & 1 & 0 & 0 & 2 & 0 & 0 & 0 & 0 & 0 & 0 & 0 & 0 \\
-1 & 2 & 0 & 0 & 2 & 1 & 0 & 0 & 0 & 0 & 0 & 0 & 0 & 0 & 0 & 0 \\
0 & -1 & 2 & 0 & 0 & 2 & 1 & 0 & 0 & 0 & 0 & 0 & 0 & 0 & 0 & 0 \\
0 & 0 & -1 & 2 & 0 & 0 & 2 & 1 & 0 & 0 & 0 & 0 & 0 & 0 & 0 & 0 \\
0 & 0 & 0 & 0 & 2 & 0 & 0 & -1 & 1 & 0 & 0 & 2 & 0 & 0 & 0 & 0 \\
0 & 0 & 0 & 0 & -1 & 2 & 0 & 0 & 2 & 1 & 0 & 0 & 0 & 0 & 0 & 0 \\
0 & 0 & 0 & 0 & 0 & -1 & 2 & 0 & 0 & 2 & 1 & 0 & 0 & 0 & 0 & 0 \\
0 & 0 & 0 & 0 & 0 & 0 & -1 & 2 & 0 & 0 & 2 & 1 & 0 & 0 & 0 & 0 \\
0 & 0 & 0 & 0 & 0 & 0 & 0 & 0 & 2 & 0 & 0 & -1 & 1 & 0 & 0 & 2 \\
0 & 0 & 0 & 0 & 0 & 0 & 0 & 0 & -1 & 2 & 0 & 0 & 2 & 1 & 0 & 0 \\
0 & 0 & 0 & 0 & 0 & 0 & 0 & 0 & 0 & -1 & 2 & 0 & 0 & 2 & 1 & 0 \\
0 & 0 & 0 & 0 & 0 & 0 & 0 & 0 & 0 & 0 & -1 & 2 & 0 & 0 & 2 & 1
\end{bmatrix}$$

Therefore:

$g = Hf =$

$$\begin{bmatrix} 1 & 0 & 0 & 2 & 0 & 0 & 0 & 0 & 0 & 0 & 0 & 0 & 2 & 0 & 0 & -1 \\ 2 & 1 & 0 & 0 & 0 & 0 & 0 & 0 & 0 & 0 & 0 & 0 & -1 & 2 & 0 & 0 \\ 0 & 2 & 1 & 0 & 0 & 0 & 0 & 0 & 0 & 0 & 0 & 0 & 0 & -1 & 2 & 0 \\ 0 & 0 & 2 & 0 & 0 & 0 & 0 & 0 & 0 & 0 & 0 & 0 & 0 & 0 & -1 & 2 \\ 2 & 0 & 0 & -1 & 1 & 0 & 0 & 2 & 0 & 0 & 0 & 0 & 0 & 0 & 0 & 0 \\ -1 & 2 & 0 & 0 & 2 & 1 & 0 & 0 & 0 & 0 & 0 & 0 & 0 & 0 & 0 & 0 \\ 0 & -1 & 2 & 0 & 0 & 2 & 1 & 0 & 0 & 0 & 0 & 0 & 0 & 0 & 0 & 0 \\ 0 & 0 & -1 & 2 & 0 & 0 & 2 & 1 & 0 & 0 & 0 & 0 & 0 & 0 & 0 & 0 \\ 0 & 0 & 0 & 0 & 2 & 0 & 0 & -1 & 1 & 0 & 0 & 2 & 0 & 0 & 0 & 0 \\ 0 & 0 & 0 & 0 & -1 & 2 & 0 & 0 & 2 & 1 & 0 & 0 & 0 & 0 & 0 & 0 \\ 0 & 0 & 0 & 0 & 0 & -1 & 2 & 0 & 0 & 2 & 1 & 0 & 0 & 0 & 0 & 0 \\ 0 & 0 & 0 & 0 & 0 & 0 & -1 & 2 & 0 & 0 & 2 & 1 & 0 & 0 & 0 & 0 \\ 0 & 0 & 0 & 0 & 0 & 0 & 0 & 0 & 2 & 0 & 0 & -1 & 1 & 0 & 0 & 2 \\ 0 & 0 & 0 & 0 & 0 & 0 & 0 & -1 & 2 & 0 & 0 & 2 & 1 & 0 & 0 \\ 0 & 0 & 0 & 0 & 0 & 0 & 0 & 0 & -1 & 2 & 0 & 0 & 2 & 1 & 0 \\ 0 & 0 & 0 & 0 & 0 & 0 & 0 & 0 & 0 & -1 & 2 & 0 & 0 & 2 & 1 \end{bmatrix} \begin{bmatrix} 22 \\ 33 \\ 32 \\ 0 \\ 23 \\ 20 \\ 18 \\ 0 \\ 16 \\ 17 \\ 21 \\ 0 \\ 0 \\ 0 \\ 0 \\ 0 \end{bmatrix} = \begin{bmatrix} 22 \\ 77 \\ 98 \\ 64 \\ 67 \\ 110 \\ 89 \\ 4 \\ 62 \\ 66 \\ 71 \\ 24 \\ 32 \\ 18 \\ 25 \\ -21 \end{bmatrix}$$

Hence:

$$g = \begin{bmatrix} 22 & 77 & 98 & 64 \\ 67 & 110 & 89 & 4 \\ 62 & 66 & 71 & 24 \\ 32 & 18 & 25 & -21 \end{bmatrix}$$

The MATLAB codes to perform 2-d convolution using matrices are shown in Tables 1.2 and 1.3.

1.3.3.3 Matrix Representation of Discrete Fourier Transform

The discrete Fourier transform (DFT) is an important transform in signal processing. It can be utilized to efficiently perform the process of convolution in the frequency domain. It is also employed in spectral estimation, where the spectrum of the signal is to be estimated from a finite number of samples. The DFT of a 1-d discrete signal of size N is defined as:

$$X(k) = \sum_{n=0}^{N-1} x(n) W_N^{nk} \quad k = 0, 1, \ldots, N-1 \tag{1.45}$$

where $W_N = \exp\left(-j\frac{2\pi}{N}\right)$, $W_N^{nk} = \exp\left(-j\frac{2\pi}{N}nk\right)$ and $j = \sqrt{-1}$

Matrices, Matrix Algebra, and Elementary Matrix Operations

TABLE 1.2

MATLAB Code to Perform 2-d Convolution Using Matrices

```
f = [22 33 32;23 20 18;16 17 21];
h = [1 2;2 -1];
[a,b] = size(f);
[c,d] = size(h);
N = a+c-1;
M = b+d-1;
fe = zeros(N,M);
fe(1:1:a,1:1:b) = f;
Y = fe';
F = Y(1:1:N*M)';
he = zeros(N,M);
he(1:1:c,1:1:d) = h;
r = [he(1,1) he(1,N:-1:2)];
Z = circmatrix(r,N-1);
eval(['H' num2str(0) '= Z;']);
for i = N:-1:2
r = [he(i,1) he(i,N:-1:2)];
eval(['H' num2str(i-1) '= circmatrix(r,N-1);']);
end
p = [0 M-1:-1:1];
for i = 1:N
WW = [];
p1 = cirshift1(p,i-1);
for j = 1:M
W = eval(['H' num2str(p1(j))]);
WW = [WW W];
end
eval(['HH' num2str(i-1) '= WW;']);
end
H = eval(['HH' num2str(0)]);
for i = 1:1:N-1
H = [H;eval(['HH' num2str(i)])];
end
G = H*F;
g = col2im(G,[1,1],[N,M])';
```

TABLE 1.3

MATLAB Code to Generate Circulant Matrix

```
function [Y] = circmatrix(X,m)
N = length(X);
Y = X;
for l = 1:m
Y = [Y;real(ifft(fft(X).*exp(-j*2*l*pi*(0:1:N-1)/N)))];
end
```

The DFT can be performed using the MATLAB® command **X = fft(x)**. Inverse DFT is obtained using MATLAB command **x = ifft(X)**. We can also compute the DFT using matrix multiplication. To do so, we can simply define the $N \times 1$ vectors x and X and the $N \times N$ matrix A as:

$$x = [x(0) \quad x(1) \quad \cdots \quad x(N-1)]^T, X = [X(0) \quad X(1) \quad \cdots \quad X(N-1)]^T \quad (1.46)$$

$$A = \begin{bmatrix} W_N^0 & W_N^0 & W_N^0 & \cdots & W_N^0 \\ W_N^0 & W_N^1 & W_N^2 & \cdots & W_N^{N-1} \\ \vdots & \vdots & \vdots & \cdots & \\ W_N^0 & W_N^{N-2} & W_N^{2(N-2)} & \cdots & W_N^{(N-2)(N-1)} \\ W_N^0 & W_N^{N-1} & W_N^{2(N-1)} & \cdots & W_N^{(N-1)(N-1)} \end{bmatrix} \quad (1.47)$$

Then:

$$X = Ax \quad (1.48)$$

The inverse DFT can be obtained by premultiplying vector X by A^{-1} as follows:

$$x = A^{-1} X \quad (1.49)$$

It can be shown that $A^{-1} = (1/N) A^H$ (see Problem 1.12). Therefore:

$$x = \frac{1}{N} A^H X \quad (1.50)$$

Example 1.14

Find the four-point DFT of $x(n) = [1\ 2\ 3\ 4]$.

Solution: Since $N=4$, then:

$$A = \begin{bmatrix} W_4^0 & W_4^0 & W_4^0 & W_4^0 \\ W_4^0 & W_4^1 & W_4^2 & W_4^3 \\ W_4^0 & W_4^2 & W_4^4 & W_4^6 \\ W_4^0 & W_4^3 & W_4^6 & W_4^9 \end{bmatrix} = \begin{bmatrix} 1 & 1 & 1 & 1 \\ 1 & -j & -1 & j \\ 1 & -1 & 1 & -1 \\ 1 & j & -1 & -j \end{bmatrix}$$

The DFT of x can be computed by premultiplying vector x by matrix A as follows:

$$X = Ax = \begin{bmatrix} 1 & 1 & 1 & 1 \\ 1 & -j & -1 & j \\ 1 & -1 & 1 & -1 \\ 1 & j & -1 & -j \end{bmatrix} \begin{bmatrix} 1 \\ 2 \\ 3 \\ 4 \end{bmatrix} = \begin{bmatrix} 10 \\ -2+j2 \\ -2 \\ -2-j2 \end{bmatrix}$$

Similarly, the 2-d DFT is an important tool in image processing. It is used for processing images in the frequency domain, image enhancement and coding. The 2-d DFT of a 2-d signal $f(n,m)$ of size $N \times N$ is defined as:

Matrices, Matrix Algebra, and Elementary Matrix Operations

$$F(k,l) = \sum_{n=0}^{N-1} \sum_{m=0}^{N-1} f(n,m) W_N^{nk} W_N^{ml} \quad k, l = 0, 1, \cdots, N-1 \qquad (1.51)$$

where $W_N = \exp(-j(2\pi/N))$.
The MATLAB command to compute the 2-d DFT of the 2-d signal $f(n,m)$ is **F=fft2(f)**.

We can compute the DFT using matrix multiplication. This is accomplished as described below. Let A represent the $N \times N$ matrix given by:

$$A = \begin{bmatrix} W_N^0 & W_N^0 & W_N^0 & \cdots & W_N^0 \\ W_N^0 & W_N^1 & W_N^2 & \cdots & W_N^{N-1} \\ \vdots & \vdots & \vdots & \cdots & \\ W_N^0 & W_N^{N-2} & W_N^{2(N-2)} & \cdots & W_N^{(N-2)(N-1)} \\ W_N^0 & W_N^{N-1} & W_N^{2(N-1)} & \cdots & W_N^{(N-1)(N-1)} \end{bmatrix} \qquad (1.52)$$

Then, Equation 1.51 can be written in matrix form as:

$$F = AfA \qquad (1.53)$$

Example 1.15

Find the four-point DFT of $f(n,m) = \begin{bmatrix} 12 & 22 & 34 & 10 \\ 10 & 11 & 18 & 23 \\ 25 & 23 & 24 & 26 \\ 20 & 15 & 17 & 18 \end{bmatrix}$

Solution: Since $N = 4$, then:

$$A = \begin{bmatrix} W_4^0 & W_4^0 & W_4^0 & W_4^0 \\ W_4^0 & W_4^1 & W_4^2 & W_4^3 \\ W_4^0 & W_4^2 & W_4^4 & W_4^6 \\ W_4^0 & W_4^3 & W_4^6 & W_4^9 \end{bmatrix} = \begin{bmatrix} 1 & 1 & 1 & 1 \\ 1 & -j & -1 & j \\ 1 & -1 & 1 & -1 \\ 1 & j & -1 & -j \end{bmatrix}$$

The 2-d DFT of f can be computed by pre and postmultiplying matrix f by matrix A.

$$F = AfA = \begin{bmatrix} 1 & 1 & 1 & 1 \\ 1 & -j & -1 & j \\ 1 & -1 & 1 & -1 \\ 1 & j & -1 & -j \end{bmatrix} \begin{bmatrix} 12 & 22 & 34 & 10 \\ 10 & 11 & 18 & 23 \\ 25 & 23 & 24 & 26 \\ 20 & 15 & 17 & 18 \end{bmatrix} \begin{bmatrix} 1 & 1 & 1 & 1 \\ 1 & -j & -1 & j \\ 1 & -1 & 1 & -1 \\ 1 & j & -1 & -j \end{bmatrix}$$

$$= \begin{bmatrix} 308 & -26+j6 & 12 & -26-j6 \\ -20+j8 & -14-j4 & 14+j10 & -32+j26 \\ 44 & -16-j24 & 16 & -16+j24 \\ -20-j8 & -32-j26 & 14-j10 & -14+j4 \end{bmatrix}$$

The inverse 2-d DFT is obtained by pre and postmultiplying the F matrix by the inverse of matrix A as follows:

$$f = A^{-1}FA^{-1} = \frac{1}{N^2} A^H F A^H \qquad (1.54)$$

1.4 Elementary Row Operations

Elementary row operations can be used to solve a simultaneous set of linear equations. There are three row operations that can be performed on the augmented matrix, constructed from the equations, so as to obtain a strictly triangular system. Back-substitution can then be applied to arrive at the solution of the system. The row operations include:

1. Interchanging two rows
2. Multiplying a row by a nonzero real number
3. Replacing a row by its sum with a multiple of another row

The augmented matrix and the usage of the above operations are illustrated in Example 1.16 to solve a linear system of equations.

Example 1.16

Solve the system of linear equations given by:

$$x_1 + 2x_2 - 2x_3 = 1$$
$$2x_1 + 5x_2 + x_3 = 9$$
$$x_1 + 3x_2 + 4x_3 = 9$$

Solution: The above equations can be expressed in matrix form as follows:

$$Ax = b \Leftrightarrow \begin{bmatrix} 1 & 2 & -2 \\ 2 & 5 & 1 \\ 1 & 3 & 4 \end{bmatrix} \begin{bmatrix} x_1 \\ x_2 \\ x_3 \end{bmatrix} = \begin{bmatrix} 1 \\ 9 \\ 9 \end{bmatrix}$$

The augmented matrix for this system is then arrived at by augmenting the matrix A with the vector b as shown below:

$$[A|b] = \begin{bmatrix} 1 & 2 & -2 & | & 1 \\ 2 & 5 & 1 & | & 9 \\ 1 & 3 & 4 & | & 9 \end{bmatrix}$$

Row operation 3 is employed twice to eliminate the two nonzero entries in the first column where the second row (R_2) is updated by taking its initial values and subtracting from them the values of the first row (R_1) after it has been multiplied

Matrices, Matrix Algebra, and Elementary Matrix Operations

by 2. This is denoted by: $R_2 \leftarrow R2 - 2R_1$. Similarly, $R_3 \leftarrow R3 - 1R_1$ indicates that the content of third row (R_3) have been replaced by the content of third row minus the content of first row (R_1). This process is illustrated visually below where the updated matrix is shown on the right hand side:

$$\begin{bmatrix} 1 & 2 & -2 & | & 1 \\ 2 & 5 & 1 & | & 9 \\ 1 & 3 & 4 & | & 9 \end{bmatrix} \xrightarrow{R_2 \leftarrow R_2 - 2R_1} \begin{bmatrix} 1 & 2 & -2 & | & 1 \\ 0 & 1 & 5 & | & 7 \\ 1 & 3 & 4 & | & 9 \end{bmatrix} \xrightarrow{R_3 \leftarrow R_3 - 1R_1} \begin{bmatrix} 1 & 2 & -2 & | & 1 \\ 0 & 1 & 5 & | & 7 \\ 0 & 1 & 6 & | & 8 \end{bmatrix}$$

Again, row operation 3 is used to eliminate the first nonzero entry in the third row (i.e. the entry 1 found in row 3 column 2) of the resulting matrix. This is done in a similar fashion to what has been described previously by using the following operation: $R_3 \leftarrow R_3 - 1R_2$. The process is illustrated below:

$$\begin{bmatrix} 1 & 2 & -2 & | & 1 \\ 0 & 1 & 5 & | & 7 \\ 0 & 1 & 6 & | & 8 \end{bmatrix} \xrightarrow{R_3 \leftarrow R_3 - 1R_2} \begin{bmatrix} 1 & 2 & -2 & | & 1 \\ 0 & 1 & 5 & | & 7 \\ 0 & 0 & 1 & | & 1 \end{bmatrix}$$

The resulting matrix, along with the corresponding set of updated equations, are shown below:

$$\begin{bmatrix} 1 & 2 & -2 & | & 1 \\ 0 & 1 & 5 & | & 7 \\ 0 & 0 & 1 & | & 1 \end{bmatrix} \Leftrightarrow \begin{matrix} x_1 + 2x_2 - 2x_3 = 1 \\ x_2 + 5x_3 = 7 \\ x_3 = 1 \end{matrix}$$

Finally, back-substitution is employed to obtain the solution set for the given system. This is done by solving the system starting with the third row (third equation) and working backwards towards the first row (first equation)—hence the name back-substitution. The process is illustrated as follows:

$$1x_3 = 1 \Rightarrow x_3 = 1$$
$$1x_3 + 5x_3 = 7 \Rightarrow x_2 = 7 - 5x_3 = 7 - 5(1) = 2$$
$$1x_1 + 2x_2 - 2x_3 = 1 \Rightarrow x_1 = 1 - 2x_2 + 2x_3 = 1 - 2(2) + 2(1) = -1$$

Hence, the solution vector is:

$$\begin{bmatrix} x_1 \\ x_2 \\ x_3 \end{bmatrix} = \begin{bmatrix} -1 \\ 2 \\ 1 \end{bmatrix}$$

1.4.1 Row Echelon Form

A matrix is said to be in row echelon form (REF) if it satisfies all of the following conditions:

1. The first nonzero entry in each nonzero row is 1.
2. If the kth row does not consist entirely of zeros, then the number of leading zero entries in the $(k+1)^{th}$ row should be greater than the number of leading zero entries in the kth row.

3. If there are any rows whose entries are all zero, they should be below the rows with nonzero entries.

The following set of examples serve to illustrate the concept of REF.

Example 1.17

The following matrices are in REF.

$$A = \begin{bmatrix} 1 & 8 & 9 \\ 0 & 1 & 6 \\ 0 & 0 & 1 \end{bmatrix}, B = \begin{bmatrix} 1 & 1 & 2 & 0 \\ 0 & 0 & 1 & 9 \\ 0 & 0 & 0 & 1 \\ 0 & 0 & 0 & 0 \end{bmatrix}, C = \begin{bmatrix} 1 & 5 & 6 \\ 0 & 0 & 1 \end{bmatrix}$$

Example 1.18

The following matrices are not in REF.

$$A = \begin{bmatrix} 2 & 8 & 1 \\ 0 & 1 & 5 \\ 0 & 0 & 7 \end{bmatrix}, B = \begin{bmatrix} 0 & 0 \\ 1 & 0 \\ 1 & 7 \end{bmatrix}, C = \begin{bmatrix} 0 & 0 & 5 \\ 1 & 5 & 8 \\ 0 & 0 & 0 \end{bmatrix}$$

1.4.2 Elementary Transformation Matrices

Elementary transformation matrices are frequently employed to make adjustments to the rows and/or columns of matrices through pre and/or postmultiplication. There are three types of elementary transformation matrices and each of these is designated by E with a subscript. Premultiplying a matrix with an elementary transformation matrix serves to change its row(s) whereas postmultiplying changes its column(s). The three types of elementary transformation matrices and their affect on general matrices are discussed below.

1.4.2.1 Type 1: Scaling Transformation Matrix (E_1)

This transformation matrix is utilized to multiply a row or column of a matrix by a constant. For example, consider the matrix A shown below:

$$A = \begin{bmatrix} 1 & 2 & 3 \\ 4 & 5 & 6 \\ 7 & 8 & 9 \end{bmatrix}$$

In order to multiply the second row by 2, premultiply A by the elementary transformation matrix

$$E_1 = \begin{bmatrix} 1 & 0 & 0 \\ 0 & 2 & 0 \\ 0 & 0 & 1 \end{bmatrix}$$

The outcome is:

$$E_1 A = \begin{bmatrix} 1 & 0 & 0 \\ 0 & 2 & 0 \\ 0 & 0 & 1 \end{bmatrix} \begin{bmatrix} 1 & 2 & 3 \\ 4 & 5 & 6 \\ 7 & 8 & 9 \end{bmatrix} = \begin{bmatrix} 1 & 2 & 3 \\ 8 & 10 & 12 \\ 7 & 8 & 9 \end{bmatrix} \quad (1.55)$$

Similarly, to multiply the third column of A by 3, postmultiply A by
$$E_1 = \begin{bmatrix} 1 & 0 & 0 \\ 0 & 1 & 0 \\ 0 & 0 & 3 \end{bmatrix}$$

$$AE_1 = \begin{bmatrix} 1 & 2 & 3 \\ 4 & 5 & 6 \\ 7 & 8 & 9 \end{bmatrix} \begin{bmatrix} 1 & 0 & 0 \\ 0 & 1 & 0 \\ 0 & 0 & 3 \end{bmatrix} = \begin{bmatrix} 1 & 2 & 9 \\ 4 & 5 & 18 \\ 7 & 8 & 27 \end{bmatrix} \quad (1.56)$$

1.4.2.2 Type 2: Interchange Transformation Matrix (E_2)

This transformation matrix is employed to accomplish a row or column switch. E_2 is derived from the identity matrix (I) by interchanging its rows or columns appropriately to accomplish the switch. This is illustrated by the following example. Consider the following matrix A:

$$A = \begin{bmatrix} 1 & -2 & 3 \\ 0 & -1 & 2 \\ 3 & 1 & 5 \end{bmatrix}$$

Row 1 and Row 2 of A can be interchanged by premultiplying A with
$$E_2 = \begin{bmatrix} 0 & 1 & 0 \\ 1 & 0 & 0 \\ 0 & 0 & 1 \end{bmatrix}$$

$$E_2 A = \begin{bmatrix} 0 & 1 & 0 \\ 1 & 0 & 0 \\ 0 & 0 & 1 \end{bmatrix} \begin{bmatrix} 1 & 2 & 3 \\ 0 & -1 & 2 \\ 3 & 1 & 5 \end{bmatrix} = \begin{bmatrix} 0 & -1 & 2 \\ 1 & 2 & 3 \\ 3 & 1 & 5 \end{bmatrix} \quad (1.57)$$

Note, E_2 is derived from the identity matrix I by interchanging the first and second rows. Similarly, to exchange columns 1 and 2 of A, A can be postmultiplied with E_2 as shown below:

$$AE_2 = \begin{bmatrix} 1 & -2 & 3 \\ 0 & -1 & 2 \\ 3 & 1 & 5 \end{bmatrix} \begin{bmatrix} 0 & 1 & 0 \\ 1 & 0 & 0 \\ 0 & 0 & 1 \end{bmatrix} = \begin{bmatrix} -2 & 1 & 3 \\ -1 & 0 & 2 \\ 1 & 3 & 5 \end{bmatrix} \quad (1.58)$$

1.4.2.3 Type 3: Combination Transformation Matrices (E_3)

This transformation matrix allows the addition of a scalar multiple of one row (or column) to another row (or column), where the row (or column) that is being multiplied remains unchanged. For example, let the matrix A be defined as follows:

$$A = \begin{bmatrix} 1 & -2 & 3 \\ 0 & -1 & 2 \\ 3 & 4 & 5 \end{bmatrix}$$

Let us assume that we prefer to subtract $4 \times R_2$ from R_3 and place the resulting values in R_3. This is denoted by the following short hand notation: $R_3 \leftarrow R_3 - 4R_2$. The above is accomplished by premultiplying A with $E_3 = \begin{bmatrix} 1 & 0 & 0 \\ 0 & 1 & 0 \\ 0 & -4 & 1 \end{bmatrix}$ as shown below:

$$E_3 A = \begin{bmatrix} 1 & 0 & 0 \\ 0 & 1 & 0 \\ 0 & -4 & 1 \end{bmatrix} \begin{bmatrix} 1 & -2 & 3 \\ 0 & -1 & 2 \\ 3 & 4 & 5 \end{bmatrix} = \begin{bmatrix} 1 & -2 & 3 \\ 0 & -1 & 2 \\ 3 & 8 & -3 \end{bmatrix} \quad (1.59)$$

where the matrix E_3 is constructed from the identity matrix except that its third row-second column element is replaced by -4 in order to accomplish the desired transformation: $R_3 \leftarrow R_3 - 4R_2$. Similarly, the operation $C_3 \leftarrow 2C_1 + C_3$ is obtained by postmultiplying A with $E_3 = \begin{bmatrix} 1 & 0 & 2 \\ 0 & 1 & 0 \\ 0 & 0 & 1 \end{bmatrix}$

$$AE_3 = \begin{bmatrix} 1 & -2 & 3 \\ 0 & -1 & 2 \\ 3 & 4 & 5 \end{bmatrix} \begin{bmatrix} 1 & 0 & 2 \\ 0 & 1 & 0 \\ 0 & 0 & 1 \end{bmatrix} = \begin{bmatrix} 1 & -2 & 5 \\ 0 & -1 & 2 \\ 3 & 4 & 11 \end{bmatrix} \quad (1.60)$$

where C_2 and C_3 denote the second and third column, respectively. Successive transformations can be done using a single composite transformation matrix T, which is made up of the concatenation of the appropriate elementary transformation matrices, i.e. $T = E_1 E_2 E_3 \ldots$. This is illustrated below:

For example, consider $A = \begin{bmatrix} 1 & -2 & 3 \\ 0 & -1 & 5 \\ 3 & 4 & 5 \end{bmatrix}$

$$E_3 E_2 E_1 A = \begin{bmatrix} 1 & 0 & 0 \\ 0 & 1 & 0 \\ 0 & -4 & 1 \end{bmatrix} \begin{bmatrix} 0 & 0 & 1 \\ 0 & 1 & 0 \\ 1 & 0 & 0 \end{bmatrix} \begin{bmatrix} 1 & 0 & 0 \\ 0 & 1 & 0 \\ 0 & 0 & 3 \end{bmatrix} \begin{bmatrix} 1 & -2 & 3 \\ 0 & -1 & 5 \\ 3 & 4 & 5 \end{bmatrix} = \begin{bmatrix} 9 & 12 & 15 \\ 0 & -1 & 5 \\ 1 & 2 & -17 \end{bmatrix} \quad (1.61)$$

where the transformation matrix E_1 is designed to multiply the third row of A by 3 ($R_3 \leftarrow 3R_3$), E_2 swaps the resultant first and third rows ($R_1 \leftrightarrow R_3$) and E_3 performs the following operation $R_3 \leftarrow R_3 - 4R_2$ on the resultant matrix $E_2 E_1 A$.

1.5 Solution of System of Linear Equations

Simultaneous sets of linear equations play an important role in many different fields of engineering. For example, the linear equations describing the behavior of a given circuit can be solved to yield currents and voltages of interest throughout the circuit. In control systems, the dynamic of linear time invariant systems are modeled using a set of first order linear differential equations. The equilibrium point of the system is obtained by solving a set of algebraic linear equations. In this section, we describe the Gaussian elimination approach employed for solving a set of linear equations. Other techniques will be covered in subsequent chapters.

1.5.1 Gaussian Elimination

Consider a set of linear equations given by:

$$Ax = b \qquad (1.62)$$

where x and b represent $n \times 1$ vectors, and A is an $n \times n$ matrix. One approach to solve the system of linear equations $Ax = b$ is by utilizing the Gaussian elimination technique. In this method, we combine matrix A and column vector b into an augmented $n \times (n+1)$ matrix as follows:

$$[A|b] = \begin{bmatrix} a_{11} & a_{12} & \cdots & a_{1n} & b_1 \\ a_{21} & a_{22} & \cdots & a_{2n} & b_2 \\ \vdots & \vdots & & \vdots & \vdots \\ a_{n1} & a_{n2} & \cdots & a_{nn} & b_n \end{bmatrix} \qquad (1.63)$$

A sequence of elementary row operations is applied to this augmented matrix to transform the matrix into an upper triangular matrix of the form:

$$\begin{bmatrix} a_{11} & a_{12} & \cdots & a_{1n} & b_1 \\ 0 & a'_{22} & \cdots & a'_{2n} & b'_2 \\ \vdots & \vdots & & \vdots & \vdots \\ 0 & 0 & \cdots & a'_{nn} & b'_n \end{bmatrix} \qquad (1.64)$$

Now, we can solve the last equation for the unknown (x_n) and substitute it back into the previous equation to obtain a solution for x_{n-1}. This process is continued using back-substitution until all of the unknowns have been found. The general solution is given by:

$$x_i = \frac{1}{a'_{ii}}\left(b'_i - \sum_{k=i+1}^{n} a'_{ik} x_k\right) \quad i = n, n-1, \cdots, 1 \tag{1.65}$$

We can extend the application of elementary matrices E_1, E_2 and E_3 to perform Gaussian elimination as shown in Examples 1.19 and 1.20. Example 1.19 illustrates their use in solving systems of linear equations without having to reduce the matrix to REF. Example 1.20 demonstrates their use by first reducing the augmented matrix to REF prior to arriving at a solution.

Example 1.19

Consider the following set of linear equations

$$2x_1 + x_2 + x_3 = 5$$
$$4x_1 - 6x_2 = -2$$
$$-2x_1 + 7x_2 + 2x_3 = 9$$

The above equations can be expressed in matrix form as: $Ax = b$

$$\begin{matrix} 2x_1 + x_2 + x_3 = 5 \\ 4x_1 - 6x_2 = -2 \\ -2x_1 + 7x_2 + 2x_3 = 9 \end{matrix} \Leftrightarrow \begin{bmatrix} 2 & 1 & 1 \\ 4 & -6 & 0 \\ -2 & 7 & 2 \end{bmatrix} \begin{bmatrix} x_1 \\ x_2 \\ x_3 \end{bmatrix} = \begin{bmatrix} 5 \\ -2 \\ 9 \end{bmatrix}$$

To solve the above system using Gaussian elimination, we first begin by forming the augmented matrix A_a as:

$$A_a = [A|b] = \begin{bmatrix} 2 & 1 & 1 & | & 5 \\ 4 & -6 & 0 & | & -2 \\ -2 & 7 & 2 & | & 9 \end{bmatrix}$$

Once A_a is formed, we proceed to eliminate the first element of the second and third row through the use of elementary matrix transformations. To eliminate the first element of the second row, we need to perform the following operation: $R_2 \leftarrow R_2 - 2R_1$. This is achieved by premultiplying A_a with E_{3-1} as follows:

$$\begin{bmatrix} 1 & 0 & 0 \\ -2 & 1 & 0 \\ 0 & 0 & 1 \end{bmatrix} \begin{bmatrix} 2 & 1 & 1 & | & 5 \\ 4 & -6 & 0 & | & -2 \\ -2 & 7 & 2 & | & 9 \end{bmatrix} = \begin{bmatrix} 2 & 1 & 1 & | & 5 \\ 0 & -8 & -2 & | & -12 \\ -2 & 7 & 2 & | & 9 \end{bmatrix}$$

the newly achieved augmented matrix

$$A_a = \begin{bmatrix} 2 & 1 & 1 & | & 5 \\ 0 & -8 & -2 & | & -12 \\ -2 & 7 & 2 & | & 9 \end{bmatrix}$$

Matrices, Matrix Algebra, and Elementary Matrix Operations

To eliminate the first element of the third row, we need to perform the following operation: $R_3 \leftarrow R_3 + R_1$. This is accomplished by premultiplying A_a with E_{3-2} as shown:

$$\begin{bmatrix} 1 & 0 & 0 \\ 0 & 1 & 0 \\ 1 & 0 & 1 \end{bmatrix} \begin{bmatrix} 2 & 1 & 1 & \vdots & 5 \\ 0 & -8 & -2 & \vdots & -12 \\ -2 & 7 & 2 & \vdots & 9 \end{bmatrix} = \begin{bmatrix} 2 & 1 & 1 & \vdots & 5 \\ 0 & -8 & -2 & \vdots & -12 \\ 0 & 8 & 3 & \vdots & 14 \end{bmatrix}$$

a newly augmented matrix

$$A_a = \begin{bmatrix} 2 & 1 & 1 & \vdots & 5 \\ 0 & -8 & -2 & \vdots & -12 \\ 0 & 8 & 3 & \vdots & 14 \end{bmatrix}$$

Finally, to eliminate the second element of the third row, we proceed to perform the following operation: $R_3 \leftarrow R_3 + R_2$. This is accomplished by premultiplying A_a with E_{3-3} as follows:

$$\begin{bmatrix} 1 & 0 & 0 \\ 0 & 1 & 0 \\ 0 & 1 & 1 \end{bmatrix} \begin{bmatrix} 2 & 1 & 1 & \vdots & 5 \\ 0 & -8 & -2 & \vdots & -12 \\ 0 & 8 & 3 & \vdots & 14 \end{bmatrix} = \begin{bmatrix} 2 & 1 & 1 & \vdots & 5 \\ 0 & -8 & -2 & \vdots & -12 \\ 0 & 0 & 1 & \vdots & 2 \end{bmatrix}$$

the following augmented matrix

$$A_a = \begin{bmatrix} 2 & 1 & 1 & \vdots & 5 \\ 0 & -8 & -2 & \vdots & -12 \\ 0 & 0 & 1 & \vdots & 2 \end{bmatrix}$$

Effectively, premultiplying A_a with $E_{3-1}, E_{3-2}, E_{3-3}$ would yield the same result.

$$A'_a = E_{3-1} E_{3-2} E_{3-3} A_a = \begin{bmatrix} 1 & 0 & 0 \\ -2 & 1 & 0 \\ 0 & 0 & 1 \end{bmatrix} \begin{bmatrix} 1 & 0 & 0 \\ 0 & 1 & 0 \\ 1 & 0 & 1 \end{bmatrix} \begin{bmatrix} 1 & 0 & 0 \\ 0 & 1 & 0 \\ 0 & 1 & 1 \end{bmatrix} \begin{bmatrix} 2 & 1 & 1 & \vdots & 5 \\ 4 & -6 & 0 & \vdots & -2 \\ -2 & 7 & 2 & \vdots & 9 \end{bmatrix}$$

Therefore, the resulting augmented matrix along with its corresponding set of equations is shown below:

$$A'_a = \begin{bmatrix} 2 & 1 & 1 & \vdots & 5 \\ 0 & -8 & -2 & \vdots & -12 \\ 0 & 0 & 1 & \vdots & 2 \end{bmatrix} \Leftrightarrow \begin{matrix} 2x_1 + x_2 + x_3 = 5 \\ -8x_2 - 2x_3 = -12 \\ x_3 = 2 \end{matrix}$$

Note that the above equations are equivalent to the ones given by Example 1.19. By utilizing the process of back-substitution, we can solve for the remaining unknowns x_1 and x_2 as follows:

$$-8x_2 - 2x_3 = -12 \Rightarrow -8x_2 - 2(2) = -12 \Rightarrow -8x_2 = -8 \Rightarrow x_2 = 1$$

Similarly,

$$2x_1 + x_2 + x_3 = 5 \Rightarrow 2x_1 = 5 - (x_2 + x_3) \Rightarrow 2x_1 = 5 - 1 - 2 \Rightarrow x_1 = 1$$

Thus, the solution is:

$$\begin{bmatrix} x_1 \\ x_2 \\ x_3 \end{bmatrix} = \begin{bmatrix} 1 \\ 1 \\ 2 \end{bmatrix}$$

Example 1.20

Solve the following set of equations using the technique of Gaussian elimination.

$$2x_1 - x_2 + x_3 = 3$$
$$4x_1 + x_2 + 3x_3 = 15$$
$$x_1 - x_2 + 5x_3 = 14$$

Solution: To solve the above set of equations using Gaussian elimination, we perform the following steps:

1. Form the augmented matrix $[A \mid b]$

$$\begin{bmatrix} 2 & -1 & 1 & \vdots & 3 \\ 4 & 1 & 3 & \vdots & 15 \\ 1 & -1 & 5 & \vdots & 14 \end{bmatrix}$$

2. Utilize the following elementary row operations to reduce the augmented matrix to REF.
   i. Interchange the rows.
   ii. Multiply a row by a nonzero real number.
   iii. Replace a row by its sum with a multiple of another row.

This is illustrated by steps shown below starting with the augmented matrix and proceeding until we reach an augmented matrix in REF:

$$\begin{bmatrix} 2 & -1 & 1 & \vdots & 3 \\ 4 & 1 & 3 & \vdots & 15 \\ 1 & -1 & 5 & \vdots & 14 \end{bmatrix} \xrightarrow{R_3 \leftrightarrow R_1} \begin{bmatrix} 1 & -1 & 5 & \vdots & 14 \\ 4 & 1 & 3 & \vdots & 15 \\ 2 & -1 & 1 & \vdots & 3 \end{bmatrix} \xrightarrow{R_2 \leftarrow R_2 - 4R_1} \begin{bmatrix} 1 & -1 & 5 & \vdots & 14 \\ 0 & 5 & -17 & \vdots & -41 \\ 2 & -1 & 1 & \vdots & 3 \end{bmatrix}$$

$$\xrightarrow{R_3 \leftarrow R_3 - 2R_1} \begin{bmatrix} 1 & -1 & 5 & \vdots & 14 \\ 0 & 5 & -17 & \vdots & -41 \\ 0 & 1 & -9 & \vdots & -25 \end{bmatrix} \xrightarrow{R_2 \leftarrow R_2/5} \begin{bmatrix} 1 & -1 & 5 & \vdots & 14 \\ 0 & 1 & -17/5 & \vdots & -41/5 \\ 0 & 1 & -9 & \vdots & -25 \end{bmatrix}$$

$$\xrightarrow{R_3 \leftarrow R_3 - R_2} \begin{bmatrix} 1 & -1 & 5 & \vdots & 14 \\ 0 & 1 & -17/5 & \vdots & -41/5 \\ 0 & 0 & -28/5 & \vdots & -84/5 \end{bmatrix} \xrightarrow{R_3 \leftarrow (-5/28)R_3} \begin{bmatrix} 1 & -1 & 5 & \vdots & 14 \\ 0 & 1 & -17/5 & \vdots & -41/5 \\ 0 & 0 & 1 & \vdots & 3 \end{bmatrix}$$

The resultant augmented matrix is in REF. The equations can now be solved using back-substitution.

$$x_3 = 3$$
$$x_2 = (-41/5)+(17/5)x_3 = (51-41)/5 = 2$$
$$x_1 = 14 + x_2 - 5x_3 = 14 + 2 - 15 = 1$$

The solution set is $[x_1 \ x_2 \ x_3]^T = [1 \ 2 \ 3]^T$.

1.5.2 Over Determined Systems

Over determined systems are systems in which the number of equations exceeds the number of unknowns. These systems are usually (not-always) inconsistent.

Example 1.21

Consider the following set of linear equations:

$$x_1 + x_2 = 1$$
$$x_1 - x_2 = 3$$
$$2x_1 + 2x_2 = 4$$

Solution: These equations can be expressed as:

$$x_2 = 1 - x_1$$
$$x_2 = x_1 - 3$$
$$x_2 = 2 - x_1$$

There are three equations with two unknowns. Plotting these three linear equations in the $x_1 - x_2$ plane as shown in Figure 1.10 reveals no common intersection point among them; therefore, the set has no solution.

The above conclusion is also arrived at by performing a Gaussian elimination as shown below starting with the augmented matrix and proceeding along until we reach the REF:

$$\begin{bmatrix} 1 & 1 & | & 1 \\ 1 & -1 & | & 3 \\ 2 & 2 & | & 4 \end{bmatrix} \xrightarrow{R_2 \leftarrow R_2 - R_1} \begin{bmatrix} 1 & 1 & | & 1 \\ 0 & -2 & | & 2 \\ 2 & 2 & | & 4 \end{bmatrix} \xrightarrow{R_3 \leftarrow R_3 - 2R_2} \begin{bmatrix} 1 & 1 & | & 1 \\ 0 & -2 & | & 2 \\ 0 & 0 & | & 2 \end{bmatrix} \xrightarrow{R_2 \leftarrow R_2/(-2)} \begin{bmatrix} 1 & 1 & | & 1 \\ 0 & 1 & | & -1 \\ 0 & 0 & | & 2 \end{bmatrix}$$

The final augmented matrix represents the following set of equations:

$$x_1 + x_2 = 1$$
$$x_2 = -1$$
$$0x_1 + 0x_2 = 2$$

Obviously, the last equation is impossible to satisfy. Hence, the above over determined system of equations is inconsistent. Note that all the coefficients of the unknown variables in the third row of the resulting augmented matrix are equal to zero indicating an inconsistent set of equations.

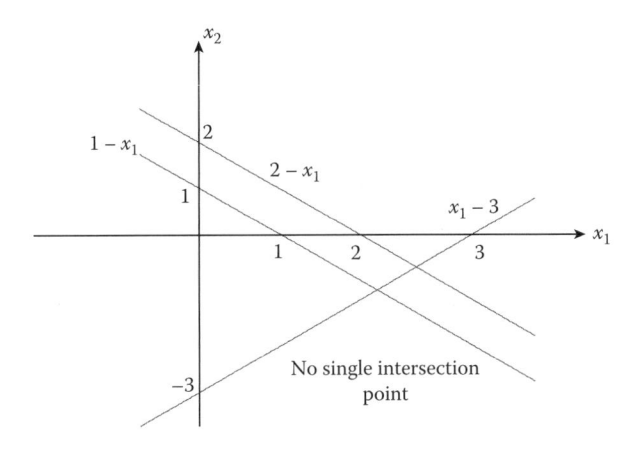

FIGURE 1.10
Graphical interpretation of equations from Example 1.21.

1.5.3 Under Determined Systems

Under determined systems are systems in which the number of equations is less than the number of unknowns. These systems are usually (not-always) consistent with infinitely many solutions.

Example 1.22

Consider the following set of equations.

$$x_1 + x_2 + x_3 = 6$$

$$2x_1 + 3x_2 + 4x_3 = 20$$

Solution: The above system of equations is simplified by utilizing the Gaussian elimination technique in a similar fashion to what has been done earlier starting with the augmented matrix and proceeding until a REF is reached. This is illustrated as follows:

$$\begin{bmatrix} 1 & 1 & 1 & | & 6 \\ 2 & 3 & 4 & | & 20 \end{bmatrix} \xrightarrow{R_2 \leftarrow R_2 - 2R_1} \begin{bmatrix} 1 & 1 & 1 & | & 6 \\ 0 & 1 & 2 & | & 8 \end{bmatrix}$$

By examining the resulting augmented matrix, it can be easily seen that the above system of equations has infinitely many solutions.

1.6 Matrix Partitions

Matrices can be divided into a number of submatrices; this process is called matrix partitioning. For example, consider the following matrix which can be partitioned into four submatrices as shown below.

Matrices, Matrix Algebra, and Elementary Matrix Operations

$$C = \begin{bmatrix} 1 & 2 & 3 & 4 & 5 \\ 6 & 7 & 8 & 9 & 10 \\ 11 & 12 & 13 & 14 & 15 \\ \hline 16 & 17 & 18 & 19 & 20 \\ 21 & 22 & 23 & 24 & 25 \end{bmatrix} = \begin{bmatrix} C_{11} & C_{12} \\ C_{21} & C_{22} \end{bmatrix} \quad (1.66)$$

where C_{11} is a 3×3 matrix, C_{12} is a 3×2 matrix, C_{21} is a 2×3 matrix, and C_{22} is a 2×2 matrix. We now consider two types of matrix partitions: row partitions and column partitions.

1.6.1 Column Partitions

Partitioning a matrix into columns is referred to as column partitioning. It is often useful in engineering applications. Let the matrix B be defined as follows:

$$B = \begin{bmatrix} -1 & 2 & 1 \\ 2 & 3 & 4 \\ 1 & 3 & 2 \end{bmatrix} \quad (1.67)$$

Matrix B can be partitioned into three column submatrices (vectors) as shown below:

$$B = [b_1 \; b_2 \; b_3]$$

where $b_1 = \begin{bmatrix} -1 \\ 2 \\ 1 \end{bmatrix}$, $b_2 = \begin{bmatrix} 2 \\ 3 \\ 3 \end{bmatrix}$ and $b_3 = \begin{bmatrix} 1 \\ 4 \\ 2 \end{bmatrix}$

Now, if A is a three column matrix, then the product of AB can be represented by:

$$AB = A(b_1 \; b_2 \; b_3) = (Ab_1, Ab_2, Ab_3) \quad (1.68)$$

Example 1.23

Let the matrix A and B be defined as follows:

$$A = \begin{bmatrix} 1 & 2 & 1 \\ 2 & 1 & -3 \end{bmatrix} \quad \text{and} \quad B = \begin{bmatrix} -1 & 2 & 1 \\ 2 & 3 & 4 \\ 1 & 3 & 2 \end{bmatrix}$$

Find the product AB and partition it into its corresponding column vectors.

Solution: As described above, AB can be written as follows: $AB = A(b_1\ b_2\ b_3) = (Ab_1, Ab_2, Ab_3)$. In order to partition AB into its corresponding column vectors, we compute each of the column vectors separately as shown below:

$$Ab_1 = \begin{bmatrix} 1 & 2 & 1 \\ 2 & 1 & -3 \end{bmatrix}\begin{bmatrix} -1 \\ 2 \\ 1 \end{bmatrix} = \begin{bmatrix} 4 \\ -3 \end{bmatrix}; Ab_2 = \begin{bmatrix} 1 & 2 & 1 \\ 2 & 1 & -3 \end{bmatrix}\begin{bmatrix} 2 \\ 3 \\ 3 \end{bmatrix} = \begin{bmatrix} 11 \\ -2 \end{bmatrix}; Ab_3 = \begin{bmatrix} 1 & 2 & 1 \\ 2 & 1 & -3 \end{bmatrix}\begin{bmatrix} 1 \\ 4 \\ 2 \end{bmatrix} = \begin{bmatrix} 11 \\ 0 \end{bmatrix}$$

It is imperative to note that the rules of matrix multiplication need to be followed in arriving at the desired partitions. Therefore:

$$A[b_1\ \ b_2\ \ b_3] = \begin{bmatrix} 4 & | & 11 & | & 11 \\ -3 & | & -2 & | & 0 \end{bmatrix}$$

In general, if A is an $m \times n$ matrix and B is an $n \times r$ matrix that has been partitioned into columns as:

$$B = [b_1\ |\ b_2\ |\ \ldots\ |\ b_r] \tag{1.69}$$

then

$$C = A_{m \times n} B_{n \times r} = A[b_1\ |\ b_2\ |\ \ldots\ |\ b_r] = [Ab_1,\ Ab_2,\ \ldots\ Ab_r]_{m \times r} \tag{1.70}$$

Thus:

$$C = [c_1\ |\ c_2\ |\ \ldots\ |\ c_r] = [Ab_1\ |\ Ab_2\ |\ \ldots\ |\ Ab_r] \tag{1.71}$$

1.6.2 Row Partitions

Similarly, partitioning a matrix into rows is referred to as row partitioning. Let A be an $m \times n$ matrix and B an $n \times r$ matrix; row partitioning of A yields:

$$A = \begin{bmatrix} a_1 \\ -- \\ a_2 \\ -- \\ \vdots \\ -- \\ a_m \end{bmatrix}_{m \times n} \tag{1.72}$$

where $a_1, a_2, \ldots, a_m$ represent row vectors of A. Then, the product AB can be expressed as:

$$AB = \begin{bmatrix} a_1 B \\ --- \\ a_2 B \\ --- \\ \vdots \\ --- \\ a_m B \end{bmatrix}_{m \times r} \tag{1.73}$$

Example 1.24

Consider $A = \begin{bmatrix} 2 & 4 \\ 3 & 1 \\ 1 & 6 \end{bmatrix}_{3\times 2}$ and $B = \begin{bmatrix} 3 & 2 & 1 \\ 1 & 2 & 4 \end{bmatrix}_{2\times 3}$. Find the row partitioned product AB.

Solution:

$$AB = \begin{bmatrix} a_1 B \\ a_2 B \\ a_3 B \end{bmatrix} = \begin{bmatrix} [2 \quad 4] \begin{bmatrix} 3 & 2 & 1 \\ 1 & 2 & 4 \end{bmatrix} \\ [3 \quad 1] \begin{bmatrix} 3 & 2 & 1 \\ 1 & 2 & 4 \end{bmatrix} \\ [1 \quad 6] \begin{bmatrix} 3 & 2 & 1 \\ 1 & 2 & 4 \end{bmatrix} \end{bmatrix} = \begin{bmatrix} 10 & 12 & 18 \\ 10 & 8 & 7 \\ 9 & 14 & 25 \end{bmatrix}$$

1.7 Block Multiplication

Let A be an $m \times n$ matrix and B an $n \times r$ matrix. The following are different ways of computing the product AB.

Case 1: Partition B into $B = \begin{bmatrix} \underbrace{B_1}_{n \times t} & | & \underbrace{B_2}_{n \times (r-t)} \end{bmatrix}_{n \times r}$. Then the product AB can be written as follows:

$$AB = A\begin{bmatrix} b_1 & b_2 & \ldots & b_t & ; & b_{t+1} & b_{t+2} & \ldots & b_{r-1} & b_r \end{bmatrix}$$
$$= \begin{bmatrix} A(b_1 \; b_2 \; \ldots \; b_t) & ; & A(b_{t+1} \; \ldots \; b_r) \end{bmatrix} \quad (1.74)$$
$$= \begin{bmatrix} AB_1 & AB_2 \end{bmatrix}$$

Therefore:

$$A\begin{bmatrix} B_1 & B_2 \end{bmatrix} = \begin{bmatrix} AB_1 & AB_2 \end{bmatrix} \quad (1.75)$$

Case 2: Partition A into $A = \begin{bmatrix} (A_1)_{k \times n} \\ \hline (A_2)_{(m-k) \times n} \end{bmatrix}_{m \times n}$. Then:

$$AB = \begin{bmatrix} A_1 \\ A_2 \end{bmatrix} B = \begin{bmatrix} a_1 \\ a_2 \\ \vdots \\ a_k \\ a_{k+1} \\ \vdots \\ a_m \end{bmatrix} B = \begin{bmatrix} a_1 B \\ a_2 B \\ \vdots \\ a_k B \\ a_{k+1} B \\ \vdots \\ a_m B \end{bmatrix} = \begin{bmatrix} \begin{pmatrix} a_1 \\ \vdots \\ a_k \end{pmatrix} B \\ \begin{pmatrix} a_{k+1} \\ \vdots \\ a_m \end{pmatrix} B \end{bmatrix} = \begin{bmatrix} A_1 B \\ A_2 B \end{bmatrix}$$

$$\begin{bmatrix} A_1 \\ A_2 \end{bmatrix} B = \begin{bmatrix} A_1 B \\ A_2 B \end{bmatrix} \qquad (1.76)$$

Case 3: Let $A = \begin{bmatrix} (A_1)_{m \times s} & (A_2)_{m \times (n-s)} \end{bmatrix}$ and $B = \begin{bmatrix} (B_1)_{s \times r} \\ (B_2)_{(n-s) \times r} \end{bmatrix}$

If $C = AB$, then:

$$C_{ij} = \sum_{l=1}^{n} a_{il} b_{lj} = \underbrace{\sum_{l=1}^{s} a_{il} b_{lj}}_{A_1 B_1} + \underbrace{\sum_{l=s+1}^{n} a_{il} b_{lj}}_{A_2 B_2}$$

Therefore:

$$AB = C = \begin{bmatrix} A_1 & A_2 \end{bmatrix} \begin{bmatrix} B_1 \\ B_2 \end{bmatrix} = A_1 B_1 + A_2 B_2 \qquad (1.77)$$

Case 4: Let A and B be partitioned as:

$$A = \begin{bmatrix} A_{11} & A_{12} \\ \hline A_{21} & A_{22} \end{bmatrix} \text{ and } B = \begin{bmatrix} B_{11} & B_{12} \\ \hline B_{21} & B_{22} \end{bmatrix}$$

For convenience, we define:

$$A_1 = \begin{bmatrix} A_{11} \\ A_{21} \end{bmatrix}, A_2 = \begin{bmatrix} A_{12} \\ A_{22} \end{bmatrix}, B_1 = \begin{bmatrix} B_{11} & B_{12} \end{bmatrix}, B_2 = \begin{bmatrix} B_{21} & B_{22} \end{bmatrix}$$

Then:

$$AB = \begin{bmatrix} A_1 & | & A_2 \end{bmatrix} \begin{bmatrix} B_1 \\ \hline B_2 \end{bmatrix} = A_1 B_1 + A_2 B_2$$

where

$$A_1 B_1 = \begin{bmatrix} A_{11} \\ A_{21} \end{bmatrix} \begin{bmatrix} B_{11} & B_{12} \end{bmatrix} = \begin{bmatrix} A_{11} B_{11} & A_{11} B_{12} \\ A_{21} B_{11} & A_{21} B_{12} \end{bmatrix}$$

and

$$A_2 B_2 = \begin{bmatrix} A_{12} \\ A_{22} \end{bmatrix} \begin{bmatrix} B_{21} & B_{22} \end{bmatrix} = \begin{bmatrix} A_{12} B_{21} & A_{12} B_{22} \\ A_{22} B_{21} & A_{22} B_{22} \end{bmatrix}$$

Therefore:

$$AB = \begin{bmatrix} A_{11} & A_{12} \\ A_{21} & A_{22} \end{bmatrix} \begin{bmatrix} B_{11} & B_{12} \\ B_{21} & B_{22} \end{bmatrix} = \begin{bmatrix} A_{11} B_{11} + A_{12} B_{21} & A_{11} B_{12} + A_{12} B_{22} \\ A_{21} B_{11} + A_{22} B_{21} & A_{21} B_{12} + A_{22} B_{22} \end{bmatrix} \qquad (1.78)$$

Matrices, Matrix Algebra, and Elementary Matrix Operations

In general, if the blocks have the proper dimensions, block multiplication is similar to matrix multiplication. For example:

$$\text{if } A = \begin{bmatrix} A_{11} & \cdots & A_{1t} \\ \vdots & \ddots & \vdots \\ A_{s1} & \cdots & A_{st} \end{bmatrix} \text{ and } B = \begin{bmatrix} B_{11} & \cdots & B_{1r} \\ \vdots & \ddots & \vdots \\ B_{t1} & \cdots & B_{tr} \end{bmatrix}, \text{ then:}$$

$$AB = C = \begin{bmatrix} C_{11} & \cdots & C_{1r} \\ \vdots & \ddots & \vdots \\ C_{s1} & \cdots & C_{sr} \end{bmatrix} \tag{1.79}$$

where

$$C_{ij} = \sum_{k=1}^{t} A_{ik} B_{kj} \tag{1.80}$$

Note that the product AB exists only if the number of columns of A_{ik} is equal to the number of rows of B_{kj}.

Example 1.25

Let $A = \begin{bmatrix} 1 & 1 & 1 & 1 \\ 2 & 2 & 1 & 1 \\ 3 & 3 & 2 & 2 \end{bmatrix}$ and $B = \begin{bmatrix} 1 & 1 & 1 & 1 \\ 1 & 2 & 1 & 1 \\ 3 & 1 & 1 & 1 \\ 3 & 2 & 1 & 2 \end{bmatrix}$

Compute AB by using block multiplication.

Solution: Partition A and B as:

$$A = \left[\begin{array}{cc|cc} 1 & 1 & 1 & 1 \\ 2 & 2 & 1 & 1 \\ 3 & 3 & 2 & 2 \end{array}\right] \quad B = \left[\begin{array}{cc|cc} 1 & 1 & 1 & 1 \\ 1 & 2 & 1 & 1 \\ \hline 3 & 1 & 1 & 1 \\ 3 & 2 & 1 & 2 \end{array}\right]$$

Then:

$$AB = \left[\begin{array}{cc|cc} 1 & 1 & 1 & 1 \\ 2 & 2 & 1 & 1 \\ 3 & 3 & 2 & 2 \end{array}\right] \left[\begin{array}{cc|cc} 1 & 1 & 1 & 1 \\ 1 & 2 & 1 & 1 \\ \hline 3 & 1 & 1 & 1 \\ 3 & 2 & 1 & 2 \end{array}\right] = \left[\begin{array}{cc|cc} 8 & 6 & 4 & 5 \\ 10 & 9 & 6 & 7 \\ 18 & 15 & 10 & 12 \end{array}\right]$$

Matrix A can also be partitioned as:

$$A = \left[\begin{array}{cc|cc} 1 & 1 & 1 & 1 \\ 2 & 2 & 1 & 1 \\ 3 & 3 & 2 & 2 \end{array}\right]$$

In this case, the product AB is computed as

$$AB = \begin{bmatrix} 1 & 1 & 1 & 1 \\ 2 & 2 & 1 & 1 \\ 3 & 3 & 2 & 2 \end{bmatrix} \begin{bmatrix} 1 & 1 & 1 & 1 \\ 1 & 2 & 1 & 1 \\ 3 & 1 & 1 & 1 \\ 3 & 2 & 1 & 2 \end{bmatrix} = \begin{bmatrix} 8 & 6 & 4 & 5 \\ 10 & 9 & 6 & 7 \\ 18 & 15 & 10 & 12 \end{bmatrix}$$

1.8 Inner, Outer, and Kronecker Products

Inner, outer, and Kronecker products are useful operations in signal processing, image processing and communication systems. They are defined as shown below.

1.8.1 Inner Product

Given two $n \times 1$ vectors x and y in R^n, the inner (scalar) product $<x,y>$ is defined as:

$$<x,y> = x^T y = \begin{bmatrix} x_1 & x_2 & \cdots & x_n \end{bmatrix} \begin{bmatrix} y_1 \\ y_2 \\ \vdots \\ y_n \end{bmatrix} = x_1 y_1 + x_2 y_2 + \ldots + x_n y_n \quad (1.81)$$

Example 1.26

Let $x = [1 \ 2 \ -3]^T$ and $y = [3 \ -2 \ 1]^T$. Find the inner product.

Solution:

$$x^T y = x_1 y_1 + x_2 y_2 + x_3 y_3 = 3 - 4 - 3 = -4$$

Example 1.27

Let $X_1, X_2, \ldots, X_N$ be N jointly Gaussian random variables. Let us define the random vector X as follows: $X = [X_1 \ X_2 \ \ldots \ X_N]^T$. The mean and covariance matrix of the random vector X are denoted by m and K and are defined as:

$$m = \begin{bmatrix} E(X_1) & E(X_2) & \cdots & E(X_N) \end{bmatrix}^T \quad (1.82)$$

$$K = E[(X-m)(X-m)^T] \quad (1.83)$$

where E is the expectation operator. The joint probability density function (PDF) of random vector X is given by:

Matrices, Matrix Algebra, and Elementary Matrix Operations 39

$$p_X(x) = \frac{1}{(2\pi)^{N/2}|K|^{1/2}} \exp\left(\frac{-1}{2}[x-m]^T K^{-1}[x-m]\right) \quad (1.84)$$

where $|K|$ stands for the determinant of matrix K and K^{-1} represents its inverse. Notice that the term in the exponential is the weighted inner product of $x-m$ with $x-m$. The computation of the determinants and inverse are discussed in detail in Chapter 2.

Example 1.28

The Euclidean distance between two vectors x and y in an N-dimensional space is given by:

$$d(x,y) = \left[(x-y)^T(x-y)\right]^{\frac{1}{2}} \quad (1.85)$$

For example, if $x = \begin{bmatrix} 1 \\ 2 \end{bmatrix}$ and $y = \begin{bmatrix} 3 \\ 4 \end{bmatrix}$ then

$$d(x,y) = \left[\begin{pmatrix} 1-3 & 2-4 \end{pmatrix}\begin{pmatrix} 1-3 \\ 2-4 \end{pmatrix}\right]^{\frac{1}{2}} = \left[\begin{pmatrix} -2 & -2 \end{pmatrix}\begin{pmatrix} -2 \\ -2 \end{pmatrix}\right]^{\frac{1}{2}} = (8)^{\frac{1}{2}} = \sqrt{8}$$

1.8.2 Outer Product

The outer product between two vectors x and y in R^n is denoted by $>x,y<$ and is defined as:

$$>x,y<= xy^T = \begin{bmatrix} x_1 \\ x_2 \\ \vdots \\ x_n \end{bmatrix} [y_1 \quad y_2 \quad \cdots \quad y_n] = \begin{bmatrix} x_1y_1 & x_1y_2 & \cdots & x_1y_n \\ \vdots & \ddots & & \vdots \\ \vdots & & \ddots & \\ x_ny_1 & x_ny_2 & & x_ny_n \end{bmatrix} \quad (1.86)$$

Example 1.29

$x = \begin{bmatrix} 1 \\ 2 \\ 3 \end{bmatrix}$ and $y = \begin{bmatrix} 4 \\ 5 \\ 6 \end{bmatrix}$. Find the outer product of x and y.

Solution:

$$xy^T = \begin{bmatrix} 1 \\ 2 \\ 3 \end{bmatrix} [4 \quad 5 \quad 6] = \begin{bmatrix} 4 & 5 & 6 \\ 8 & 10 & 12 \\ 12 & 15 & 18 \end{bmatrix}$$

The covariance matrix $K = E[(x-m)(x-m)^T]$ and the correlation matrix $R = E(xx^T)$ of the random vector x represent examples of outer product of vectors.

1.8.3 Kronecker Products

The Kronecker product of two matrices A and B is defined as:

$$A \otimes B = \begin{bmatrix} a_{11}B & a_{12}B & \cdots & a_{1m}B \\ a_{21}B & a_{22}B & \cdots & a_{2n}B \\ \vdots & \vdots & & \\ a_{n1}B & a_{n2}B & \cdots & a_{nm}B \end{bmatrix} \qquad (1.87)$$

If A is an $n \times m$ matrix and B is a $p \times q$ matrix, then $A \times B$ is an $np \times mq$ matrix. The MATLAB® command to find the Kronecker product of A and B is **kron(A, B)**.

Example 1.30

Find the Kronecker product of matrices A and B.

$$A = \begin{bmatrix} -1 & 2 & 3 \\ 1 & 4 & 2 \end{bmatrix} \quad B = \begin{bmatrix} 2 & 3 \\ -1 & 4 \end{bmatrix}$$

Solution:

$$A \otimes B = \begin{bmatrix} a_{11}B & a_{12}B & \cdots & a_{1m}B \\ a_{21}B & a_{22}B & \cdots & a_{2n}B \\ \vdots & \vdots & & \\ a_{n1}B & a_{n2}B & \cdots & a_{nm}B \end{bmatrix} = \begin{bmatrix} a_{11}B & a_{12}B & a_{13}B \\ a_{21}B & a_{22}B & a_{23}B \end{bmatrix} = \begin{bmatrix} -B & 2B & 3B \\ +B & 4B & 2B \end{bmatrix}$$

$$= \begin{bmatrix} -2 & -3 & 4 & 6 & 6 & 9 \\ 1 & -4 & -2 & 8 & -3 & 12 \\ 2 & 3 & 8 & 12 & 4 & 6 \\ -1 & 4 & -4 & 16 & -2 & 8 \end{bmatrix}$$

Problems

PROBLEM 1.1

For each given matrix, state whether it is square, upper triangular, lower triangular or diagonal.

a. $A = \begin{bmatrix} 1 & 7 & 6 \\ 0 & 8 & 9 \\ 0 & 0 & 6 \end{bmatrix}$;

b. $B = \begin{bmatrix} 4 & 0 & 0 & 0 \\ 6 & 3 & 0 & 0 \\ -1 & -2 & 5 & 0 \\ 3 & 4 & 7 & 8 \\ 9 & 11 & -1 & -2 \end{bmatrix}$;

Matrices, Matrix Algebra, and Elementary Matrix Operations

c. $C = \begin{bmatrix} 3 & 2 & 1 \\ -2 & -1 & 0 \\ 2 & 8 & 10 \end{bmatrix}$; d. $D = \begin{bmatrix} 1 & 0 \\ 0 & 2 \end{bmatrix}$; e. $E = \begin{bmatrix} 1 & -4 & 7 & 8 & 9 \\ 0 & 2 & -1 & 3 & 5 \\ 0 & 0 & 3 & 2 & 1 \end{bmatrix}$

PROBLEM 1.2

If $A_1 = \begin{bmatrix} 1 & 2 \\ 3 & 4 \end{bmatrix}$ and $A_2 = \begin{bmatrix} 1+j2 & 2+j3 \\ 3+j4 & 4+j5 \end{bmatrix}$ compute

A_1^T, $(A_1 + A_2^H)^H$ and A_2^H

PROBLEM 1.3

Given that $A = \begin{bmatrix} 3 & 4 & 5 \\ 3 & -6 & 7 \\ 2 & 1 & 4 \end{bmatrix}$ and $B = \begin{bmatrix} 2 & 1 & 6 \\ -2 & -1 & 4 \\ 5 & 3 & 9 \end{bmatrix}$ compute the following:

a. $A+B$ b. $A-B$ c. AB d. $B-A$
e. BA f. Trace(A) g. Trace(B) h. Trace(A^2B)

PROBLEM 1.4

If $A = \begin{bmatrix} 8 & 0 & -3 & 4 \\ 5 & 5 & 8 & -7 \\ 3 & -2 & -1 & 0 \\ 0 & 7 & 5 & 4 \end{bmatrix}$ and $B = \begin{bmatrix} -3 & 1 & 1 & 10 \\ 3 & 4 & 15 & 12 \\ 0 & -4 & -8 & 9 \\ 2 & 1 & 1 & 2 \end{bmatrix}$ compute

a. $2A$; b. $A+B$; c. $2A-3B$; d. $(2A)^T-(3B)^T$; e. AB;
f. BA; g. A^TB^T; h. $(BA)^T$

PROBLEM 1.5
Prove that:
a. $(A^T)^T = A$ and $(A^H)^H = A$
b. $(AB)^T = B^TA^T$ and $(AB)^H = B^HA^H$

PROBLEM 1.6
Which of the following matrices are in REF?

a. $\begin{bmatrix} 1 & -1 & 0 & 8 & 9 \\ 0 & 1 & 3 & 0 & 0 \end{bmatrix}$; b. $\begin{bmatrix} 1 & 0 & 0 \\ 0 & 0 & 0 \\ 0 & 1 & 0 \end{bmatrix}$; c. $\begin{bmatrix} 1 & 2 & 3 \\ 0 & 0 & 1 \\ 0 & 0 & 0 \end{bmatrix}$;

d. $\begin{bmatrix} 1 & -8 & -7 & 0 & 0 \\ 0 & 0 & 1 & 2 & 3 \\ 0 & 0 & 0 & 1 & 1 \end{bmatrix}$; e. $\begin{bmatrix} 0 & 0 \\ 1 & 0 \end{bmatrix}$; f. $\begin{bmatrix} 1 & 0 & 9 \\ 0 & 0 & 1 \end{bmatrix}$

PROBLEM 1.7
For the following matrices, show that Trace(AB)=Trace(BA)

$$A = \begin{bmatrix} -6 & 2 & -5 \\ 2 & 3 & -1 \end{bmatrix} \quad B = \begin{bmatrix} 2 & 3 \\ -3 & -4 \\ 2 & -6 \end{bmatrix}$$

PROBLEM 1.8
a. Find AB and BA:

$$A = \begin{bmatrix} 1 & 2 & 3 \\ 4 & 5 & 6 \\ 7 & 8 & 9 \end{bmatrix}, \quad B = \begin{bmatrix} 1 & 0 & -1 \\ 0 & 1 & 2 \\ 1 & 2 & -1 \end{bmatrix}$$

b. Is $AB = BA$?

PROBLEM 1.9
Given matrices A and B, compute:
a. AB
b. BA
c. $AB - B^T A^T$

$$A = \begin{bmatrix} 1 & 2 & 3 \\ 0 & 1 & 2 \\ 0 & 0 & 1 \end{bmatrix}, \quad B = \begin{bmatrix} 1 & 0 & 0 \\ 2 & 1 & 0 \\ 3 & 2 & 1 \end{bmatrix}$$

PROBLEM 1.10
Given that

$$A = \begin{bmatrix} 3 & 2 & 8 & -12 \\ 9 & 0 & 0 & 3 \\ 2 & 6 & 5 & 8 \\ -2 & -4 & 2 & 1 \end{bmatrix}, \quad B = \begin{bmatrix} -4 & 0 & 0 & 0 \\ 3 & 3 & 7 & -2 \\ 1 & 7 & 6 & -5 \\ 2 & -1 & 0 & 9 \end{bmatrix} \text{ and }$$

$$C = \begin{bmatrix} 6 & 1 & -10 & 0 \\ 0 & 5 & -5 & 4 \\ 0 & 0 & 2 & 1 \\ 3 & 7 & 6 & 1 \end{bmatrix} \text{ show that:}$$

a. $A(B+C) = AB + AC$
b. $A(BC) = (AB)C$

PROBLEM 1.11

Let $A = \begin{bmatrix} \frac{1}{2} & -\frac{1}{2} \\ -\frac{1}{2} & \frac{1}{2} \end{bmatrix}$. Compute A^2 and A^3. What can be said about A^n?

PROBLEM 1.12
Show that the inverse of the DFT matrix A is $A^{-1}=(1/N)A^H$, where:

$$A = \begin{bmatrix} W_N^0 & W_N^0 & W_N^0 & \cdots & W_N^0 \\ W_N^0 & W_N^1 & W_N^2 & \cdots & W_N^{N-1} \\ \vdots & \vdots & \vdots & \cdots & \\ W_N^0 & W_N^{N-2} & W_N^{2(N-2)} & \cdots & W_N^{(N-2)(N-1)} \\ W_N^0 & W_N^{N-1} & W_N^{2(N-1)} & \cdots & W_N^{(N-1)(N-1)} \end{bmatrix}$$

PROBLEM 1.13
For each of the following systems of equations, use the technique of Gaussian elimination to solve the system.

a. $\begin{aligned} 3x_1 - 6x_2 &= 9 \\ 4x_1 - 2x_2 &= 18 \end{aligned}$; b. $\begin{aligned} 3x_1 + 6x_2 + 3x_3 &= 6 \\ -x_1 - x_2 + 2x_3 &= 3; \\ 4x_1 + 6x_2 &= 0 \end{aligned}$ c. $\begin{aligned} -2x_1 + 4x_2 - 2x_3 &= 4 \\ -2x_1 + 2x_2 + x_3 &= 4 \\ 6x_1 + 4x_2 + 4x_3 &= 10 \\ -3x_1 + 8x_2 + 5x_3 &= 17 \end{aligned}$

PROBLEM 1.14
Write the transformation matrices for each of the following transformations.
a. $R_2 \to R_2 - 4R_1$
b. $R_3 \to -16R_3$
c. $C_1 \to C_1 - 2C_3$
d. $C_2 \to 3C_2$

PROBLEM 1.15
Show that if a matrix is skew-symmetric ($A^T = -A$), then its diagonal entries must all be zero.

PROBLEM 1.16
Let A be an $n \times n$ matrix and let $B = A + A^T$ and $C = A - A^T$. Show that:

a. B is symmetric and C is skew-symmetric.
b. Every $n \times n$ matrix can be represented as a sum of a symmetric and skew-symmetric matrix.

PROBLEM 1.17
For each of the following pairs of matrices, find an elementary matrix E such that $EA = B$.

a. $A = \begin{bmatrix} 3 & 8 \\ 5 & 16 \end{bmatrix}$, $B = \begin{bmatrix} 12 & 32 \\ 5 & 16 \end{bmatrix}$

b. $A = \begin{bmatrix} 7 & 6 & 3 \\ 2 & 0 & 9 \\ -8 & 7 & 4 \end{bmatrix}$, $B = \begin{bmatrix} -8 & 7 & 4 \\ 2 & 0 & 9 \\ 7 & 6 & 3 \end{bmatrix}$

c. $A = \begin{bmatrix} 1 & 4 & 5 & 9 \\ 0 & 6 & 2 & 1 \\ -4 & 5 & 2 & -3 \\ 3 & -3 & 9 & 0 \end{bmatrix}$, $B = \begin{bmatrix} 1 & 4 & 5 & 9 \\ 6 & 0 & 20 & 1 \\ -4 & 5 & 2 & -3 \\ 3 & -3 & 9 & 0 \end{bmatrix}$

PROBLEM 1.18

For each of the following pairs of matrices, find an elementary matrix E such that $AE=B$.

a. $A = \begin{bmatrix} 5 & 2 \\ 3 & -6 \end{bmatrix}$, $B = \begin{bmatrix} 5 & -8 \\ 3 & -12 \end{bmatrix}$

b. $A = \begin{bmatrix} 2 & 4 & 0 \\ -1 & -3 & 2 \\ -5 & 6 & 10 \end{bmatrix}$, $B = \begin{bmatrix} 4 & 2 & 0 \\ -3 & -1 & 2 \\ 6 & -5 & 10 \end{bmatrix}$

c. $A = \begin{bmatrix} 3 & 8 & 7 & 3 \\ 2 & -1 & 0 & 6 \\ 3 & 2 & -4 & -5 \\ 7 & -9 & 6 & 5 \end{bmatrix}$, $B = \begin{bmatrix} 10 & 8 & 7 & 3 \\ 2 & -1 & 0 & 6 \\ -1 & 2 & -4 & -5 \\ 13 & -9 & 6 & 5 \end{bmatrix}$

PROBLEM 1.19

Is the transpose of an elementary matrix the same type of elementary matrix? Is the product of two elementary matrices an elementary matrix?

PROBLEM 1.20

Let $B = A^T A$. Show that $b_{ij} = a_i^T a_j$.

PROBLEM 1.21

Let
$$A = \begin{bmatrix} 3 & 4 & 5 \\ 2 & -1 & -7 \\ 4 & 6 & -8 \end{bmatrix} \text{ and } B = \begin{bmatrix} 5 & 7 & -1 & -2 \\ 0 & 8 & 3 & -6 \\ 4 & 2 & -1 & 0 \end{bmatrix}.$$

a. Calculate Ab_1 and Ab_2.
b. Calculate AB and verify that the column vectors are as obtained in (a).

PROBLEM 1.22

Let

$I = \begin{bmatrix} 1 & 0 \\ 0 & 1 \end{bmatrix}$, $E = \begin{bmatrix} 0 & 1 \\ 1 & 0 \end{bmatrix}$, $O = \begin{bmatrix} 0 & 0 \\ 0 & 0 \end{bmatrix}$, $A = \begin{bmatrix} 4 & 3 \\ 2 & -6 \end{bmatrix}$, $B = \begin{bmatrix} 3 & -1 \\ -2 & 8 \end{bmatrix}$ and

$C = \begin{bmatrix} C_{11} & C_{12} \\ C_{21} & C_{22} \end{bmatrix} = \begin{bmatrix} 1 & 3 & 3 & 5 \\ 1 & 4 & -1 & 3 \\ 1 & 1 & 0 & -4 \\ 2 & 2 & 5 & 6 \end{bmatrix}$

Matrices, Matrix Algebra, and Elementary Matrix Operations

Perform each of the following block multiplications:

a. $\begin{bmatrix} O & I \\ I & O \end{bmatrix} \begin{bmatrix} C_{11} & C_{12} \\ C_{21} & C_{22} \end{bmatrix}$

b. $\begin{bmatrix} A & O \\ O & A \end{bmatrix} \begin{bmatrix} C_{11} & C_{12} \\ C_{21} & C_{22} \end{bmatrix}$

c. $\begin{bmatrix} B & O \\ O & I \end{bmatrix} \begin{bmatrix} C_{11} & C_{12} \\ C_{21} & C_{22} \end{bmatrix}$

d. $\begin{bmatrix} E & O \\ O & E \end{bmatrix} \begin{bmatrix} C_{11} & C_{12} \\ C_{21} & C_{22} \end{bmatrix}$

PROBLEM 1.23

Write each of the following systems of equations as a matrix equation:

a. $\begin{cases} 5x_1 + 6x_2 = 1 \\ 8x_1 - 3x_2 = 9 \end{cases}$

b. $\begin{cases} x_1 + x_3 = 4 \\ 3x_2 + 8x_3 = 10 \\ x_1 - 5x_2 - 7x_3 = 0 \end{cases}$

c. $\begin{cases} x_1 + x_2 + x_3 + x_4 = 6 \\ x_3 + 7x_4 = 9 \\ x_2 + 5x_3 - 2x_4 = 4 \\ x_1 - 2x_2 - 5x_3 = 7 \end{cases}$

PROBLEM 1.24

Use back-substitution to solve the following systems of equations:

a. $\begin{cases} x_1 + 5x_2 = 10 \\ 2x_2 = 4 \end{cases}$

b. $\begin{cases} x_1 + 3x_2 - 6x_3 = 12 \\ x_2 - 9x_3 = 8 \\ x_3 = 5 \end{cases}$

c. $\begin{cases} 5x_1 + 4x_2 - 8x_3 + 7x_4 - 5x_5 = 6 \\ x_2 - 10x_3 + 2x_4 - x_5 = 9 \\ x_3 + 16x_4 + 6x_5 = 6 \\ 3x_4 - 4x_5 = 0 \\ x_5 = 9 \end{cases}$

PROBLEM 1.25
Find the inner product for the following vectors:

a. $x=[1\ 8\ 9\ 0]^T$ and $y=[3\ 9\ 10\ -5]^T$
b. $x=[8\ -9\ -2\ 0\ 0\ -1\ 3\ 4]^T$ and $y=[6\ 6\ -5\ 0\ 2\ 2\ -1\ 0]^T$

PROBLEM 1.26
Compute the outer product for each of the given combinations of vectors:

a. $x=[34\ 5\ 6]^T$ and $y=[-1\ 0\ 5]^T$
b. $x=[4\ 3\ -7\ 0\ 6]^T$ and $y=[2\ 1\ 0\ -9\ 3]^T$

PROBLEM 1.27
Solve the following set of linear equations:

a. $\begin{cases} 2x_1+3x_2+x_3=11 \\ -2x_1+x_2+x_3=3 \\ 4x_1-3x_2+4x_3=10 \end{cases}$;
b. $\begin{cases} 2x_1+3x_2+4x_3+x_4=24 \\ x_2+x_3-3x_4=18 \\ 4x_3+5x_4=10 \\ x_1-x_3=7 \end{cases}$;
c. $\begin{cases} 3x_1+2x_2+x_3=0 \\ -2x_1+x_2-x_3=2 \\ 2x_1-x_2+2x_3=-1 \end{cases}$

d. $\begin{cases} x_1+2x_2-2x_3=9 \\ 2x_1+5x_2+x_3=9 \\ x_1+3x_2+4x_3=-2 \end{cases}$;
e. $\begin{cases} 3x_1-4x_2+4x_3=-15 \\ 3x_1+2x_2+37x_3=0 \\ -4x_1+6x_2-5x_3=25 \end{cases}$;
f. $\begin{cases} 3x_1+x_2+x_3=0 \\ x_1+2x_2-3x_3=-4 \\ -4x_2-6x_3=26 \end{cases}$

PROBLEM 1.28
a. Write the PDF of a Gaussian random vector of size 3×1 having a mean of $E(X)=[1\ -1\ 2]^T$ and covariance matrix $K=\begin{bmatrix} 2 & 0 & 0 \\ 0 & 2 & 0 \\ 0 & 0 & 1 \end{bmatrix}$

b. Repeat part (a) if the mean vector is zero and the covariance matrix is

$$K=\begin{bmatrix} 2 & 1 & 0 \\ 1 & 2 & -1 \\ 0 & -1 & 1 \end{bmatrix}$$

PROBLEM 1.29
a. Find a 2×2 nonzero matrix such that $A^2=0$.
b. Find a 3×3 nonzero matrix such that $A^3=0$.

PROBLEM 1.30
Let

$$B = \begin{bmatrix} 0 & 1 & 0 & 0 \\ 0 & 0 & 1 & 0 \\ 0 & 0 & 0 & 1 \\ 0 & 0 & 0 & 0 \end{bmatrix} \quad \text{and} \quad C = \begin{bmatrix} 0 & 1 & 0 & 0 & 0 \\ 0 & 0 & 1 & 0 & 0 \\ 0 & 0 & 0 & 1 & 0 \\ 0 & 0 & 0 & 0 & 1 \\ 0 & 0 & 0 & 0 & 0 \end{bmatrix}$$

Show that $B^4 = 0$ and $C^5 = 0$.

PROBLEM 1.31
Using the results of Problem 1.30, construct a nonzero 6×6 matrix A such that $A^6 = 0$.

PROBLEM 1.32 (MATLAB)
Find the Kronecker product of matrices A and B where:

$$A = \begin{bmatrix} 2 & -1 & 3 \\ 0 & 4 & -2 \\ 6 & 5 & -2 \end{bmatrix}, \quad B = \begin{bmatrix} 1 & 0 \\ -1 & 2 \end{bmatrix}$$

PROBLEM 1.33 (MATLAB)
a. Use MATLAB command $f = \text{round}(255*\text{round}(16,16))$ to generate a random 8-bit 16×16 image. Use the convolution command $g = \text{conv2}(f, h)$ to filter the image using FIR filter h given by $h = \begin{bmatrix} 1 & -1 \\ 1 & 1 \end{bmatrix}$

b. Use the approach of Section 1.3.3.2 to find g. Compare your approach with result of part (a).

c. Repeat parts (a) and (b) for the same image and filter $h = \begin{bmatrix} 1 & 1 \\ 1 & 1 \end{bmatrix}$

PROBLEM 1.34 (MATLAB)
a. Use MATLAB command $f = \text{imread}(\text{'cameraman.tif'});$ to read the cameraman image into MATLAB. Take the Fourier transform of the image and display its magnitude as an intensity image using log scale. You can use the following MATLAB code

```
f=imread('cameraman.tif');
f=double(f)/255;
F=abs(fft2(f));
F=fftshift(F);
a=max(max(F));
b=min(min(F));
Flog=log(1+F-b)/log(1+a-b);
imshow(Flog)
```

b. Use the approach of Section 1.3.3.3 to find F. Compare the result with part (a).

PROBLEM 1.35 (MATLAB)

Let A be a 8×8 symmetric matrix with $a_{ij}=2(0.9)^{|i-j|}$ and B be another 8×8 symmetric matrix with $b_{ij}=3(0.8)^{|i-j|}$. Find:

a. c_{ij} and d_{ij} if $C=AB$ and $D=BA$.
b. Trace(A), Trace(B), and Trace(AB)

PROBLEM 1.36 (MATLAB)

Let A be a 16×16 symmetric matrix with $a_{ij}=2\cos 0.4(i-j)\pi$ and let B be another 16×16 symmetric matrix with $b_{ij}=4\cos 0.5(i-j)\pi$. Find:

a. c_{ij} and d_{ij} if $C=A^2$ and $D=BA$.
b. Trace(A), Trace(B), Trace(AB), and Trace(BA).

PROBLEM 1.37 (MATLAB)

The following matrix equation is known as the Lypunov equation, where A is an $n\times n$ matrix, Q is an $n\times n$ symmetric matrix, and P is an unknown $n\times n$ matrix:

$$A^T P + PA = -Q$$

a. Show that the unknown matrix P is symmetric.
b. Solve the equation for P if $Q=I$ and $A = \begin{bmatrix} 1 & 1 \\ 2 & 3 \end{bmatrix}$

c. Repeat part (b) if $A = \begin{bmatrix} 0 & 1 & 0 \\ 0 & 0 & 1 \\ -2 & -3 & -4 \end{bmatrix}$ and $Q=I$. Hint: The MATLAB command to solve the Lypunov equation is $P=\text{lyap}(A,Q)$.

PROBLEM 1.38 (MATLAB)

The following matrix equation is known as the Riccati equation where A is an $n\times n$ matrix, Q and R are $n\times n$ symmetric matrices, B is $n\times 1$, and P is an unknown $n\times n$ matrix:

$$A^T P + PA - PBRB^T P = -Q$$

a. Show that the unknown matrix P is symmetric.
b. Solve the Riccati equation for P if $Q=R=I$, $B = \begin{bmatrix} 1 \\ -1 \end{bmatrix}$ and $A = \begin{bmatrix} 1 & 1 \\ 2 & 3 \end{bmatrix}$.

c. Repeat part (b) if $Q=R=I$, $B = \begin{bmatrix} 0 \\ 0 \\ 1 \end{bmatrix}$ and $A = \begin{bmatrix} 0 & 1 & 0 \\ 0 & 0 & 1 \\ -2 & -3 & -4 \end{bmatrix}$

Hint: The MATLAB command to solve the Riccati equation is $[K,P,E]=\text{lqr}(A,B,Q,R,0)$.

2

Determinants, Matrix Inversion and Solutions to Systems of Linear Equations

2.1 Introduction

Systems of linear equations arise frequently in many different engineering disciplines. For example, in electrical engineering, direct current (DC) and alternating current (AC) circuits are analyzed by solving a set of linear algebraic equations with real (DC circuits) or complex (AC circuits) coefficients. In mechanical engineering, finite element analysis is used to convert a set of partial differential equations into a set of linear equations. Most of the differential-integral equations encountered in the field of electromagnetics are solved by using the method of moments, which converts these equations into linear algebraic equations. In this chapter, we consider linear equations of the form $Ax=b$, where A is an $n \times n$ square matrix with real or complex coefficients and x and b are $n \times 1$ vectors in R^n or C^n. The existence and uniqueness of the solution as well as different methodologies for solving a set of linear equations will be thoroughly discussed.

2.2 Determinant of a Matrix

Let A be a square matrix of size $n \times n$. We associate with A a scalar quantity known as the *determinant* of A and denoted as det(A). The value of det(A) is an indication of the singularity or invertability of matrix A. A square matrix A is said to be nonsingular if it has an inverse; That is if there exists a matrix denoted by A^{-1} such that:

$$AA^{-1}=A^{-1}A=I \qquad (2.1)$$

where I is the $n \times n$ identity matrix. The existence of A^{-1} is directly related to the determinant of A. To define the determinant of a matrix, we present a

49

case by case analysis by first studying matrices of sizes 1×1, 2×2, and 3×3 and then extending it to matrices of size $n\times n$ when $n\geq 4$.

Case 1: Matrix A is 1×1

If $A=[a_{11}]$, then $\det(A)=a_{11}$. If $\det(A)\neq 0$, then matrix A has an inverse and A is said to be nonsingular. The inverse in this case is simply $A^{-1}=[1/a_{11}]$ and the solution to $Ax=b$ is unique and is given by $x=b/a_{11}$. If $\det(A)=a_{11}=0$ then, the algebraic equation $Ax=b$ has no solution.

Case 2: Matrix A is 2×2.

Let the matrix A be defined as follows: $A = \begin{bmatrix} a_{11} & a_{12} \\ a_{21} & a_{22} \end{bmatrix}$. To establish the condition of singularity of this matrix, we consider a set of linear equations of the form $Ax=b$. For simplicity, assume that $b=0$, then A is nonsingular if $Ax=0$ has one solution, which is the trivial solution $x=0$. In matrix form, we have:

$$\begin{bmatrix} a_{11} & a_{12} \\ a_{21} & a_{22} \end{bmatrix}\begin{bmatrix} x_1 \\ x_2 \end{bmatrix} = \begin{bmatrix} 0 \\ 0 \end{bmatrix} \qquad (2.2)$$

The above matrix equation can be written as two linear coupled equations given as:

$$a_{11}x_1 + a_{12}x_2 = 0 \qquad (2.3)$$

and

$$a_{21}x_2 + a_{22}x_2 = 0 \qquad (2.4)$$

Multiplying both sides of Equation 2.3 by a_{22} and Equation 2.4 by a_{12} results in:

$$a_{22}\,a_{11}x_1 + a_{22}\,a_{12}x_2 = 0 \qquad (2.5)$$

$$a_{12}\,a_{21}x_1 + a_{22}\,a_{12}x_2 = 0 \qquad (2.6)$$

Subtracting Equation 2.6 from Equation 2.5 yields:

$$(a_{11}a_{22} - a_{12}a_{21})x_1 = 0 \qquad (2.7)$$

Equation 2.7 has a unique solution $x_1=0$ if the coefficient of x_1 is nonzero, that is:

$$a_{11}a_{22} - a_{12}a_{21} \neq 0 \qquad (2.8)$$

Therefore, equation $Ax=0$ has a unique solution $x=0$ if $a_{11}a_{22}-a_{12}a_{21}\neq 0$. Hence, the determinant of the 2×2 matrix A is defined as:

Determinants, Matrix Inversion and Solutions 51

$$\det(A) = a_{11}a_{22} - a_{12}a_{21} \qquad (2.9)$$

Therefore matrix A is nonsingular if $\det(A) = a_{11}a_{22} - a_{12}a_{21} \neq 0$.

Case 3: Matrix A is 3×3

$$\text{Let } A = \begin{bmatrix} a_{11} & a_{12} & a_{13} \\ a_{21} & a_{22} & a_{23} \\ a_{31} & a_{32} & a_{33} \end{bmatrix}$$

and consider a set of linear equations of the form $Ax=0$. Then A is nonsingular if $Ax=0$ has a unique solution $x=0$. Expanding the matrix equation $Ax=0$ yields:

$$a_{11}x_1 + a_{12}x_2 + a_{13}x_3 = 0 \qquad (2.10)$$

$$a_{21}x_1 + a_{22}x_2 + a_{23}x_3 = 0 \qquad (2.11)$$

$$a_{31}x_1 + a_{32}x_2 + a_{33}x_3 = 0 \qquad (2.12)$$

Eliminating x_2 and x_3 from the above equations yields:

$$[a_{11}(a_{22}a_{33} - a_{23}a_{32}) - a_{12}(a_{21}a_{33} - a_{23}a_{31}) + a_{13}(a_{21}a_{32} - a_{22}a_{31})]x_1 = 0 \qquad (2.13)$$

Equation 2.13 has a unique solution $x_1 = 0$ if the coefficient of x_1 is nonzero, that is:

$$a_{11}(a_{22}a_{33} - a_{23}a_{32}) - a_{12}(a_{21}a_{33} - a_{23}a_{31}) + a_{13}(a_{21}a_{32} - a_{22}a_{31}) \neq 0 \qquad (2.14)$$

Therefore, equation $Ax=0$ has a unique solution $x=0$ if Equation 2.14 is satisfied. Hence, the determinant of the 3×3 matrix A is defined as:

$$\det(A) = a_{11}(a_{22}a_{33} - a_{23}a_{32}) - a_{12}(a_{21}a_{33} - a_{23}a_{31}) + a_{13}(a_{21}a_{32} - a_{22}a_{31}) \qquad (2.15)$$

Therefore, matrix A is nonsingular if $\det(A) \neq 0$. This approach can be extended to matrices of size 4×4 and higher.

In general, the determinant of an $n \times n$ matrix A is defined as:

$$\det(A) = \sum_{j=1}^{n} (-1)^{i+j} a_{ij} M_{ij} \text{ for any } i=1,2,\ldots,n \qquad (2.16)$$

where M_{ij} is the minor corresponding to a_{ij} and is the determinant of the matrix obtained by deleting the row and column containing a_{ij}. Let $A_{ij} = (-1)^{i+j} M_{ij}$, where A_{ij} is defined as the cofactor of a_{ij}. Then:

$$\det(A) = \sum_{j=1}^{n} (-1)^{i+j} a_{ij} M_{ij} = \sum_{j=1}^{n} a_{ij} A_{ij} \qquad (2.17)$$

Note that applying Equation 2.16 with $i=1$ to the 3×3 matrix A defined in Case 3 results in Equation 2.15 (see Example 2.1).

2.2.1 Properties of Determinant

Let A and B be $n \times n$ matrices and let α be a scalar, then:

(a) $\det(AB) = \det(A)\det(B)$.
(b) $\det(AB) = \det(BA)$.
(c) $\det(A^T) = \det(A)$.
(d) $\det(\alpha A) = \alpha^n \det(A)$.
(e) If A is an $n \times n$ triangular matrix then, $\det(A)$ equals to the product of the diagonal elements of A.
(f) If A has a row or column of zeros then, $\det(A) = 0$.
(g) If A has two identical rows or columns then, $\det(A) = 0$.

The proofs of the above properties are left as an exercise to the reader (see Problem 2.4).

Example 2.1

Find the determinant of the 3×3 matrix A

$$A = \begin{bmatrix} 1 & 2 & 1 \\ 3 & 1 & 2 \\ -1 & 2 & 4 \end{bmatrix}.$$

Solution: Let $i=1$ in Equation 2.16 then:

$$\det(A) = \sum_{j=1}^{n}(-1)^{i+j}a_{ij}M_{ij} = \sum_{j=1}^{n}(-1)^{1+j}a_{1j}M_{1j}$$

Therefore:

$$|A| = \det(A) = (-1)^{1+1}(1) \times \begin{vmatrix} 1 & 2 \\ 2 & 4 \end{vmatrix} + (-1)^{2+1}(2) \times \begin{vmatrix} 3 & 2 \\ -1 & 4 \end{vmatrix} + (-1)^{3+1}(1) \times \begin{vmatrix} 3 & 1 \\ -1 & 2 \end{vmatrix}$$

$$= 1 \times (4-4) - 2 \times (12+2) + 1 \times (6+1) = 1 \times (0) - 2 \times (14) + 1 \times (7)$$

$$= -21$$

Since $\det(A) = -21 \neq 0$, matrix A is nonsingular. The MATLAB® command to compute determinant of matrix A is **det(A)**.

Determinants, Matrix Inversion and Solutions

Example 2.2

Use MATLAB to compute the determinant of the 4×4 matrix B

$$B = \begin{bmatrix} 1 & -2 & 3 & -4 \\ 1 & 2 & -5 & 3 \\ 2 & 0 & 2 & 3 \\ -4 & -2 & 6 & 7 \end{bmatrix}.$$

Solution:

$$B = [1\ -2\ 3\ -4;\ 1\ 2\ -5\ 3;\ 2\ 0\ 2\ 3;\ -4\ -2\ 6\ 7];$$

$$d = \det(B)$$

$$d = 168$$

2.2.2 Row Operations and Determinants

In this section, we describe certain common row operations and their effects on the determinant of a matrix.

2.2.2.1 Interchange of Two Rows

Let A be the 2×2 matrix defined as: $A = \begin{bmatrix} a_{11} & a_{12} \\ a_{21} & a_{22} \end{bmatrix}$ and $E = \begin{bmatrix} 0 & 1 \\ 1 & 0 \end{bmatrix}$ (recall that E is a type 2 matrix formed by interchanging the rows of the identity matrix I), then:

$$EA = \begin{bmatrix} a_{21} & a_{22} \\ a_{11} & a_{12} \end{bmatrix} \qquad (2.18)$$

and

$$\det(EA) = \begin{vmatrix} a_{21} & a_{22} \\ a_{11} & a_{12} \end{vmatrix} = a_{21}a_{12} - a_{22}a_{11} = -\det(A) \qquad (2.19)$$

Therefore, interchanging two rows of a matrix results in a matrix whose determinant equal to the negative of the determinant of the original matrix. The proof is shown by mathematical induction. Consider the matrix A of size 3×3 shown in Case 3 above and let E_{13} be an elementary matrix that interchanges row one and row three of A. Then:

$$\det(E_{13}A) = \det\left(\begin{bmatrix} 0 & 0 & 1 \\ 0 & 1 & 0 \\ 1 & 0 & 0 \end{bmatrix} \begin{bmatrix} a_{11} & a_{12} & a_{13} \\ a_{21} & a_{22} & a_{23} \\ a_{31} & a_{32} & a_{33} \end{bmatrix}\right) = \det\begin{bmatrix} a_{31} & a_{32} & a_{33} \\ a_{21} & a_{22} & a_{23} \\ a_{11} & a_{12} & a_{13} \end{bmatrix} \qquad (2.20)$$

If we proceed to compute the determinant by expanding along the second row, we get:

$$\det(E_{13}A) = -a_{21}\begin{vmatrix} a_{32} & a_{33} \\ a_{12} & a_{13} \end{vmatrix} + a_{22}\begin{vmatrix} a_{31} & a_{33} \\ a_{11} & a_{13} \end{vmatrix} - a_{23}\begin{vmatrix} a_{31} & a_{32} \\ a_{11} & a_{12} \end{vmatrix}$$

$$= a_{21}\begin{vmatrix} a_{12} & a_{13} \\ a_{32} & a_{33} \end{vmatrix} - a_{22}\begin{vmatrix} a_{11} & a_{13} \\ a_{31} & a_{33} \end{vmatrix} + a_{23}\begin{vmatrix} a_{11} & a_{12} \\ a_{31} & a_{32} \end{vmatrix} \quad (2.21)$$

$$= -\det(A)$$

The above approach can be generalized to an $n \times n$ matrix A. If E_{ij} is the $n \times n$ elementary matrix formed by interchanging the ith and j^{th} row of the identity matrix I, then:

$$\det(E_{ij}A) = -\det(A) \quad (2.22)$$

Therefore, for any elementary matrix of Type 2, we have:

$$\det(EA) = -\det(A) = \det(E)\det(A) \quad (2.23)$$

2.2.2.2 Multiplying a Row of A by a Nonzero Constant

Let E be a Type 1 elementary matrix formed by multiplying the ith row of the $n \times n$ identity matrix I by a nonzero constant α. That is:

$$E = \begin{bmatrix} 1 & 0 & 0 & 0 & 0 & 0 & 0 & 0 \\ 0 & 1 & 0 & 0 & 0 & 0 & 0 & 0 \\ 0 & 0 & \ddots & 0 & 0 & 0 & 0 & 0 \\ 0 & 0 & 0 & 1 & 0 & 0 & 0 & 0 \\ 0 & 0 & 0 & 0 & \alpha & 0 & 0 & 0 \\ 0 & 0 & 0 & 0 & 0 & 1 & 0 & 0 \\ 0 & 0 & 0 & 0 & 0 & 0 & \ddots & 0 \\ 0 & 0 & 0 & 0 & 0 & 0 & 0 & 1 \end{bmatrix} \quad (2.24)$$

Note that since E is diagonal, then the determinant of E is: $\det(E) = \alpha$. The matrix generated by the product of E and A is given as:

$$EA = \begin{bmatrix} a_{11} & a_{12} & \cdots & a_{1n-1} & a_{1n} \\ \vdots & \vdots & & \vdots & \vdots \\ \alpha a_{i1} & \alpha a_{i2} & & \alpha a_{in-1} & \alpha a_{in} \\ \vdots & & & & \\ a_{n1} & a_{n2} & & a_{nn-1} & a_{nn} \end{bmatrix} \quad (2.25)$$

Determinants, Matrix Inversion and Solutions

To find det(EA), we utilize Equation 2.17 and expand along the ith row. Then:

$$\det(EA) = \alpha a_{i1}\det(A_{i1}) + \alpha a_{i2}\det(A_{i2}) + \cdots + \alpha a_{in}\det(A_{in})$$
$$= \alpha\left(a_{i1}\det(A_{i1}) + a_{i2}\det(A_{i2}) + \cdots + a_{in}\det(A_{in})\right)$$
$$= \alpha\det(A) \qquad (2.26)$$

Another approach would be to use the fact that the determinant of the product of two matrices is equal to the product of the determinant of the individual matrices. Hence:

$$\det(EA) = \det(E)\det(A) = \alpha\det(A) \qquad (2.27)$$

2.2.2.3 Adding a Multiple of One Row to Another Row

Let E be a Type 3 elementary matrix that is designed to add a multiple of one row to another row. In this case, E is a triangular matrix and $\det(E)$ is the product of its diagonal elements which are all equal to 1. As a result:

$$\det(EA) = \det(E)\det(A) = 1 \times \det(A) = \det(A) \qquad (2.28)$$

The above operations can be summarized as follows. If E is an elementary matrix of Type 1, 2 or 3, then:

$$\det(EA) = \det(E)\det(A) \qquad (2.29)$$

where

$$\det(E) = \begin{cases} \alpha \neq 0 & \text{If } E \text{ is of Type 1} \\ -1 & \text{If } E \text{ is of Type 2} \\ 1 & \text{If } E \text{ is of Type 3} \end{cases} \qquad (2.30)$$

Similar results can be derived for column operations since:

$$\det(AE) = \det((AE)^T) = \det(E^T A^T) = \det(E^T)\det(A^T) = \det(E)\det(A) \qquad (2.31)$$

2.2.3 Singular Matrices

Theorem 2.1: An $n \times n$ matrix A is singular if $\det(A) = 0$

Proof: To determine that matrix A is singular, it suffices to show that $Ax = 0$ has a nontrivial solution $x \neq 0$. This can be demonstrated by reducing matrix A to row echelon form with a finite number of row operations. That is:

$$B = E_k E_{k-1} \ldots E_1 A \qquad (2.32)$$

where B represents the row echelon form matrix deduced from A and E_k, $E_{k-1},...,E_1$ are the corresponding elementary matrices. For $Bx=0$ to have a nontrivial solution, the last row of B must be a zero row vector. Recall that elementary matrices are nonsingular therefore:

$$\det(B)=\det(E_k E_{k-1}...E_1 A)=\det(E_k)\det(E_{k-1})...\det(E_1)\det(A) \quad (2.33)$$

But since $\det(E_i) \neq 0$ (E_i's are not singular), hence:

$$\det(A)=0 \quad \text{if and only if} \quad \det(B)=0 \quad (2.34)$$

Hence, if A is singular, B has a row full of zeros and $\det(B)=0$. The above can also be utilized to compute the determinant of A by first reducing it to REF and then employing the properties of determinants. This is outlined using the following steps:

To compute $\det(A)$, do the following:

(a) Reduce A to row echelon form through the use of elementary matrices. This yields the equation:

$$B=E_k E_{k-1}...E_1 A \quad (2.35)$$

(b) If B has a row of zeros (last row), then A must be singular and $\det(A)=0$.

(c) Otherwise, using the property of determinants, we have:

$$\det(B)=\det(E_k)\det(E_{k-1})...\det(E_1)\det(A) \quad (2.36)$$

$\det(A)$ can then be computed as follows:

$$\det(A) = \frac{\det(B)}{\det(E_k)\det(E_{k-1})...\det(E_1)} \quad (2.37)$$

Note that if A is nonsingular, it can be reduced to triangular form using Type 2 and 3 elementary matrices. This yields the following:

$$T=E_m E_{m-1}...E_1 A \quad (2.38)$$

where

$$T = \begin{bmatrix} t_{11} & & & \text{nonzero} \\ & t_{22} & & \text{elements} \\ & & \ddots & \\ & \text{zeros} & & \\ & & & t_{nn} \end{bmatrix} \quad (2.39)$$

Determinants, Matrix Inversion and Solutions 57

T represents the triangular matrix arrived at in Equation 2.38 above. Then:

$$\det(A) = \frac{1}{\det(E_m)\det(E_{m-1})\ldots\det(E_1)}\det(T) = \pm\det(T) = \pm t_{11}t_{12}\ldots t_{nn} \quad (2.40)$$

The $\pm$ sign depends on the number of times a Type 2 matrix has been employed to arrive at the REF. The sign is positive, if a Type 2 matrix has been used an even number of times, otherwise it is negative.

Example 2.3

Consider the 3×3 matrix A defined as:

$$A = \begin{bmatrix} 1 & 2 & 1 \\ 3 & 1 & 2 \\ -1 & 2 & 4 \end{bmatrix}$$

Find its determinant using the method introduced above.

Solution: We proceed by first reducing matrix A to triangular form and then computing its determinant by utilizing Equation 2.40. This is done by utilizing Type 2 and 3 elementary matrices as follows:

$$\begin{bmatrix} 1 & 2 & 1 \\ 3 & 1 & 2 \\ -1 & 2 & 4 \end{bmatrix} \xrightarrow{R_2 \leftarrow R_2 - 3R_1} \begin{bmatrix} 1 & 2 & 1 \\ 0 & -5 & -1 \\ -1 & 2 & 4 \end{bmatrix} \xrightarrow{R_3 \leftarrow R_3 - (-1)R1} \begin{bmatrix} 1 & 2 & 1 \\ 0 & -5 & -1 \\ 0 & 4 & 5 \end{bmatrix}$$

$$\xrightarrow{R_3 \leftrightarrow R_2} \begin{bmatrix} 1 & 2 & 1 \\ 0 & 4 & 5 \\ 0 & -5 & -1 \end{bmatrix} \xrightarrow{R_3 \leftarrow R_3 + (5/4)R_2} \begin{bmatrix} 1 & 2 & 1 \\ 0 & 4 & 5 \\ 0 & 0 & \frac{21}{4} \end{bmatrix}$$

Therefore:

$$T = \begin{bmatrix} 1 & 2 & 1 \\ 0 & 4 & 5 \\ 0 & 0 & \frac{21}{4} \end{bmatrix}$$

Hence, the triangular matrix T can be expressed mathematically as:

$$T = E_4 E_3 E_2 E_1 A = \begin{bmatrix} 1 & 0 & 0 \\ 0 & 1 & 0 \\ 0 & \frac{5}{4} & 1 \end{bmatrix} \begin{bmatrix} 1 & 0 & 0 \\ 0 & 0 & 1 \\ 0 & 1 & 0 \end{bmatrix} \begin{bmatrix} 1 & 0 & 0 \\ 0 & 1 & 0 \\ 1 & 0 & 1 \end{bmatrix} \begin{bmatrix} 1 & 0 & 0 \\ -3 & 1 & 0 \\ 0 & 0 & 1 \end{bmatrix} \begin{bmatrix} 1 & 2 & 1 \\ 3 & 1 & 2 \\ -1 & 2 & 4 \end{bmatrix}$$

Therefore:

$$\det(A) = -\det(T) = -(1)(4)\left(\frac{21}{4}\right) = -21$$

Note that the sign is negative, since only one elementary matrix of Type 2 matrix (E_3) was utilized.

2.3 Matrix Inversion

Let A be $n \times n$ matrix. The inverse of matrix A, denoted by A^{-1}, is an $n \times n$ matrix that satisfies the following:

$$AA^{-1} = A^{-1}A = I \qquad (2.41)$$

where I is the $n \times n$ identity matrix. If A has an inverse then matrix A is said to be nonsingular. A necessary and sufficient condition for matrix A to be nonsingular is that the determinant of A be nonzero. Under this condition, A^{-1} is computed using the following equation:

$$A^{-1} = \frac{\text{adj}(A)}{\det(A)} \qquad (2.42)$$

where adj(A) stands for the adjoint matrix of A which is defined to be the $n \times n$ matrix of cofactors given by:

$$\text{adj}(A) = \begin{bmatrix} C_{11} & C_{12} & \cdots & C_{1n} \\ C_{21} & C_{22} & \cdots & C_{2n} \\ \vdots & \vdots & \vdots & \vdots \\ C_{n1} & C_{n2} & \cdots & C_{nn} \end{bmatrix}^T \qquad (2.43)$$

where

$$C_{ij} = (-1)^{i+j} \det(M_{ij}) \qquad (2.44)$$

and M_{ij} is the minor corresponding to a_{ij} and is defined as the $(n-1) \times (n-1)$ matrix obtained by eliminating the ith row and j^{th} column of A. If A is a 2×2 matrix shown below:

$$A = \begin{bmatrix} a_{11} & a_{12} \\ a_{21} & a_{22} \end{bmatrix} \qquad (2.45)$$

Then A^{-1} is given by:

$$A^{-1} = \left(\frac{1}{\det(A)}\right) \text{adj}(A) \qquad (2.46)$$

Determinants, Matrix Inversion and Solutions 59

Hence:

$$A^{-1} = \left(\frac{1}{a_{11}a_{22} - a_{12}a_{21}}\right)\begin{bmatrix} a_{22} & -a_{12} \\ -a_{21} & a_{11} \end{bmatrix} \quad (2.47)$$

If A is a diagonal matrix and $\det(A) \neq 0$. Then, A is nonsingular and A^{-1} exists. Note since $\det(A)$ is the product of the diagonal elements of A, this implies that all its diagonal elements are nonzero, i.e. if:

$$A = \begin{bmatrix} d_1 & 0 & \cdots & 0 \\ 0 & d_2 & & 0 \\ \vdots & \vdots & \ddots & \vdots \\ 0 & 0 & \cdots & d_n \end{bmatrix}_{n \times n} \quad (2.48)$$

Then $\det(A) = d_1 d_2 \ldots d_n \neq 0$ only if $d_i \neq 0 \; \forall i$. A^{-1} is then computed as follows:

$$A^{-1} = \begin{bmatrix} \frac{1}{d_1} & 0 & \cdots & 0 & 0 \\ 0 & \frac{1}{d_2} & 0 & \cdots & 0 \\ \vdots & 0 & \ddots & & \vdots \\ 0 & \vdots & & \ddots & 0 \\ 0 & 0 & \cdots & 0 & \frac{1}{d_n} \end{bmatrix} \quad (2.49)$$

Example 2.4

Find the inverse of the matrix

$$A = \begin{bmatrix} 3 & 2 & 4 \\ 1 & -3 & -2 \\ 4 & 6 & 2 \end{bmatrix}$$

Solution: First, we compute determinant of A to find out whether A is singular or nonsingular.

$$\det(A) = 3\begin{vmatrix} -3 & -2 \\ 6 & 2 \end{vmatrix} - 2\begin{vmatrix} 1 & -2 \\ 4 & 2 \end{vmatrix} + 4\begin{vmatrix} 1 & -3 \\ 4 & 6 \end{vmatrix} = 3(-6+12) - 2(2+8) + 4(6+12) = 70$$

Since $\det(A) = 70 \neq 0$, matrix A is nonsingular and its inverse is computed as follows:

$$\text{adj}(A) = \begin{bmatrix} C_{11} & C_{12} & C_{13} \\ C_{21} & C_{22} & C_{23} \\ C_{31} & C_{32} & C_{33} \end{bmatrix}^T = \begin{bmatrix} 6 & -10 & 18 \\ 20 & -10 & -10 \\ 8 & 10 & -11 \end{bmatrix}^T = \begin{bmatrix} 6 & 20 & 8 \\ -10 & -10 & 10 \\ 18 & -10 & -11 \end{bmatrix}$$

and

$$A^{-1} = \frac{\text{adj}(A)}{\det(A)} = \frac{1}{70}\begin{bmatrix} 6 & 20 & 8 \\ -10 & -10 & 10 \\ 18 & -10 & -11 \end{bmatrix} = \begin{bmatrix} \frac{6}{70} & \frac{2}{7} & \frac{8}{70} \\ -\frac{1}{7} & -\frac{1}{7} & \frac{1}{7} \\ -\frac{18}{70} & -\frac{1}{7} & -\frac{11}{70} \end{bmatrix}$$

The MATLAB® command to compute inverse of matrix A is **inv(A)**.

2.3.1 Properties of Matrix Inversion

If A and B are nonsingular $n \times n$ square matrices, then:

(a) $AA^{-1} = A^{-1}A = I$.
(b) $(AB)^{-1} = B^{-1}A^{-1}$.
(c) $\det(A^{-1}) = \dfrac{1}{\det(A)}$.

The proofs are left as an exercise (see Problem 2.7)

2.3.2 Gauss–Jordan Method for Calculating Inverse of a Matrix

Instead of computing the inverse of A by utilizing Equation 2.42, A^{-1} can be derived by utilizing the Gauss–Jordan method. This requires that A be reduced to its reduced row echelon form (RREF). Once in this form, its inverse is then available. The process is illustrated below.

Definition: A matrix is in RREF if:

(a) It is in row echelon form.
(b) The first nonzero entry in each row is the only nonzero entry in the corresponding column.

The following matrices are in RREF:

$$\begin{bmatrix} 1 & 0 & 0 \\ 0 & 1 & 0 \\ 0 & 0 & 1 \end{bmatrix}; \begin{bmatrix} 1 & 0 & 0 & | & 2 \\ 0 & 1 & 0 & | & -1 \\ 0 & 0 & 1 & | & 3 \end{bmatrix} \quad (2.50)$$

The process of transforming a matrix into its RREF is called the Gauss–Jordan elimination.

Let A^{-1} be partitioned as: $A^{-1} = [x_1 \ x_2 \ \ldots \ x_n]$, where $\{x_i\}_{i=1}^{n}$ are $n \times 1$ column vectors. Since $AA^{-1} = I$, we have:

$$AA^{-1} = A[x_1 \ x_2 \ \ldots \ x_n] = [e_1 \ e_2 \ \ldots \ e_n] \quad (2.51)$$

Determinants, Matrix Inversion and Solutions 61

where $e_i = [0 \ 0 \ \ldots \ 1 \ 0 \ \ldots \ 0]^T$ is a $n \times 1$ vector of all zero elements except for the ith row which is one. Using Equation 2.51 above, we can generate n systems of linear equations as follows:

$$Ax_1 = e_1, \ Ax_2 = e2, \ \ldots, \ \text{and} \ Ax_n = e_n \qquad (2.52)$$

In order to solve each of the above systems of equations, we perform Gauss–Jordan elimination on all the systems simultaneously by constructing a single augmented matrix as shown in the equation below:

$$[A \ | \ e_1 \ | \ e_2 \ | \ \cdots \ | \ e_n] \qquad (2.53)$$

Once the solutions for $\{x_i\}_{i=1}^n$ are obtained, then inverse of matrix A is formed by concatenating the solution vectors as follows: $A^{-1} = [x_1 \ x_2 \ \ldots \ x_n]$. The above steps are illustrated by the following example.

Example 2.5

Compute the inverse of the following matrix using the Gauss–Jordan method.

$$A = \begin{bmatrix} 2 & 1 & 1 \\ 4 & -6 & 0 \\ -2 & 7 & 2 \end{bmatrix}$$

Solution:

Let $A^{-1} = [x_1 \ | \ x_2 \ | \ x_3]$ and $I = \begin{bmatrix} 1 & 0 & 0 \\ 0 & 1 & 0 \\ 0 & 0 & 1 \end{bmatrix} = [e_1 \ | \ e_2 \ | \ e_3]$, then:

$$AA^{-1} = I \Leftrightarrow \begin{bmatrix} 2 & 1 & 1 \\ 4 & -6 & 0 \\ -2 & 7 & 2 \end{bmatrix} [x_1 \ | \ x_2 \ | \ x_3] = [e_1 \ | \ e_2 \ | \ e_3]$$

or

$$Ax_1 = e_1$$

$$Ax_2 = e_2$$

$$Ax_3 = e_3$$

There are now three systems of equations, namely:

System 1: $[A \ | \ e_1]$
System 2: $[A \ | \ e_2]$
System 3: $[A \ | \ e_3]$

Construct the augmented matrix:

$$[A \mid e_1 \mid e_2 \mid e_3] = \begin{bmatrix} 2 & 1 & 1 & | & 1 & 0 & 0 \\ 4 & -6 & 0 & | & 0 & 1 & 0 \\ -2 & 7 & 2 & | & 0 & 0 & 1 \end{bmatrix}$$

and proceed to perform elementary row operations on the above augmented matrix until it is in RREF. That is:

$$\begin{bmatrix} 2 & 1 & 1 & | & 1 & 0 & 0 \\ 4 & -6 & 0 & | & 0 & 1 & 0 \\ -2 & 7 & 2 & | & 0 & 0 & 1 \end{bmatrix} \xrightarrow{R_2 \leftarrow R_2 - 2R_1} \begin{bmatrix} 2 & 1 & 1 & | & 1 & 0 & 0 \\ 0 & -8 & -2 & | & -2 & 1 & 0 \\ -2 & 7 & 2 & | & 0 & 0 & 1 \end{bmatrix}$$

$$\xrightarrow{R_3 \leftarrow R_3 + R_1} \begin{bmatrix} 2 & 1 & 1 & | & 1 & 0 & 0 \\ 0 & -8 & -2 & | & -2 & 1 & 0 \\ 0 & 8 & 3 & | & 1 & 0 & 1 \end{bmatrix} \xrightarrow{R_3 \leftarrow R_3 + R_2} \begin{bmatrix} 2 & 1 & 1 & | & 1 & 0 & 0 \\ 0 & -8 & -2 & | & -2 & 1 & 0 \\ 0 & 0 & 1 & | & -1 & 1 & 1 \end{bmatrix}$$

$$\xrightarrow{R_2 \leftarrow R_2 + 2R_3} \begin{bmatrix} 2 & 1 & 1 & | & 1 & 0 & 0 \\ 0 & -8 & 0 & | & -4 & 3 & 2 \\ 0 & 0 & 1 & | & -1 & 1 & 1 \end{bmatrix} \xrightarrow{R_1 \leftarrow R_1 + \tfrac{1}{8}R_2} \begin{bmatrix} 2 & 0 & 1 & | & \tfrac{1}{2} & \tfrac{3}{8} & \tfrac{1}{4} \\ 0 & -8 & 0 & | & -4 & 3 & 2 \\ 0 & 0 & 1 & | & -1 & 1 & 1 \end{bmatrix}$$

$$\xrightarrow{R_1 \leftarrow R_1 - R_3} \begin{bmatrix} 2 & 0 & 0 & | & \tfrac{3}{2} & \tfrac{-5}{8} & \tfrac{-6}{8} \\ 0 & -8 & 0 & | & -4 & 3 & 2 \\ 0 & 0 & 1 & | & -1 & 1 & 1 \end{bmatrix} \xrightarrow{R_1 \leftarrow R_1/2} \begin{bmatrix} 1 & 0 & 0 & | & \tfrac{3}{4} & \tfrac{-5}{16} & \tfrac{-6}{16} \\ 0 & -8 & 0 & | & -4 & 3 & 2 \\ 0 & 0 & 1 & | & -1 & 1 & 1 \end{bmatrix}$$

$$\xrightarrow{R_2 \leftarrow -\tfrac{1}{8}R_2} \begin{bmatrix} 1 & 0 & 0 & | & \tfrac{12}{16} & \tfrac{-5}{16} & \tfrac{-6}{16} \\ 0 & 1 & 0 & | & \tfrac{4}{8} & \tfrac{-3}{8} & \tfrac{-2}{8} \\ 0 & 0 & 1 & | & -1 & 1 & 1 \end{bmatrix} \Leftrightarrow [I \mid x_1 \mid x_2 \mid x_3]$$

At this time, the above augmented matrix is in the form: $[I \mid x_1 \mid x_2 \mid x_3]$, where I represents the left hand side 3×3 identity matrix and x_1, x_2, x_3 encompass the column vectors shown on the right hand side. These represent the solution vectors to the systems of equations shown above. Therefore:

$$A^{-1} = [x_1 \mid x_2 \mid x_3] = \begin{bmatrix} \tfrac{3}{4} & \tfrac{-5}{16} & \tfrac{-3}{8} \\ \tfrac{1}{2} & \tfrac{-3}{8} & \tfrac{-1}{4} \\ -1 & 1 & 1 \end{bmatrix}$$

2.3.3 Useful Formulas for Matrix Inversion

(a) Let A be a nonsingular $n \times n$ matrix and u and v be vectors of size $n \times 1$ such that the $n \times n$ matrix $A + uv^T$ is nonsingular, then:

$$(A + uv^T)^{-1} = A^{-1} - \frac{A^{-1}uv^T A^{-1}}{1 + v^T A^{-1} u} \qquad (2.54)$$

Equation 2.54 suggests that if A^{-1} is known, then the inverse of $A + uv^T$ can be obtained in terms of A^{-1}. The above property can be generalized yielding the following matrix inversion lemma.

Matrix Inversion Lemma: Let A and B be $n \times n$ and $m \times m$ nonsingular matrices, respectively. Then, for any matrices C and D of appropriate sizes, we have:

$$(A + CBD)^{-1} = A^{-1} - A^{-1}C(DA^{-1}C + B^{-1})^{-1}DA^{-1} \qquad (2.55)$$

Proof: To prove the matrix inversion lemma, we show that product of $(A + CBD)$ and $A^{-1} - A^{-1}C(DA^{-1}C + B^{-1})^{-1}DA^{-1}$ is equal to the identity matrix.

$(A + CBD)[A^{-1} - A^{-1}C(DA^{-1}C + B^{-1})^{-1}DA^{-1}] =$

$= AA^{-1} - AA^{-1}C(DA^{-1}C + B^{-1})^{-1}DA^{-1} + CBDA^{-1} - CBDA^{-1}C(DA^{-1}C + B^{-1})^{-1}DA^{-1}$

$= I - C(DA^{-1}C + B^{-1})^{-1}DA^{-1} + CBDA^{-1} - CBDA^{-1}C(DA^{-1}C + B^{-1})^{-1}DA^{-1}$

$= I - C[(DA^{-1}C + B^{-1})^{-1} - B + BDA^{-1}C(DA^{-1}C + B^{-1})^{-1}]DA^{-1}$

$= I - C[I + BDA^{-1}C)(DA^{-1}C + B^{-1})^{-1} - B]DA^{-1}$

$= I - C[(BB^{-1} + BDA^{-1}C)(DA^{-1}C + B^{-1})^{-1} - B]DA^{-1}$

$= I - C[(B(B^{-1} + DA^{-1}C)(DA^{-1}C + B^{-1})^{-1} - B]DA^{-1} = I - C[B - B]DA^{-1} = I \qquad (2.56)$

(b) If A and B are nonsingular matrices, then:

$$\begin{bmatrix} A & 0 \\ C & B \end{bmatrix}^{-1} = \begin{bmatrix} A^{-1} & 0 \\ -B^{-1}CA^{-1} & B^{-1} \end{bmatrix} \qquad (2.57)$$

and

$$\begin{bmatrix} A & C \\ 0 & B \end{bmatrix}^{-1} = \begin{bmatrix} A^{-1} & -A^{-1}CB^{-1} \\ 0 & B^{-1} \end{bmatrix} \qquad (2.58)$$

The proof is left as an exercise (see Problem 2.15)

2.3.4 Recursive Least Square (RLS) Parameter Estimation

The matrix inversion lemma has many applications in estimation theory. As an example, consider the recursive least square (RLS) parameter estimation where measurement $y(k)$ at time k is related to the parameter vector θ and the measurement noise $v(k)$ through the linear regression equation:

$$y(k) = h^T(k)\theta + v(k) \tag{2.59}$$

where θ and $h(k)$ are $n \times 1$ vectors. Given N samples of $y(k)$, the least square error estimate of the parameter vector θ is given by:

$$\hat{\theta}(N) = \left[\sum_{k=1}^{N} h(k)h^T(k)\right]^{-1} \sum_{k=1}^{N} y(k)h(k) \tag{2.60}$$

To derive the recursive least square estimator, we define $P(N) = \sum_{k=1}^{N} h(k)h^T(k)$, then:

$$\hat{\theta}(N) = P^{-1}(N) \sum_{k=1}^{N} y(k)h(k) \tag{2.61}$$

From the definition of $P(N)$, we have:

$$P(N+1) = \sum_{k=1}^{N+1} h(k)h^T(k) = \sum_{k=1}^{N} h(k)h^T(k) + h(N+1)h^T(N+1)$$
$$= P(N) + h(N+1)h^T(N+1) \tag{2.62}$$

Hence:

$$\sum_{k=1}^{N+1} y(k)h(k) = \sum_{k=1}^{N} y(k)h(k) + y(N+1)h(N+1)$$
$$= P(N)\hat{\theta}(N) + y(N+1)h(N+1) \tag{2.63}$$
$$= P(N)\hat{\theta}(N) + y(N+1)h(N+1)$$
$$\quad + h(N+1)h^T(N+1)\hat{\theta}(N) - h(N+1)h^T(N+1)\hat{\theta}(N)$$
$$= [P(N) + h(N+1)h^T(N+1)]\hat{\theta}(N) + h(N+1)[y(N+1) - h^T(N+1)\hat{\theta}(N)]$$

Therefore:

$$\sum_{k=1}^{N+1} y(k)h(k) = [P(N) + h(N+1)h^T(N+1)]\hat{\theta}(N)$$
$$+ h(N+1)[y(N+1) - h^T(N+1)\hat{\theta}(N)] \tag{2.64}$$
$$= P(N+1)\hat{\theta}(N) + h(N+1)[y(N+1) - h^T(N+1)\hat{\theta}(N)]$$

Determinants, Matrix Inversion and Solutions 65

or

$$P(N+1)\hat{\theta}(N+1) = P(N+1)\hat{\theta}(N) + h(N+1)[y(N+1) - h^T(N+1)\hat{\theta}(N)] \quad (2.65)$$

Pre-multiplying both sides of Equation 2.65 by $P^{-1}(N+1)$ yields:

$$\hat{\theta}(N+1) = \hat{\theta}(N) + P^{-1}(N+1)h(N+1)[y(N+1) - h^T(N+1)\hat{\theta}(N)] \quad (2.66)$$

From Equation 2.56, we have:

$$P^{-1}(N+1) = [P(N) + h(N+1)h^T(N+1)]^{-1} \quad (2.67)$$

If we utilize the matrix inversion lemma given by Equation 2.55, with $A = P(N)$, $B = I$, $C = h(N+1)$, and $D = h^T(N+1)$ to modify the recursive equation shown above Equation 2.67. The resulting recursive equation is:

$$P^{-1}(N+1) = P^{-1}(N) - P^{-1}(N)h(N+1)$$
$$\times [h^T(N+1)P^{-1}(N)h(N+1) + 1]^{-1}h^T(N+1)P^{-1}(N) \quad (2.68)$$

Let $K(N) = P^{-1}(N)h(N)$, then:

$$\hat{\theta}(N+1) = \hat{\theta}(N) + K(N+1)[y(N+1) - h^T(N+1)\hat{\theta}(N)] \quad (2.69)$$

where

$$K(N+1) = K(N) - K(N)h(N+1)[h^T(N+1)K(N)h(N+1) + 1]^{-1} \quad (2.70)$$

The gain vector $K(\cdot)$ is known as the Kalman gain and can be computed recursively using Equation 2.70. The advantage of this recursive equation is that there is no need for matrix inversion since $h^T(N+1)K(N)h(N+1) + 1$ is a scalar quantity.

Example 2.6

As an example of the application of RLS algorithm, consider estimating the parameter vector $\theta = \begin{bmatrix} \theta_1 \\ \theta_2 \end{bmatrix}$ from the noisy observation y. Let the measurement y at time k be related to the parameter vector θ by:

$$y(k) = [h_1(k) \ h_2(k)]\theta + v(k) \quad k = 1, 2, \ldots, 200$$

where $v(k)$ is the measurement noise assumed to be zero mean Gaussian with variance $\sigma_v^2 = 0.1$. To simulate the above, we assumed that $\theta = \begin{bmatrix} 2 \\ 3 \end{bmatrix}$. The measurement gain vector $[h_1(k) \ h_2(k)]$ is generated randomly. The plots of $\theta_1(k)$ and $\theta_2(k)$ as a function of iteration index k are shown in Figure 2.1. The initial estimate of vector θ is assumed to be $\hat{\theta}_1(0) = \hat{\theta}_2(0) = 0$. The MATLAB simulation code is shown in Table 2.1.

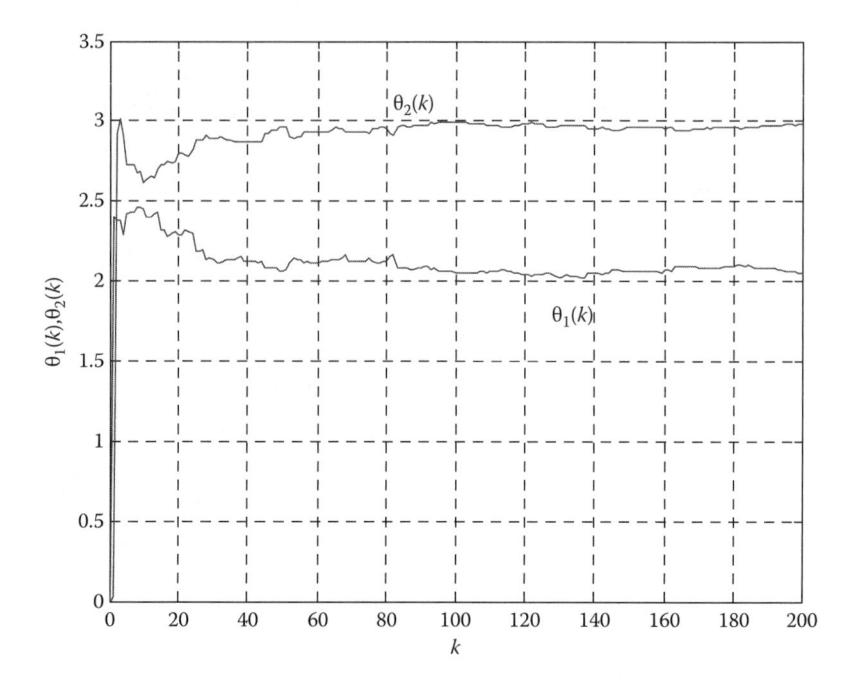

FIGURE 2.1
Convergence of the RLS algorithm.

TABLE 2.1

MATLAB® Code for Example 2.6

```
teta = [2 3]';
N = 200;
h = rand(N,2);
sigma = sqrt(0.1);
y = h*teta + sigma*randn(N,1);
teth = [0 0]';
p = 10^(6)*eye(2);
P = p;
K = [];
x = teth;
for i = 1:N
  h1 = h(i,:);
  k = p*h1'/(h1*p*h1' + 1);
  x = x + k*(y(i) - h1*x);
  p = (p - k*h1*p);
  P = [P p];
  K = [K k];
  teth = [teth x];
end
plot(0:1:N, teth(1,:))
hold on
grid on
plot(0:1:N, teth(2,:))
xlabel('k')
ylabel('\theta_1(k),\theta_2(k)')
```

2.4 Solution of Simultaneous Linear Equations

Consider the set of simultaneous linear equation given by:

$$\begin{aligned} a_{11}x_1 + a_{12}x_2 + \ldots + a_{1n}x_n &= b_1 \\ a_{21}x_1 + a_{22}x_2 + \ldots + a_{2n}x_n &= b_2 \\ \vdots \quad \vdots \quad \vdots \quad \vdots& \\ a_{n1}x_1 + a_{n2}x_2 + \ldots + a_{nn}x_n &= b_n \end{aligned} \qquad (2.71)$$

This set of simultaneous linear equations can also be expressed in matrix form as shown below:

$$\begin{bmatrix} a_{11} & a_{12} & \cdots & a_{1n} \\ a_{21} & a_{22} & \cdots & a_{2n} \\ \vdots & & \ddots & \\ a_{n1} & a_{n2} & & a_{nn} \end{bmatrix} \begin{bmatrix} x_1 \\ x_2 \\ \vdots \\ x_n \end{bmatrix} = \begin{bmatrix} b_1 \\ b_2 \\ \vdots \\ b_n \end{bmatrix} \qquad (2.72)$$

or in a more compact form as:

$$Ax = b \qquad (2.73)$$

where A is the $n \times n$ coefficient matrix, b is a known $n \times 1$ column vector and x represents the $n \times 1$ unknown vector. The system has either: (i) no solution, (ii) one solution or (iii) infinitely many solutions. As an example, consider the following system of equations:

$$a_{11}x_1 + a_{12}x_2 = b_1 \qquad (2.74)$$

$$a_{21}x_1 + a_{22}x_2 = b_2 \qquad (2.75)$$

The above two equations represent equations of two straight lines in the two dimensional x_1–x_2 plane. (x_1, x_2) is the unique solution to the above set of equations if it lies on both lines.

Example 2.7

Consider the set of simultaneous linear equations

$$2x_1 + 2x_2 = 2$$

$$2x_1 - 2x_2 = 2$$

The above set of equations has a unique solution given by $[x_1 \ x_2] = [1 \ 0]$. The coefficient matrix for this set of equation is given by:

$$A = \begin{bmatrix} 2 & 2 \\ 2 & -2 \end{bmatrix}$$

Since det(A)=$-8\neq 0$, then A is a nonsingular matrix. Figure 2.2 shows the plots of the two lines and their corresponding intersection point. The equations of the two lines are: $x_2=1-x_1$, and $x_2=-1+x_1$.

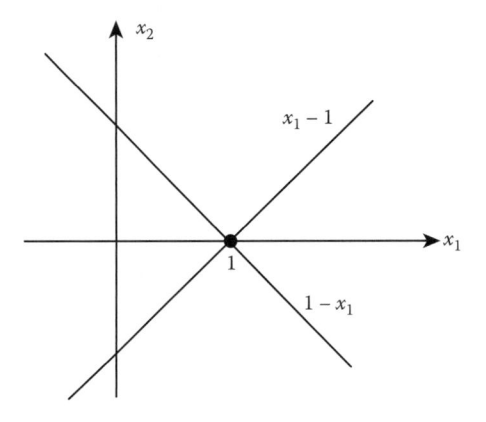

FIGURE 2.2
The two lines and their intersection.

Example 2.8

Consider a set of simultaneous linear equations

$$3x_1+3x_2=3$$

$$3x_1+3x_2=6$$

Solution: The above equations can be expressed as $x_2=1-x_1$ and $x_2=2-x_1$. The plots are shown in Figure 2.3. As can be seen from the figure, the lines are parallel and hence the set of equations have no solution. The coefficient matrix for this set of equations is given by:

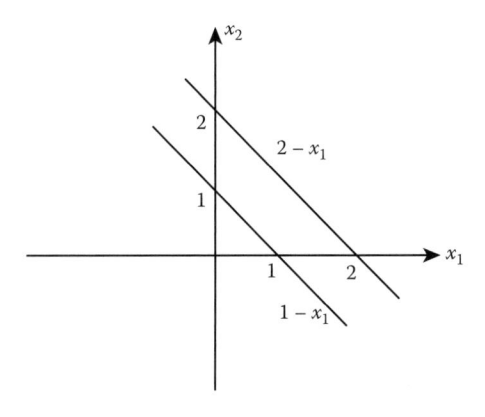

FIGURE 2.3
Plots of system of equations of Example 2.8.

Determinants, Matrix Inversion and Solutions

$$A = \begin{bmatrix} 3 & 3 \\ 3 & 3 \end{bmatrix}$$

Note that since det(A)=9−9=0, matrix A is singular.

Example 2.9

Consider the set of linear equations

$$2x_1 + 2x_2 = 2 \Rightarrow x_2 = 1 - x_1$$
$$4x_1 + 4x_2 = 4 \Rightarrow x_2 = 1 - x_1$$

As can be easily seen, both equations are representation of the same line and hence the set of equations has infinitely many solutions.

2.4.1 Equivalent Systems

Definition: Two systems of equations are said to be equivalent if they possess the same solution set.

Example 2.10

The following sets are equivalent since they have same solution $x_1 = 1$ and $x_2 = 0$

$$\begin{cases} x_1 + x_2 = 1 \\ x_1 - x_2 = 1 \end{cases}, \begin{cases} 4x_1 + 4x_2 = 4 \\ 4x_1 - 4x_2 = 4 \end{cases}, \text{ and } \begin{cases} 2x_1 + 2x_2 = 2 \\ -x_2 = 0 \end{cases}$$

2.4.2 Strict Triangular Form

Definition: A system is said to be in "strict triangular form" if the coefficients of the first $k-1$ variables in the kth equation are all zero and the coefficient of x_k is nonzero for all $k=1, 2, ..., n$. Systems in this form are solved by utilizing a technique known as back substitution. The above concepts are illustrated in Example 2.11.

Example 2.11

Consider the set of linear equations with three unknowns x_1, x_2, and x_3 given as

$$-2x_1 + 2x_2 + 3x_3 = 1$$
$$2x_2 - 2x_3 = 4$$
$$4x_3 = 12$$

The above set of equations is in strict-triangular form since in the third equation, the coefficients of all the other variables are zero except for the last one. We can use back-substitution to solve the above system.

$$4x_3 = 12 \Rightarrow x_3 = 3$$

$$2x_2 - 2x_3 = 4 \Rightarrow 2x_2 = 4 + 2x_3 = 4 + 2(3) = 10 \Rightarrow x_2 = 5$$

$$-2x_1 + 2x_2 + 3x_3 = 1 \Rightarrow -2x_1 = 1 - 2x_2 - 3x_3 = 1 - 2(5) - 3(3) = 1 - 10 - 9 = -18 \Rightarrow x_1 = 9$$

The solution to the above set of equations is: $[x_1 \ x_2 \ x_3] = [9 \ 5 \ 3]$.

Elementary matrix transformations are employed to convert a set of linear equations of the form $Ax = b$ to the strict triangular form given by $Tx = \tilde{b}$ as discussed in Chapter 1. Back substitution is then utilized to solve for x as shown in this example. In the next section, we explain another common approach called Cramer's rule for solving simultaneous set of linear equations.

2.4.3 Cramer's Rule

Let A be an $n \times n$ matrix and let $b \in R^n$, then the solution to $Ax = b$ is unique if and only if matrix A is nonsingular, i.e. $\det(A) \neq 0$. The solution is given by:

$$Ax = b \Rightarrow x = A^{-1}b = \frac{adj(A)}{\det(A)} b = \frac{1}{\det(A)} \begin{bmatrix} C_{11} & C_{21} & \cdots & C_{n1} \\ C_{12} & C_{22} & \cdots & C_{n2} \\ \vdots & \vdots & \vdots & \vdots \\ C_{1n} & C_{2n} & \cdots & C_{nn} \end{bmatrix} \begin{bmatrix} b_1 \\ b_2 \\ \vdots \\ b_n \end{bmatrix} \quad (2.76)$$

Therefore:

$$x_i = \frac{C_{1i} b_1 + C_{2i} b_2 + \cdots + C_{ni} b_n}{\det(A)} \quad (2.77)$$

Now let A_i be equal to the matrix obtained by replacing the ith column of A by vector b, then the determinant of A_i is given by:

$$|A_i| = C_{1i} b_1 + C_{2i} b_2 + \cdots + C_{ni} b_n \quad (2.78)$$

Note that the determinant of A_i shown in Equation 2.78 is equal to the numerator of Equation 2.77. As a result:

$$x_i = \frac{\det(A_i)}{\det(A)} \quad i = 1, 2, \ldots, n \quad (2.79)$$

This is known as Cramer's rule. The process of utilizing Cramer's rule to solve systems of linear equations is illustrated in Example 2.12.

Example 2.12

Use Cramer's rule to solve the following set of linear equations

$$\begin{bmatrix} 2 & 3 & 1 \\ 3 & 2 & -4 \\ 1 & -1 & 5 \end{bmatrix} \begin{bmatrix} x_1 \\ x_2 \\ x_3 \end{bmatrix} = \begin{bmatrix} 6 \\ 12 \\ -4 \end{bmatrix} \Leftrightarrow Ax = b$$

Determinants, Matrix Inversion and Solutions

Solution: Since $\det(A) = -50 \neq 0$, A is nonsingular and the above systems of equations has a unique solution. Let A_1, A_2 and A_3 represent the matrices formed by respectively replacing the first, second and third columns of matrix A by the vector b. Then, the solution for the above system of equations is arrived at as follows:

$$x_1 = \frac{\det(A_1)}{\det(A)} = \frac{\det\begin{bmatrix} 6 & 3 & 1 \\ 12 & 2 & -4 \\ -4 & -1 & 5 \end{bmatrix}}{\det\begin{bmatrix} 2 & 3 & 1 \\ 3 & 2 & -4 \\ 1 & -1 & 5 \end{bmatrix}} = \frac{-100}{-50} = 2$$

$$x_2 = \frac{\det(A_2)}{\det(A)} = \frac{\det\begin{bmatrix} 2 & 6 & 1 \\ 3 & 12 & -4 \\ 1 & -4 & 5 \end{bmatrix}}{\det\begin{bmatrix} 2 & 3 & 1 \\ 3 & 2 & -4 \\ 1 & -1 & 5 \end{bmatrix}} = \frac{-50}{-50} = 1$$

$$x_3 = \frac{\det(A_3)}{\det(A)} = \frac{\det\begin{bmatrix} 2 & 3 & 5 \\ 3 & 2 & 12 \\ 1 & -1 & -4 \end{bmatrix}}{\det\begin{bmatrix} 2 & 3 & 1 \\ 3 & 2 & -4 \\ 1 & -1 & 5 \end{bmatrix}} = \frac{50}{-50} = -1$$

2.4.4 LU Decomposition

The lower–upper (LU) decomposition is yet another approach for solving systems of linear equations by first factoring the $n \times n$ matrix A as:

$$A = LU \tag{2.80}$$

where L is a lower triangular matrix with ones on the main diagonal and U is an upper triangular matrix. Hence, the systems of equations $Ax = b$ can be expressed as:

$$Ax = b \Rightarrow (LU)x = b \Rightarrow L(Ux) = b$$

If we let $Ux = c$, then $Lc = b$, then the solution vector x can be arrived at by solving both systems using forward elimination followed by back substitution, respectively. Hence, given a system of equations, the following general steps are done sequentially to find the solution using LU decomposition:

(1) Compute U by utilizing Gaussian elimination.
(2) Derive L by inspection or by deduction.

(3) Solve the system of equations: $Lc=b$ for c using forward elimination.
(4) Solve the system $Ux=c$ for x using back substitution.

The above process is illustrated in Examples 2.13 and 2.14.

Example 2.13

Consider the system of equations given by:

$$\begin{bmatrix} 2 & 1 & 1 \\ 4 & -6 & 0 \\ -2 & 7 & 2 \end{bmatrix} \begin{bmatrix} x_1 \\ x_2 \\ x_3 \end{bmatrix} = \begin{bmatrix} 5 \\ -2 \\ 9 \end{bmatrix} \Leftrightarrow Ax = b$$

Solve for x using LU decomposition.

Solution: We first begin by using Gaussian elimination in order to determine the upper triangular matrix U. This is done as follows:

$$\begin{bmatrix} 2 & 1 & 1 \\ 4 & -6 & 0 \\ -2 & 7 & 2 \end{bmatrix} \xrightarrow{R_2 \leftarrow R_2 - 2R_1} \begin{bmatrix} 2 & 1 & 1 \\ 0 & -8 & -2 \\ -2 & 7 & 2 \end{bmatrix} \xrightarrow{R_3 \leftarrow R_3 + R_1} \begin{bmatrix} 2 & 1 & 1 \\ 0 & -8 & -2 \\ 0 & 8 & 3 \end{bmatrix}$$

$$\xrightarrow{R_3 \leftarrow R_3 + R_2} \begin{bmatrix} 2 & 1 & 1 \\ 0 & -8 & -2 \\ 0 & 0 & 1 \end{bmatrix}$$

From the above, matrix U is given by:

$$U = \begin{bmatrix} 2 & 1 & 1 \\ 0 & -8 & -2 \\ 0 & 0 & 1 \end{bmatrix}$$

The same result can be arrived at by utilizing the elementary transformation matrices as described in Chapter 1. Hence, matrix U can be derived by:

$$U = E_3 E_2 E_1 A = \begin{bmatrix} 1 & 0 & 0 \\ 0 & 1 & 0 \\ 0 & 1 & 1 \end{bmatrix} \begin{bmatrix} 1 & 0 & 0 \\ 0 & 1 & 0 \\ 1 & 0 & 1 \end{bmatrix} \begin{bmatrix} 1 & 0 & 0 \\ -2 & 1 & 0 \\ 0 & 0 & 1 \end{bmatrix} \begin{bmatrix} 2 & 1 & 1 \\ 4 & -6 & 0 \\ -2 & 7 & 2 \end{bmatrix} = \begin{bmatrix} 2 & 1 & 1 \\ 0 & -8 & -2 \\ 0 & 0 & 1 \end{bmatrix}$$

Since $A = LU$, then:

$$A = LU \Rightarrow A = LE_3 E_2 E_1 A \Rightarrow I = LE_3 E_2 E_1 \Rightarrow L = (E_3 E_2 E_1)^{-1}$$

Therefore:

$$L = (E_3 E_2 E_1)^{-1} = E_1^{-1} E_2^{-1} E_3^{-1} = \begin{bmatrix} 1 & 0 & 0 \\ -2 & 1 & 0 \\ 1 & 0 & 1 \end{bmatrix}^{-1} \begin{bmatrix} 1 & 0 & 0 \\ 0 & 1 & 0 \\ 1 & 0 & 1 \end{bmatrix}^{-1} \begin{bmatrix} 1 & 0 & 0 \\ 0 & 1 & 0 \\ 0 & 1 & 1 \end{bmatrix}^{-1}$$

$$= \begin{bmatrix} 1 & 0 & 0 \\ 2 & 1 & 0 \\ -1 & 0 & 1 \end{bmatrix} \begin{bmatrix} 1 & 0 & 0 \\ 0 & 1 & 0 \\ -1 & 0 & 1 \end{bmatrix} \begin{bmatrix} 1 & 0 & 0 \\ 0 & 1 & 0 \\ 0 & -1 & 1 \end{bmatrix} = \begin{bmatrix} 1 & 0 & 0 \\ 2 & 1 & 0 \\ -1 & -1 & 1 \end{bmatrix}$$

Determinants, Matrix Inversion and Solutions

Note that matrix L can also be obtained directly by inspection from the steps utilized to reduce the original matrix A into the upper triangular form U. This is done as follows. During the reduction process shown above, we have performed the following elementary row operations: $R_2 \leftarrow R_2 - 2R_1$, $R_3 \leftarrow R_3 - (-1) R_1$, and $R_3 \leftarrow R_3 - (-1) R_2$. Hence, matrix L can be arrived at by placing the coefficients from each of the above operation in their corresponding locations in the lower diagonal matrix L. Also note that the diagonal entries are all equal to one and the upper diagonal entries are all equal to zero. To this effect, the operations: $R_2 \leftarrow R_2 - 2R_1$, $R_3 \leftarrow R_3 - (-1) R_1$, and $R_3 \leftarrow R_3 - (-1) R_2$ yield the coefficients of 2, -1, -1 in the $l_{21} = 2$, $l_{31} = -1$ and $l_{32} = -1$, respectively. Hence, matrix L is given by:

$$L = \begin{bmatrix} 1 & 0 & 0 \\ 2 & 1 & 0 \\ -1 & -1 & 1 \end{bmatrix}$$

Now that we have computed L and U, we can utilize forward elimination to solve for c using $Lc = b$ as follows:

$$\begin{bmatrix} 1 & 0 & 0 \\ 2 & 1 & 0 \\ -1 & -1 & 1 \end{bmatrix} \begin{bmatrix} c_1 \\ c_2 \\ c_3 \end{bmatrix} = \begin{bmatrix} 5 \\ -2 \\ 9 \end{bmatrix} \Rightarrow \begin{matrix} c_1 = 5 \\ 2c_1 + c_2 = -2 \Rightarrow c_2 = -12 \\ -c_1 - c_2 + c_3 = 9 \Rightarrow c_3 = 9 + c_1 + c_2 = 2 \end{matrix} \rightarrow \begin{bmatrix} c_1 \\ c_2 \\ c_3 \end{bmatrix} = \begin{bmatrix} 5 \\ -12 \\ 2 \end{bmatrix}$$

Finally, we utilize $Ux = c$ and proceed to solve for x by back substitution as follows:

$$\begin{bmatrix} 2 & 1 & 1 \\ 0 & -8 & -2 \\ 0 & 0 & 1 \end{bmatrix} \begin{bmatrix} x_1 \\ x_2 \\ x_3 \end{bmatrix} = \begin{bmatrix} 5 \\ -12 \\ 2 \end{bmatrix}$$

$1x_3 = 2 \Rightarrow x_3 = 2$

$-8x_2 - 2x_3 = -12 \Rightarrow -8x_2 = -12 + 2x_3 \Rightarrow x_2 = (-12 + 4)/(-8) = 1$

$2x_1 + x_2 + x_3 = 5 \Rightarrow x_1 = (5 - x_2 - x_3)/2 = (5 - 1 - 2)/2 = 1$

$2x_1 + x_2 + x_3 = 5 \Rightarrow x_1 = (5 - x_2 - x_3)/2 = (5 - 1 - 2)/2 = 1$

Hence, the solution is: $x_1 = 1$, $x_2 = 1$, and $x_3 = 2$.

Example 2.14

Solve the following systems of equations using LU decomposition.

$$2x_1 + 3x_2 + 3x_3 = 2$$

$$5x_2 + 7x_3 = 2$$

$$6x_1 + 9x_2 + 8x_3 = 5$$

Solution: The above system of equations can be written in the $Ax=b$ form as:

$$\begin{bmatrix} 2 & 3 & 3 \\ 0 & 5 & 7 \\ 6 & 9 & 8 \end{bmatrix} \begin{bmatrix} x_1 \\ x_2 \\ x_3 \end{bmatrix} = \begin{bmatrix} 2 \\ 2 \\ 5 \end{bmatrix}$$

Factorizing A into its L,U factors yields:

$$A = \begin{bmatrix} 2 & 3 & 3 \\ 0 & 5 & 7 \\ 6 & 9 & 8 \end{bmatrix} \xrightarrow{R_3 \leftarrow R_3 - 3R_1} \begin{bmatrix} 2 & 3 & 3 \\ 0 & 5 & 7 \\ 0 & 0 & -1 \end{bmatrix} = U$$

The elementary transformation matrix that represents the above operation is:

$$E_1 = \begin{bmatrix} 1 & 0 & 0 \\ 0 & 1 & 0 \\ -3 & 0 & 1 \end{bmatrix}$$

Therefore:

$$L = E_1^{-1} = \begin{bmatrix} 1 & 0 & 0 \\ 0 & 1 & 0 \\ -3 & 0 & 1 \end{bmatrix}^{-1} = \begin{bmatrix} 1 & 0 & 0 \\ 0 & 1 & 0 \\ 3 & 0 & 1 \end{bmatrix}$$

Utilize forward elimination to solve for c in the system of equations: $Lc=b$. This yields:

$$\begin{bmatrix} 1 & 0 & 0 \\ 0 & 1 & 0 \\ 3 & 0 & 1 \end{bmatrix} \begin{bmatrix} c_1 \\ c_2 \\ c_3 \end{bmatrix} = \begin{bmatrix} 2 \\ 2 \\ 5 \end{bmatrix} \rightarrow \begin{bmatrix} c_1 \\ c_2 \\ c_3 \end{bmatrix} = \begin{bmatrix} 2 \\ 2 \\ -1 \end{bmatrix}$$

Utilize back substitution to solve for x using the system of equations: $Ux=c$. This results in:

$$\begin{bmatrix} 2 & 3 & 3 \\ 0 & 5 & 7 \\ 0 & 0 & -1 \end{bmatrix} \begin{bmatrix} x_1 \\ x_2 \\ x_3 \end{bmatrix} = \begin{bmatrix} 2 \\ 2 \\ -1 \end{bmatrix}$$

$$\Rightarrow x_3 = 1$$

$$5x_2 + 7x_3 = 2 \Rightarrow x_2 = (2 - 7x_3)/5 = -1$$

$$2x_1 + 3x_2 + 3x_3 = 2 \Rightarrow x_1 = (2 - 3x_2 - 3x_3)/2 = 1$$

The solution to the set of equations is:

$$\begin{bmatrix} x_1 \\ x_2 \\ x_3 \end{bmatrix} = \begin{bmatrix} 1 \\ -1 \\ 1 \end{bmatrix}$$

2.5 Applications: Circuit Analysis

Direct current (DC) electric circuits are analyzed by solving a set of linear equations with real coefficients. The unknowns are usually currents and/or voltage. Alternative current (AC) circuits are similar to DC circuits except that the coefficients are complex numbers. In this section, we provide several examples of DC and AC circuits and demonstrate the use of matrices to solve the underlying equations.

Example 2.15 (DC circuit)

Consider the DC circuit shown in Figure 2.4 where there are two independent and one dependent source. Solve the circuit and compute the power supplied by each source. Is the dependent source an active or passive source?

Let I_1, I_2, and I_3 represent the currents through the three branches of the circuit as shown in Figure 2.4. By utilizing Kirchhoff voltage law (KVL), we can write the following equations for each of the meshes:

$$\text{Mesh 1: } 3I_1 + 3(I_1 - I_2) + 10 + (I_1 - I_3) - 23 = 0$$
$$\text{Mesh 2: } 4.4I_2 + 4I_1 + 0.4(I_2 - I_3) - 10 + 3(I_2 - I_1) = 0$$
$$\text{Mesh 3: } 0.4(I_3 - I_2) + 0.2I_3 + 0.1I_3 + (I_3 - I_1) = 0$$

The above equations can be simplified to yield:

$$7I_1 - 3I_2 - I_3 = 13$$
$$I_1 + 7.8I_2 - 0.4I_3 = 10$$
$$-I_1 - 0.4I_2 - 1.7I_3 = 0$$

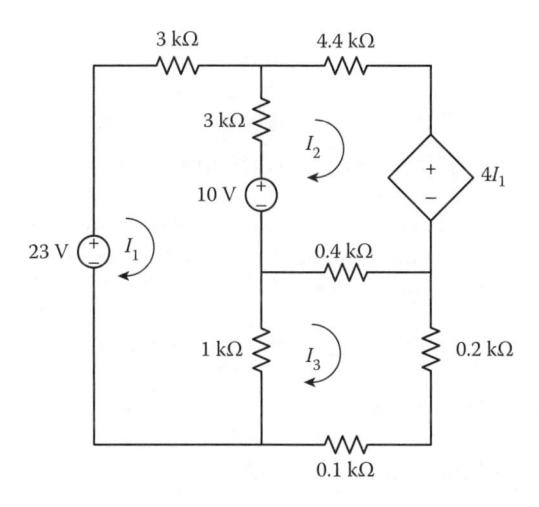

FIGURE 2.4
DC circuit.

and then expressed in matrix form as:

$$\begin{bmatrix} 7 & -3 & -1 \\ 1 & 7.8 & -0.4 \\ -1 & -0.4 & 1.7 \end{bmatrix} \begin{bmatrix} I_1 \\ I_2 \\ I_3 \end{bmatrix} = \begin{bmatrix} 13 \\ 10 \\ 0 \end{bmatrix} \Leftrightarrow Ax = b$$

Using Cramer's rule, we can easily solve for the unknown current I_1, I_2, and I_3 as follows:

$$I_1 = \frac{\begin{vmatrix} 13 & -3 & -1 \\ 10 & 7.8 & -0.4 \\ 0 & -0.4 & 1.7 \end{vmatrix}}{\begin{vmatrix} 7 & -3 & -1 \\ 1 & 7.8 & -0.4 \\ -1 & -0.4 & 1.7 \end{vmatrix}} = \frac{225.3}{88.2} = 2.554 \text{ mA}$$

$$I_2 = \frac{\begin{vmatrix} 7 & 13 & -1 \\ 1 & 10 & -0.4 \\ -1 & 0 & 1.7 \end{vmatrix}}{\begin{vmatrix} 7 & -3 & -1 \\ 1 & 7.8 & -0.4 \\ -1 & -0.4 & 1.7 \end{vmatrix}} = \frac{92.10}{88.2} = 1.0442 \text{ mA}$$

and

$$I_3 = \frac{\begin{vmatrix} 7 & -3 & 13 \\ 1 & 7.8 & 10 \\ -1 & -0.4 & 0 \end{vmatrix}}{\begin{vmatrix} 7 & -3 & -1 \\ 1 & 7.8 & -0.4 \\ -1 & -0.4 & 1.7 \end{vmatrix}} = \frac{154.2}{88.2} = 1.7483 \text{ mA}$$

The power supplied by the 23 V and the 10 V sources are:

$$P_{E1} = E_1 I_1 = 23 \times 2.5544 = 58.7512 \text{ mW}$$

$$P_{E2} = E_2 I_2 = 10 \times 1.0442 = 10.442 \text{ mW}$$

The voltage across the dependent source is $4I_1$ and the current through it is I_2 and since I_2 is positive and entering the positive terminal of the source, it is a passive source receiving power. The amount of received power is:

$$P = VI = 4I_1 I_2 = 4 \times 2.5544 \times 1.0442 = 10.6692 \text{ mW}$$

where W stands for the units of Watts.

Example 2.16 (AC circuit)

In the AC circuit shown in Figure 2.5, assume that all voltage values are root mean square (RMS) values and the impedance values are in Ohms. Find:

Determinants, Matrix Inversion and Solutions

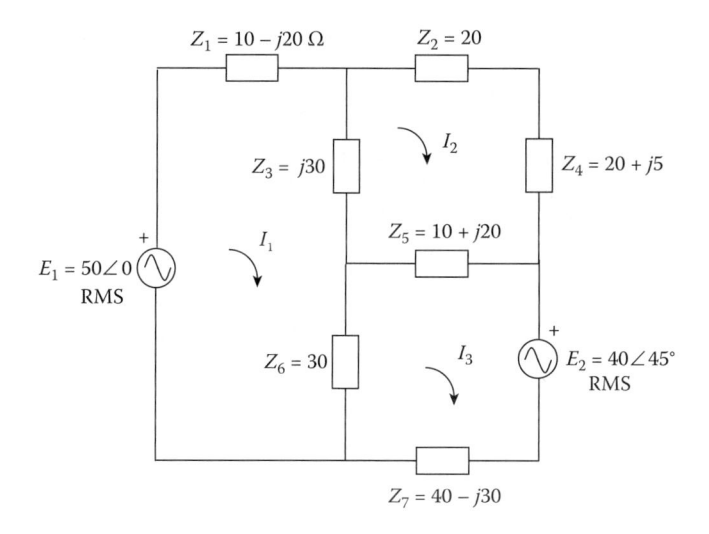

FIGURE 2.5
AC circuit.

a. The mesh currents I_1, I_2 and I_3.
b. A time domain expression for the voltage drop across Z_4 if the frequency of the two sources is 50 Hz.
c. The power dissipated by impedance Z_7.

Solution: The three mesh currents are assigned as indicated in Figure 2.5. By applying KVL around each mesh, we can write the following equations:

(a) $Z_1 I_1 + Z_3(I_1 - I_2) + Z_6(I_1 - I_3) = E_1$

$(Z_2 + Z_4)I_2 + Z_5(I_2 - I_3) + Z_3(I_2 - I_1) = 0$

$Z_5(I_3 - I_2) + Z_7 I_3 + Z_6(I_3 - I_1) = -E_2$

Simplifying and putting the equations in matrix form results in:

$$\begin{bmatrix} Z_1 + Z_3 + Z_6 & -Z_3 & -Z_6 \\ -Z_3 & Z_2 + Z_3 + Z_4 + Z_5 & -Z_5 \\ -Z_6 & -Z_5 & Z_5 + Z_6 + Z_7 \end{bmatrix} \begin{bmatrix} I_1 \\ I_2 \\ I_3 \end{bmatrix} = \begin{bmatrix} E_1 \\ 0 \\ -E_2 \end{bmatrix}$$

Substituting numerical values, we have:

$$\begin{bmatrix} 40 + j10 & -j30 & -30 \\ -j30 & 50 + j55 & -10 - j20 \\ -30 & -10 - j20 & 80 - j10 \end{bmatrix} \begin{bmatrix} I_1 \\ I_2 \\ I_3 \end{bmatrix} = \begin{bmatrix} 50 \\ 0 \\ -40\angle 45 \end{bmatrix} = \begin{bmatrix} 50 \\ 0 \\ -20\sqrt{2} - j20\sqrt{2} \end{bmatrix}$$

Using Cramer's rule, we solve for the three loop currents as follows:

$$I_1 = \frac{\begin{vmatrix} 50 & -j30 & -30 \\ 0 & 50+j55 & -10-j20 \\ -20\sqrt{2}-j20\sqrt{2} & -10-j20 & 80-j10 \end{vmatrix}}{\begin{vmatrix} 40+j10 & -j30 & -30 \\ -j30 & 50+j55 & -10-j20 \\ -30 & -10-j20 & 80-j10 \end{vmatrix}} = 0.1023 + j0.3136 = 0.3299 e^{j0.3996\pi} \text{ A}$$

$$I_2 = \frac{\begin{vmatrix} 40+j10 & 50 & -30 \\ -j30 & 0 & -10-j20 \\ -30 & -20\sqrt{2}-j20\sqrt{2} & 80-j10 \end{vmatrix}}{\begin{vmatrix} 40+j10 & -j30 & -30 \\ -j30 & 50+j55 & -10-j20 \\ -30 & -10-j20 & 80-j10 \end{vmatrix}} = -0.5521 + j0.2944 = 0.5994 e^{j0.8365\pi} \text{ A}$$

$$I_3 = \frac{\begin{vmatrix} 40+j10 & -j30 & -30 \\ -j30 & 50+j55 & -10-j20 \\ -30 & -10-j20 & 80-j10 \end{vmatrix}}{\begin{vmatrix} 40+j10 & -j30 & 50 \\ -j30 & 50+j55 & 0 \\ -30 & -10-j20 & -20\sqrt{2}-j20\sqrt{2} \end{vmatrix}} = -1.3403 + j0.9744 = 1.6571 e^{j0.7999\pi} \text{ A}$$

where A stands for the units of Amperes.

(b) $V = Z_4 I_2 = (20+j5)(-0.5521+j0.2944) = -27.5775 + j12.1119 = 30.12 e^{j0.8683\pi}$. V

Hence, the time-domain equation is:

$$v(t) = 30.12\sqrt{2} \cos(120\pi t + 0.8683\pi) = 42.5961 \cos(120\pi t + 0.8683\pi)$$

(c) The power dissipated in Z_7 is:

$$P = \text{real}(V\bar{I}) = \text{real}(Z_7 I_3 \bar{I}_3) = \text{real}(Z_7|I_3|^2) = \text{real}(Z_7)|I_3|^2 = 40(1.6571)^2 = 109.837 \text{ W}$$

2.6 Homogeneous Coordinates System

The homogeneous coordinates of a point $p = [x \; y \; z]^T$ in 3-d space are defined as $p_h = [kx \; ky \; kz \; k]$ where k is a nonzero constant. Without loss of generality, we can assume that $k = 1$. Therefore, a point in 3-d space can be expressed with Cartesian or homogenous coordinates, respectively as:

$$p = \begin{bmatrix} x \\ y \\ z \end{bmatrix} \text{ or } p_h = \begin{bmatrix} x \\ y \\ z \\ 1 \end{bmatrix} \quad (2.81)$$

2.6.1 Applications of Homogeneous Coordinates in Image Processing

The homogeneous coordinate system is useful in image processing and computer vision for the representation of image transformations such as translation, scaling and rotation of 2-d and 3-d objects. For example, translation of a point with coordinates $(x\ y\ z)$ by a displacement $(x_0\ y_0\ z_0)$ will produce a new point whose coordinates are given by:

$$x' = x + x_0$$
$$y' = y + y_0 \qquad (2.82)$$
$$z' = z + z_0$$

These equations can be put in matrix form using the homogenous coordinate system as:

$$\begin{bmatrix} x' \\ y' \\ z' \\ 1 \end{bmatrix} = \begin{bmatrix} 1 & 0 & 0 & x_0 \\ 0 & 1 & 0 & y_0 \\ 0 & 0 & 1 & z_0 \\ 0 & 0 & 0 & 1 \end{bmatrix} \begin{bmatrix} x \\ y \\ z \\ 1 \end{bmatrix} \qquad (2.83)$$

or

$$p'_k = A p_k \qquad (2.84)$$

where p_k and p'_k are the homogenous coordinates before and after the translation and A is a 4×4 nonsingular transformation. Similarly, other transformations such as rotation and scaling can be defined. Different transformations for both 2-d and 3-d objects are shown in Tables 2.2 and 2.3, respectively.

TABLE 2.2

2-d transformations

Transformation Type	Transformation Matrix
Translation	$A = \begin{bmatrix} 1 & 0 & x_0 \\ 0 & 1 & y_0 \\ 0 & 0 & 1 \end{bmatrix}$
Rotation by θ	$A = \begin{bmatrix} \cos\theta & -\sin\theta & 0 \\ \sin\theta & \cos\theta & 0 \\ 0 & 0 & 1 \end{bmatrix}$
Scaling	$A = \begin{bmatrix} k_x & 0 & 0 \\ 0 & k_y & 0 \\ 0 & 0 & 1 \end{bmatrix}$
x-Shearing	$A = \begin{bmatrix} 1 & \beta & 0 \\ 0 & 1 & 0 \\ 0 & 0 & 1 \end{bmatrix}$

TABLE 2.3
3-d Transformations

Transformation Type	Transformation Matrix
Translation	$A = \begin{bmatrix} 1 & 0 & 0 & x_0 \\ 0 & 1 & 0 & y_0 \\ 0 & 0 & 1 & z_0 \\ 0 & 0 & 0 & 1 \end{bmatrix}$
Rotation about x-axis by θ	$A = \begin{bmatrix} 1 & 0 & 0 & 0 \\ 0 & \cos\theta & -\sin\theta & 0 \\ 0 & \sin\theta & \cos\theta & 0 \\ 0 & 0 & 0 & 1 \end{bmatrix}$
Rotation about y-axis by φ	$A = \begin{bmatrix} \cos\psi & 0 & \sin\psi & 0 \\ 0 & 1 & 0 & 0 \\ -\sin\psi & 0 & \cos\psi & 0 \\ 0 & 0 & 0 & 1 \end{bmatrix}$
Rotation about z-axis by ϕ	$A = \begin{bmatrix} \cos\phi & -\sin\phi & 0 & 0 \\ \sin\phi & \cos\phi & 0 & 0 \\ 0 & 0 & 1 & 0 \\ 0 & 0 & 0 & 1 \end{bmatrix}$
Scaling (zoom in and zoom out)	$A = \begin{bmatrix} k_x & 0 & 0 & 0 \\ 0 & k_y & 0 & 0 \\ 0 & 0 & k_z & 0 \\ 0 & 0 & 0 & 1 \end{bmatrix}$
x-Shearing	$A = \begin{bmatrix} 1 & \beta & 0 & 0 \\ 0 & 1 & 0 & 0 \\ 0 & 0 & 1 & 0 \\ 0 & 0 & 0 & 1 \end{bmatrix}$

Example 2.17

Consider the two dimensional ellipsoid object O_1 defined by:

$$x^2 + \frac{y^2}{4} = 1$$

Find the transformation A that would scale the object by 0.4 and 0.6 in x and y directions respectively, rotate it counterclockwise about the x axis by 45° and then translate the object to location $[x_0 \ y_0] = [1 \ 1]$.

Solution:

$$A = A_T \times A_\theta \times A_s = \begin{bmatrix} 1 & 0 & x_0 \\ 0 & 1 & y_0 \\ 0 & 0 & 1 \end{bmatrix} \begin{bmatrix} \cos\theta & -\sin\theta & 0 \\ \sin\theta & \cos\theta & 0 \\ 0 & 0 & 1 \end{bmatrix} \begin{bmatrix} k_x & 0 & 0 \\ 0 & k_y & 0 \\ 0 & 0 & 1 \end{bmatrix}$$

Determinants, Matrix Inversion and Solutions

$$A = \begin{bmatrix} 1 & 0 & 1 \\ 0 & 1 & 1 \\ 0 & 0 & 1 \end{bmatrix} \begin{bmatrix} \cos\dfrac{\pi}{4} & -\sin\dfrac{\pi}{4} & 0 \\ \sin\dfrac{\pi}{4} & \cos\dfrac{\pi}{4} & 0 \\ 0 & 0 & 1 \end{bmatrix} \begin{bmatrix} 0.4 & 0 & 0 \\ 0 & 0.6 & 0 \\ 0 & 0 & 1 \end{bmatrix} = \begin{bmatrix} 0.2828 & 0.4243 & 1 \\ -0.2828 & 0.4243 & 1 \\ 0 & 0 & 1 \end{bmatrix}$$

The original object O_1 and the transformed object O_2 are shown in Figure 2.6. A simple MATLAB® implementation is given in Table 2.4.

TABLE 2.4

MATLAB code for Example 2.17

```
TR= [1 0 1;0 1 1;0 0 1];
R= [cos(pi/4)  sin(pi/4)  0 ;-sin(pi/4)
 cos(pi/4)  0;0 0 1];
S= [0.4 0 0;0 0.6 0;0 0 1];
T=TR*R*S;
theta=linspace(0,2*pi,512);
x=cos(theta);
y=2*sin(theta);
object1= [x;y;ones(size(x))];
object2 =T*image1;
plot(x,y)
hold on
plot(object(1,:),object2(2,:))
hold off
axis([-2 2 -3 3])
grid on
```

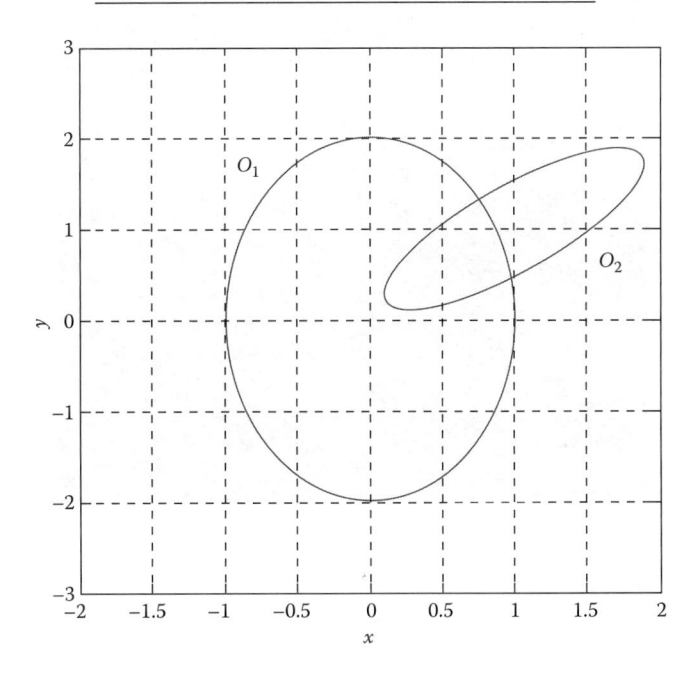

FIGURE 2.6
Original and transformed 2-d object.

Example 2.18

Consider a 256×256 gray scale image. The original and the transformed images are shown in Figures 2.7 and 2.8. The transformation is composed of a scaling factor of 1.2 and 0.5 in x and y direction, respectively followed by a counter-clockwise rotation of 5° followed by a translation of 20 and 30 pixels in x and y direction, respectively. The overall transform is given by:

$$A = A_T \times A_\theta \times A_s = \begin{bmatrix} 1 & 0 & x_0 \\ 0 & 1 & y_0 \\ 0 & 0 & 1 \end{bmatrix} \begin{bmatrix} \cos\theta & -\sin\theta & 0 \\ \sin\theta & \cos\theta & 0 \\ 0 & 0 & 1 \end{bmatrix} \begin{bmatrix} k_x & 0 & 0 \\ 0 & k_y & 0 \\ 0 & 0 & 1 \end{bmatrix}$$

$$A = \begin{bmatrix} 1 & 0 & 20 \\ 0 & 1 & 30 \\ 0 & 0 & 1 \end{bmatrix} \begin{bmatrix} \cos\frac{\pi}{36} & -\sin\frac{\pi}{36} & 0 \\ \sin\frac{\pi}{36} & \cos\frac{\pi}{36} & 0 \\ 0 & 0 & 1 \end{bmatrix} \begin{bmatrix} 1.2 & 0 & 0 \\ 0 & 0.5 & 0 \\ 0 & 0 & 1 \end{bmatrix} = \begin{bmatrix} 1.1954 & 0.0436 & 20 \\ -0.1046 & 0.4981 & 30 \\ 0 & 0 & 1 \end{bmatrix}$$

The MATLAB implementation codes are shown in Tables 2.5 and 2.6. Table 2.5 provides the MATLAB code for the function file called **affine.m** that accept an input image and transform it to an output image through scaling, rotation, and

FIGURE 2.7
Original image.

Determinants, Matrix Inversion and Solutions 83

FIGURE 2.8
Transformed image.

translation. Table 2.6 provides the main code utilized to perform the transformation and produces the output for the cameraman image.

The application of several transformations can be represented by a single 4×4 transformation. For example, if the application requires a translation followed by

TABLE 2.5

Main MATLAB Code

```
function [imageout]=affine(imagein,x0,y0,tet,sx,sy,type)
TR=[1 0 x0;0 1 y0;0 0 1];
tet=tet*pi/180;
R=[cos(tet) sin(tet) 0;-sin(tet) cos(tet) 0; 0 0  1];
S=[sx 0 0;0 sy 0;0 0 1];
switch type
    case 'RTS'
        Tin=inv(S*TR*R);
    case 'RST'
        Tin=inv(TR*S*R);
    case 'SRT'
        Tin=inv(TR*R*S);
    case 'STR'
        Tin=inv(R*TR*S);
    case 'TRS'
        Tin=inv(S*R*TR);
    case 'TSR'
        Tin=inv(R*S*TR);
    otherwise
end
```

(continued)

TABLE 2.5 (Continued)

```
[n1,m1,k]=size(imagein);
n2=round(n1*sx);
m2=round(m1*sy);
n=max(n2,m2);
m=n;
XY=ones(n*m,2);
x1=(0:1:n1-1)';
y1=(0:1:m1-1)';
x2=(0:1:n-1)';
y2=(0:1:m-1)';
XY(:,2)=kron(ones(n,1),y2);
XY(:,1)=kron(x2,ones(m,1));
XY0=[XY ones(n*m,1)]*Tin';
XY0=round(XY0);
[XX,YY]=meshgrid(y1,x1);
if k==1
z=interp2(XX,YY,imagein,XY0(:,2),XY0(:,1),'cubic');
z(isnan(z))=1;
imageout=zeros(n,m);
imageout(1:n*m)=z(1:n*m);
imageout=imageout';
imageout=imageout(1:n2,1:m2);
else
    r=imagein(:,:,1);
    g=imagein(:,:,2);
    b=imagein(:,:,3);
    z1=interp2(XX,YY,r,XY0(:,2),XY0(:,1),'cubic');
    z2=interp2(XX,YY,g,XY0(:,2),XY0(:,1),'cubic');
    z3=interp2(XX,YY,b,XY0(:,2),XY0(:,1),'cubic');
z1(isnan(z1))=1;
z2(isnan(z2))=1;
z3(isnan(z3))=1;
imageout=zeros(n,m);
imageout(1:n*m)=z1(1:n*m);
imageout=imageout';
rout=imageout(1:n2,1:m2);
imageout=zeros(n,m);
imageout(1:n*m)=z2(1:n*m);
imageout=imageout';
gout=imageout(1:n2,1:m2);
imageout=zeros(n,m);
imageout(1:n*m)=z3(1:n*m);
imageout=imageout';
bout=imageout(1:n2,1:m2);
imageout=zeros(n2,m2,k);
imageout(:,:,1)=rout;
imageout(:,:,2)=gout;
imageout(:,:,3)=bout;
end
```

TABLE 2.6

MATLAB code for Example 2.18

```
f = imread('cameraman.tif');
f = double(f)/255;
[g] = affine(f,20,30,5,0.5,1.2),'TRS';
figure(1)
imshow(f)
figure(2)
imshow(g)
```

a rotation followed by a scaling of a point X_a where $X_a = [x \ y \ z \ 1]^T$, then the following concatenated transformation can be utilized:

$$X_b = S(R(TX_a)) = AX_a \qquad (2.85)$$

where R, S and T represent the rotation, scaling and translations about the respective axes. This can be performed using a single matrix A generated by the concatenation of the R, S and T matrices as follows: $A = SRT$. Note that this operation is not, in general, commutative. The inverse transformation is given by:

$$X_a = A^{-1}X_b = (SRT)^{-1} X_b = T^{-1}R^{-1}S^{-1}X_b \qquad (2.86)$$

2.7 Rank, Null Space and Invertibility of Matrices

2.7.1 Null Space $N(A)$

Let A be an $m \times n$ matrix. The set of vectors in R^n for which $Ax = 0$ is called the null space of A. It is denoted by $N(A)$. Therefore:

$$N(A) = \{x \in R^n \text{ such that } Ax = 0\} \qquad (2.87)$$

The null space of A is a subspace of the vector space R^n. It can be easily derived by utilizing the Gauss–Jordan technique in order to identify the solution space. This is accomplished by first generating the corresponding augmented matrix that represents the given set of equations. The augmented matrix is then simplified to a RREF or a suitable form that allows for the identification of the appropriate pivot and free variables. These, in turn, are employed to derive the solution space. The procedure is illustrated in the example given below.

Example 2.19

Find the null space of the 2×3 matrix A

$$A = \begin{bmatrix} 1 & 3 & 2 \\ 2 & -3 & 4 \end{bmatrix}$$

Solution: The null space of A is obtained by solving the system $Ax=0$. We first start by generating the augmented matrix A_a and then simplifying it by utilizing the Gauss–Jordan technique. This is shown as:

$$Ax = \begin{bmatrix} 1 & 3 & 2 \\ 2 & -3 & -5 \end{bmatrix} \begin{bmatrix} x_1 \\ x_2 \\ x_3 \end{bmatrix} = \begin{bmatrix} x_1 + 3x_2 + 2x_3 \\ 2x_1 - 3x_2 - 5x_3 \end{bmatrix} = \begin{bmatrix} 0 \\ 0 \end{bmatrix}$$

$$A_a = \begin{bmatrix} 1 & 3 & 2 & 0 \\ 2 & -3 & -5 & 0 \end{bmatrix} \xrightarrow{R_2 \leftarrow R_2 - 2R_1} \begin{bmatrix} 1 & 3 & 2 & 0 \\ 0 & -9 & -9 & 0 \end{bmatrix} \xrightarrow{R_2 \leftarrow R_2/-9} \begin{bmatrix} 1 & 3 & 2 & 0 \\ 0 & 1 & 1 & 0 \end{bmatrix}$$

The value of 1 in the (1, 1) position indicates that the variable x_1 is a "pivot" variable. Similarly, the value of 1 in the (2, 2) position indicates that the variable x_2 is also a "pivot" variable. On the other hand, x_3 is a "free" variable. The justification for the above becomes clear by examining the resulting equations that are generated from the simplified matrix, where the "pivot" variable(s) are written as a function of the "free" variable(s) as shown:

$$x_2 + x_3 = 0 \Rightarrow x_2 = -x_3$$
$$1x_1 + 3x_2 + 2x_3 = 0 \Rightarrow x_1 = -3x_2 - 2x_3 = -3(-x_3) - 2x_3 = x_3$$

This yields: $x_1 = x_3$ and $x_2 = -x_3$. So x_3, the "free" variable, can take on any value, while x_1 and x_2 are fixed once x_3 has been assigned. The final solution can be formatted as follows:

$$\begin{bmatrix} x_1 \\ x_2 \\ x_3 \end{bmatrix} = x_3 \begin{bmatrix} 1 \\ -1 \\ 1 \end{bmatrix}$$

Therefore, the null space of A is the set of vectors in R^3 given by $x = [\alpha \ -\alpha \ \alpha]^T$. Notice that the null space of A is also a vector space which is a subspace of R^3. The solution can be validated by selecting a value for x_3, deriving the corresponding values for x_1 and x_2 and ensuring that $Ax=0$. To this effect, let $x_3 = 2$, then $x = [2 \ -2 \ 2]^T$ and Ax equals:

$$Ax = \begin{bmatrix} 1 & 3 & 2 \\ 2 & -3 & -5 \end{bmatrix} \begin{bmatrix} 2 \\ -2 \\ 2 \end{bmatrix} = \begin{bmatrix} 0 \\ 0 \end{bmatrix}$$

Example 2.20

Find the null space of the 3×3 matrix A

$$A = \begin{bmatrix} 1 & 2 & -1 \\ 1 & 0 & 2 \\ -1 & 1 & 3 \end{bmatrix}$$

Solution: The null space of A is obtained by solving the equation $Ax=0$. Since, in this case, $\det(A) = 1 \neq 0$, matrix A is nonsingular and $Ax=0$ has a unique solution

Determinants, Matrix Inversion and Solutions 87

$x=0$ since $x=A^{-1}0=0$. Therefore, the null space of matrix A is empty. That is $N(A)=\{\phi\}$.

2.7.2 Column Space C(A)

The space spanned by the columns of the $m \times n$ matrix A is called the column space of A. It is also called the range of matrix A and is denoted by $C(A)$. Therefore:

$$C(A) = \{y \in R^m \text{ such that } y = Ax \text{ for } x \in R^n\} \tag{2.88}$$

Given that A is an $m \times n$ matrix, the range of A is a subspace of R^m. If $y_1 \in C(A)$, then y_1 is a linear combinations of the independent columns of A.

Example 2.21

Find the range of the 2×3 matrix A

$$A = \begin{bmatrix} 2 & -3 & 4 \\ 1 & 3 & -2 \end{bmatrix}$$

Solution: The range of A is the subspace spanned by the columns of A and is obtained by:

$$y = Ax = \begin{bmatrix} 2 & -3 & 4 \\ 1 & 3 & -2 \end{bmatrix} \begin{bmatrix} a \\ b \\ c \end{bmatrix} = \begin{bmatrix} 2a - 3b + 4c \\ a + 3b - 2c \end{bmatrix}$$

$$= (2a - 3b + 4c)\begin{bmatrix} 1 \\ 0 \end{bmatrix} + (a + 3b - 2c)\begin{bmatrix} 0 \\ 1 \end{bmatrix} = \alpha \begin{bmatrix} 1 \\ 0 \end{bmatrix} + \beta \begin{bmatrix} 0 \\ 1 \end{bmatrix}$$

Hence, $C(A) = R^2$, which means that any arbitrarily vector in R^2 is in the range space of matrix A.

2.7.3 Row Space R(A)

The row space of an $m \times n$ matrix A is the column space (range) of A^T. The dimension of the column space of A represents the rank of matrix A and is less than or equal to the minimum of n and m. Similarly, the row space of matrix A is the linear combination of the rows of A. The dimension of the row space of A is also equal to the rank of matrix A.

Example 2.22

Find the row space of the 2×3 matrix A

$$A = \begin{bmatrix} 1 & -2 & 3 \\ 1 & 3 & 4 \end{bmatrix}$$

Solution: The row space of A is the subspace spanned by the columns of A^T and is obtained by:

$$y = A^T x = \begin{bmatrix} 1 & 1 \\ -2 & 3 \\ 3 & 4 \end{bmatrix} \begin{bmatrix} a \\ b \end{bmatrix} = \begin{bmatrix} a+b \\ -2a+3b \\ 3a+4b \end{bmatrix}$$

$$= a \begin{bmatrix} 1 \\ -2 \\ 3 \end{bmatrix} + b \begin{bmatrix} 1 \\ 3 \\ 4 \end{bmatrix}$$

Therefore, the row space of A is:

$$R(A) = \left\{ \begin{bmatrix} 1 \\ -2 \\ 3 \end{bmatrix}, \begin{bmatrix} 1 \\ 3 \\ 4 \end{bmatrix} \right\}$$

Example 2.23

Find the row and column spaces of matrix A.

$$A = \begin{bmatrix} 1 & 2 & 2 \\ 2 & 3 & 4 \\ 2 & -3 & 4 \\ -1 & 2 & -2 \end{bmatrix}$$

Solution: The column space of A is the linear combination of the columns of A. Thus:

$$C(A) = \alpha_1 \begin{bmatrix} 1 \\ 2 \\ 2 \\ -1 \end{bmatrix} + \alpha_2 \begin{bmatrix} 2 \\ 3 \\ -3 \\ 2 \end{bmatrix} + \alpha_3 \begin{bmatrix} 2 \\ 4 \\ 4 \\ -2 \end{bmatrix} = \alpha_1 \begin{bmatrix} 1 \\ 2 \\ 2 \\ -1 \end{bmatrix} + \alpha_2 \begin{bmatrix} 2 \\ 3 \\ -3 \\ 2 \end{bmatrix} + 2\alpha_3 \begin{bmatrix} 1 \\ 2 \\ 2 \\ -1 \end{bmatrix}$$

$$= (\alpha_1 + 2\alpha_3) \begin{bmatrix} 1 \\ 2 \\ 2 \\ -1 \end{bmatrix} + \alpha_2 \begin{bmatrix} 2 \\ 3 \\ -3 \\ 2 \end{bmatrix} = \beta_1 \begin{bmatrix} 1 \\ 2 \\ 2 \\ -1 \end{bmatrix} + \beta_2 \begin{bmatrix} 2 \\ 3 \\ -3 \\ 2 \end{bmatrix}$$

Therefore:

$$C(A) = \operatorname{span} \left(\begin{bmatrix} 1 \\ 2 \\ 2 \\ -1 \end{bmatrix}, \begin{bmatrix} 2 \\ 3 \\ -3 \\ 2 \end{bmatrix} \right)$$

Similarly, the row space of A is given by:

$$R(A) = \operatorname{span} \left(\begin{bmatrix} 1 \\ 2 \\ 2 \end{bmatrix}, \begin{bmatrix} 2 \\ 3 \\ 4 \end{bmatrix} \right)$$

Determinants, Matrix Inversion and Solutions

Notice that the dimension of both the row space as well as the column space of A is equal to 2.

2.7.4 Rank of a Matrix

The dimension of the column space (or the row space) of the matrix A is called the rank of A and is denoted by r. Thus:

$$r = \dim(R(A)) = \dim(R(A^T)) \quad (2.89)$$

The rank of a matrix represents the number of independent rows or columns in the matrix. Therefore, for an $m \times n$ matrix:

$$r \leq \min(m,n) \quad (2.90)$$

Definition: An $m \times n$ matrix is said to be full rank if

$$r = \min(m,n) \quad (2.91)$$

A matrix which is not full rank is said to be rank deficient.

Example 2.24

Find the rank of matrix A

$$A = \begin{bmatrix} 2 & -3 & 4 \\ 1 & 3 & -2 \end{bmatrix}$$

Solution: The dimension of the column space of A is 2, therefore the rank of matrix A is 2. That is:

$$r = 2$$

This matrix is full rank since $r = \min(m,n) = \min(2,3) = 2$

Example 2.25

Find the rank of matrix A

$$A = \begin{bmatrix} 1 & 2 & 4 & 1 & -3 \\ 2 & 4 & 0 & 0 & -6 \\ 3 & 6 & 4 & -2 & -9 \end{bmatrix}$$

Solution: The dimension of the column space of A is 2, therefore the rank of A is 2, i.e. $r = 2$. Thus, there are two independent columns. This matrix is rank deficient since $r = 2 < \min(m,n) = \min(3,4) = 3$.

Example 2.26

Find the rank of matrix A given by:

$$A = \begin{bmatrix} 1 & 2 & 4 \\ 2 & 0 & 1 \\ -2 & -1 & 2 \end{bmatrix}$$

Solution: This matrix is square. Its determinant is $\det(A) = -19 \neq 0$. Therefore the matrix is full rank and $r = n = 3$. Generally speaking, a nonsingular square matrix of size $n \times n$ is full rank that is $r = n$.

An $n \times n$ matrix A is invertible or nonsingular if any of these conditions holds. These represent different tests for invertibility of an $n \times n$ matrix:

a. $\det(A) \neq 0$.
b. $\text{rank}(A) = n$.
c. $Ax = 0$ implies that $x = 0$ is the unique solution. The null space of A is empty.
d. The rows (columns) of A are linearly independent.

Properties of matrix rank. The following properties hold for the rank of the product of two matrices. It is assumed that A and B are matrices of appropriate sizes to allow for the product to take place. The proof is left as an exercise (see Problem 2.16).

a. $\text{rank}(AB) \leq \min[\text{rank}(A), \text{rank}(B)]$
b. If matrix A is $m \times n$ and of full rank, then $B = A^T A$ is full rank or nonsingular.
c. If matrix A is $m \times n$ and B is $n \times l$, then

$\text{rank}(A) + \text{rank}(B) - n \leq \text{rank}(AB) \leq \min[\text{rank}(A), \text{rank}(B)]$
This inequality is known as Sylvester's inequality.

2.8 Special Matrices with Applications

Certain matrices of special structures appear frequently in the field of signal processing, communications and controls. Here, we briefly introduce some of these matrices along with their potential applications.

2.8.1 Vandermonde Matrix

An $n \times p$ matrix of the form:

$$V = \begin{bmatrix} 1 & 1 & \cdots & 1 \\ z_1 & z_2 & \cdots & z_p \\ z_1^2 & z_2^2 & \cdots & z_p^2 \\ \vdots & \vdots & \cdots & \vdots \\ z_1^{n-1} & z_2^{n-1} & \cdots & z_p^{n-1} \end{bmatrix} \quad (2.92)$$

Determinants, Matrix Inversion and Solutions 91

is called a Vandermonde matrix. This matrix appears in applications such as harmonic retrieval, system identification and decoding of Bose, Chaudhuri and Hocquenghem (BCH) block codes. Here is an example of a 3×3 Vandermonde matrix:

$$V = \begin{bmatrix} 1 & 1 & 1 \\ 2 & 3 & 4 \\ 4 & 9 & 16 \end{bmatrix}$$

The MATLAB® commands to generate a Vandermonde matrix whose second row is $r = [z_1 \ z_2 \ \ldots \ z_p]$ is V=**vander**(r)′; V=V(end:-1:1,:).

2.8.2 Hankel Matrix

An $(n-p) \times p$ matrix is a Hankel matrix if the elements that belong to a given reverse diagonal of the matrix are equal. The general structure is:

$$H = \begin{bmatrix} x_0 & x_1 & \cdots & x_{p-1} \\ x_1 & x_2 & \cdots & x_p \\ x_2 & x_3 & \cdots & x_{p+1} \\ \vdots & \vdots & \cdots & \vdots \\ x_{n-1-p} & x_{n-p} & \cdots & x_{n-2} \end{bmatrix} \qquad (2.93)$$

An example of a 3×4 Hankel matrix:

$$H = \begin{bmatrix} 1 & 2 & 3 & 4 \\ 2 & 3 & 4 & -6 \\ 3 & 4 & -6 & 5 \end{bmatrix}$$

Hankel matrices are used in signal decomposition and signal parameters estimation. The MATLAB command to produce a square Hankel matrix whose first column is defined by the elements found in a given vector C is H=**hankel**(C).

2.8.3 Toeplitz Matrices

A Toeplitz matrix is a matrix with equal values along the diagonals. A general form of an $n \times n$ Toeplitz matrix is shown below. Toeplitz matrices may not be square matrices

$$T = \begin{bmatrix} x_0 & x_1 & x_2 & \cdots & x_{n-1} \\ x_{-1} & x_0 & x_1 & \ddots & \vdots \\ x_{-2} & x_{-1} & x_0 & \ddots & x_2 \\ \vdots & \ddots & \ddots & \ddots & x_1 \\ x_{-n+1} & \cdots & x_{-2} & x_{-1} & x_0 \end{bmatrix} \qquad (2.94)$$

An example of a 4×4 Toeplitz matrix:

$$T = \begin{bmatrix} 1 & 3 & 4 & 6 \\ 2 & 1 & 3 & 4 \\ -3 & 2 & 1 & 3 \\ 8 & -3 & 2 & 1 \end{bmatrix}$$

The MATLAB command to create a Toeplitz matrix having C as its first column and R as its first row is *T=toeplitz(C, R)*.

2.8.4 Permutation Matrix

A $n \times n$ permutation matrix P is a matrix whose elements are only either zero or one with exactly one 1 in each row and column. For example the following 3×3 matrix is a permutation matrix:

$$P = \begin{bmatrix} 0 & 0 & 1 \\ 1 & 0 & 0 \\ 0 & 1 & 0 \end{bmatrix}$$

Permutation matrices are used to interchange rows and columns of a matrix. If P is a permutation matrix, then PA is a row-permuted version of matrix A and AP is a column-permuted version of matrix A. As an example consider

$$A = \begin{bmatrix} 2 & 3 & 4 \\ -1 & 4 & -5 \\ 3 & 6 & 7 \end{bmatrix} \text{ and } P = \begin{bmatrix} 0 & 0 & 1 \\ 1 & 0 & 0 \\ 0 & 1 & 0 \end{bmatrix}$$

Then

$$PA = \begin{bmatrix} 3 & 6 & 7 \\ 2 & 3 & 4 \\ -1 & 4 & -5 \end{bmatrix} \text{ and } AP = \begin{bmatrix} 3 & 4 & 2 \\ 4 & -5 & -1 \\ 6 & 7 & 3 \end{bmatrix}$$

It can be shown that the inverse of any permutation matrix is equal to its transpose that is $P^{-1}=P^T$. Furthermore, the product of permutation matrices is a permutation matrix.

2.8.5 Markov Matrices

The $n \times n$ matrix A is called a Markov matrix if $a_{ij}>0$ and $\sum_{i=1}^{n} a_{ij} = 1$. For example the following matrices are Markov matrices:

$$A = \begin{bmatrix} \rho & 1-\rho \\ 1-\rho & \rho \end{bmatrix} \quad B = \begin{bmatrix} 0.3 & 0.2 & 0.5 \\ 0.2 & 0.4 & 0.2 \\ 0.5 & 0.4 & 0.3 \end{bmatrix}$$

Determinants, Matrix Inversion and Solutions

Markov matrices are used in modeling discrete communication channels and Markov random sources.

2.8.6 Circulant Matrices

The $n \times n$ matrix A is a circulant matrix if each row is generated by a one sample circular shift of the previous row. The general form for an $n \times n$ circulant matrix is:

$$A = \begin{bmatrix} c_0 & c_1 & c_2 & \cdots & c_{n-2} & c_{n-1} \\ c_{n-1} & c_0 & c_1 & \cdots & c_{n-3} & c_{n-2} \\ c_{n-2} & c_{n-1} & c_0 & \cdots & c_{n-4} & c_{n-3} \\ \vdots & \vdots & \vdots & \vdots & \vdots & \vdots \\ c_2 & c_3 & c_4 & \cdots & c_0 & c_1 \\ c_1 & c_2 & c_3 & \cdots & c_{n-1} & c_0 \end{bmatrix} \quad (2.95)$$

Circulant and block circulant matrices are used in 1-d and 2-d deconvolutions. Image restoration and communication channel equalization are examples of 2-d and 1-d deconvolution. The matrices shown below provide examples of circulant (matrix A) and a block circulant matrix (matrix B).

$$A = \begin{bmatrix} 1 & 2 & 4 & 3 \\ 3 & 1 & 2 & 4 \\ 4 & 3 & 1 & 2 \\ 2 & 4 & 3 & 1 \end{bmatrix}$$

$$B = \begin{bmatrix} B_1 & B_2 & B_3 \\ B_3 & B_1 & B_2 \\ B_2 & B_3 & B_1 \end{bmatrix} = \begin{bmatrix} 1 & 4 & -1 & 2 & 5 & 6 \\ 4 & 1 & 2 & -1 & 6 & 5 \\ 5 & 6 & 1 & 4 & -1 & 2 \\ 6 & 5 & 4 & 1 & 2 & -1 \\ -1 & 2 & 5 & 6 & 1 & 4 \\ 2 & -1 & 6 & 5 & 4 & 1 \end{bmatrix}$$

2.8.7 Hadamard Matrices

The Hadamard matrix of order 2 is given by:

$$H_2 = \begin{bmatrix} 1 & 1 \\ 1 & -1 \end{bmatrix} \quad (2.96)$$

The Hadamard matrix of order N is generated by utilizing the recursive expression:

$$H_N = \begin{bmatrix} H_{\frac{N}{2}} & H_{\frac{N}{2}} \\ H_{\frac{N}{2}} & -H_{\frac{N}{2}} \end{bmatrix} \qquad (2.97)$$

where $H_{N/2}$ is the Hadamard matrix of order $N/2$. It is assumed that $N=2^m$. For example the Hadamard matrix of order 4 is given by:

$$H_4 = \begin{bmatrix} H_2 & H_2 \\ H_2 & -H_2 \end{bmatrix} = \begin{bmatrix} 1 & 1 & 1 & 1 \\ 1 & -1 & 1 & -1 \\ 1 & 1 & -1 & -1 \\ 1 & -1 & -1 & 1 \end{bmatrix} \qquad (2.98)$$

The columns of the Hadamard matrix form an orthogonal set. Hence the inverse of Hadamard matrix is:

$$H_N^{-1} = \frac{1}{N} H_N \qquad (2.99)$$

Hadamard matrices form the basis for the Hadamard transform. The Hadamard transform is used in image data compression. The 1-d Hadamard transform of a N point sequence is defined as:

$$F = H_N f \qquad (2.100)$$

The inverse Hadamard transform is:

$$f = H_N^{-1} = \frac{1}{N} H_N F \qquad (2.101)$$

Example 2.27

Find the Hadamard transform of the 1-d sequence $f = [12 \ \ 14 \ \ 10 \ \ 11]^T$.

Solution:

$$F = H_4 f = \begin{bmatrix} 1 & 1 & 1 & 1 \\ 1 & -1 & 1 & -1 \\ 1 & 1 & -1 & -1 \\ 1 & -1 & -1 & 1 \end{bmatrix} \begin{bmatrix} 12 \\ 14 \\ 10 \\ 11 \end{bmatrix} = \begin{bmatrix} 47 \\ -3 \\ 5 \\ -1 \end{bmatrix}$$

2.8.8 Nilpotent Matrices

Matrix A is said to be nilpotent if there exists an integer $r > 1$ such that:

$$A^r = 0 \qquad (2.102)$$

Nilpotent matrices are generally found in the design of state feedback control systems with deadbeat response. Examples of 2×2, 3×3, and 4×4 Nilpotent matrices are:

$$A = \begin{bmatrix} 0 & 1 \\ 0 & 0 \end{bmatrix}, \quad B = \begin{bmatrix} 0 & 1 & 0 \\ 0 & 0 & 1 \\ 0 & 0 & 0 \end{bmatrix}, \quad \text{and} \quad C = \begin{bmatrix} 0 & 1 & 0 & 0 \\ 0 & 0 & 1 & 0 \\ 0 & 0 & 0 & 1 \\ 0 & 0 & 0 & 0 \end{bmatrix}$$

Notice that $A^2=0$, $B^3=0$, and $C^4=0$.

2.9 Derivatives and Gradients

In a variety of engineering optimization problems, it is often necessary to differentiate a function with respect to a vector. In this section, we examine derivatives or gradients of functions with respect to vectors.

2.9.1 Derivative of Scalar with Respect to a Vector

Let $f(x)$ be a scalar function of vector $x \in R^n$. Then by definition, the derivative or gradient of $f(x)$ with respect to vector x is an $n \times 1$ vector defined as:

$$\frac{\partial f(x)}{\partial x} = \nabla_x f(x) = \begin{bmatrix} \frac{\partial f(x)}{\partial x_1} \\ \frac{\partial f(x)}{\partial x_2} \\ \vdots \\ \frac{\partial f(x)}{\partial x_n} \end{bmatrix} \quad (2.103)$$

The second derivative of $f(x)$ with respect to $x \in R^n$ is an $n \times n$ matrix is defined by:

$$\frac{\partial^2 f(x)}{\partial x^2} = \nabla_x^2 f(x) = \begin{bmatrix} \frac{\partial^2 f(x)}{\partial x_1^2} & \frac{\partial^2 f(x)}{\partial x_1 \partial x_2} & \cdots & \frac{\partial^2 f(x)}{\partial x_1 \partial x_n} \\ \frac{\partial^2 f(x)}{\partial x_2 \partial x_1} & \frac{\partial^2 f(x)}{\partial x_2^2} & \cdots & \frac{\partial^2 f(x)}{\partial x_2 \partial x_n} \\ \vdots & \vdots & \vdots & \vdots \\ \frac{\partial^2 f(x)}{\partial x_n \partial x_1} & \frac{\partial^2 f(x)}{\partial x_n \partial x_2} & \cdots & \frac{\partial^2 f(x)}{\partial x_n^2} \end{bmatrix} \quad (2.104)$$

Example 2.28

Let $x \in R^3$ and $f(x) = 2x_1^2 + x_2 + 3x_2x_3 + 4$, then:

$$\frac{\partial f(x)}{\partial x} = \nabla_x f(x) = \begin{bmatrix} \frac{\partial f(x)}{\partial x_1} \\ \frac{\partial f(x)}{\partial x_2} \\ \frac{\partial f(x)}{\partial x_3} \end{bmatrix} = \begin{bmatrix} 4x_1 \\ 1 + 3x_3 \\ 3x_2 \end{bmatrix}$$

and

$$\frac{\partial^2 f(x)}{\partial x^2} = \begin{bmatrix} 4 & 0 & 0 \\ 0 & 0 & 3 \\ 0 & 3 & 0 \end{bmatrix}$$

Example 2.29

Let $x \in R^n$ and $f(x) = c^T x$, where c is an $n \times 1$ column vector, then:

$$f(x) = c^T_x = c_1 x_1 + c_2 x_2 + \cdots + c_n x_n$$

and

$$\frac{\partial f(x)}{\partial x} = \begin{bmatrix} \frac{\partial f(x)}{\partial x_1} \\ \frac{\partial f(x)}{\partial x_2} \\ \vdots \\ \frac{\partial f(x)}{\partial x_n} \end{bmatrix} = \begin{bmatrix} c_1 \\ c_2 \\ \vdots \\ c_n \end{bmatrix}$$

Hence:

$$\frac{\partial c^T x}{\partial x} = c$$

Similarly

$$\frac{\partial x^T c}{\partial x} = c$$

2.9.2 Quadratic Functions

Let $x \in R^n$ and $f(x) = x^T A x$, where A is an $n \times n$ symmetric matrix, then:

$$f(x) = x^T A x = \sum_{i=1}^{n} \sum_{j=1}^{n} a_{ij} x_i x_j \qquad (2.105)$$

Determinants, Matrix Inversion and Solutions

is a quadratic function of its argument. Consider the first component of $\partial f(x)/\partial x$, that is:

$$\frac{\partial f(x)}{\partial x_1} = \frac{\partial}{\partial x_1} \sum_{i=1}^{n}\sum_{j=1}^{n} a_{ij} x_i x_j$$

$$= \frac{\partial}{\partial x_1}\sum_{i=2}^{n}\sum_{j=2}^{n} a_{ij}x_ix_j + \frac{\partial}{\partial x_1} a_{11}x_1x_1 + \frac{\partial}{\partial x_1}\sum_{j=2}^{n} a_{1j}x_1x_j + \frac{\partial}{\partial x_1}\sum_{i=2}^{n} a_{i1}x_ix_1 \quad (2.106)$$

$$= \frac{\partial}{\partial x_1}\sum_{i=2}^{n}\sum_{j=2}^{n} a_{ij}x_ix_j + \frac{\partial}{\partial x_1} a_{11}x_1^2 + \frac{\partial}{\partial x_1} x_1 \sum_{j=2}^{n} a_{1j}x_j + \frac{\partial}{\partial x_1} x_1 \sum_{i=2}^{n} a_{i1}x_i$$

The first term in the above expansion is independent of x_1, therefore it does not contribute to the derivative. Hence:

$$\frac{\partial f(x)}{\partial x_1} = 0 + 2a_{11}x_1 + \sum_{j=2}^{n} a_{1j}x_j + \sum_{i=2}^{n} a_{i1}x_i = \sum_{j=1}^{n} a_{1j}x_j + \sum_{i=1}^{n} a_{i1}x_i = 2\sum_{j=1}^{n} a_{1j}x_j \quad (2.107)$$

This is equal to the first element of the product $2Ax$. Similarly it can be shown that $\partial f(x)/\partial x_i$ is equal to the ith element of the product $2Ax$, hence:

$$\frac{\partial x^T A x}{\partial x} = 2Ax \quad (2.108)$$

Similarly, it can be shown that:

$$\frac{\partial^2 x^T A x}{\partial x^2} = 2A \quad (2.109)$$

Example 2.30

Let $x \in R^3$ and $f(x) = 2x_1^2 + +3x_2^2 + 4x_3^2 - 2x_1x_2 + 4x_1x_3 + 2x_2x_3$.
Find $\partial f(x)/\partial x$ and $\partial^2 f(x)/\partial x^2$.

Solution:

$$f(x) = 2x_1^2 + +3x_2^2 + 4x_3^2 - 2x_1x_2 + 4x_1x_3 + 2x_2x_3$$

$$= \begin{bmatrix} x_1 & x_2 & x_3 \end{bmatrix} \begin{bmatrix} 2 & -1 & 2 \\ -1 & 3 & 1 \\ 2 & 1 & 4 \end{bmatrix} \begin{bmatrix} x_1 \\ x_2 \\ x_3 \end{bmatrix} = x^T A x$$

The first derivative of $f(x)$ is:

$$\frac{\partial f(x)}{\partial x} = 2Ax = 2\begin{bmatrix} 2 & -1 & 2 \\ -1 & 3 & 1 \\ 2 & 1 & 4 \end{bmatrix}\begin{bmatrix} x_1 \\ x_2 \\ x_3 \end{bmatrix} = \begin{bmatrix} 4x_1 - 2x_2 + 4x_3 \\ -2x_1 + 6x_2 + 2x_3 \\ 4x_1 + 2x_2 + 8x_3 \end{bmatrix}$$

The second derivative of $f(x)$ is:

$$\frac{\partial^2 f(x)}{\partial x^2} = \frac{\partial 2Ax}{\partial x} = 2A = \begin{bmatrix} 4 & -2 & 4 \\ -2 & 6 & 2 \\ 4 & 2 & 8 \end{bmatrix}$$

Example 2.31

Let $x \in R^n$, A be an $n \times n$ symmetric matrix, b be an $n \times 1$ column vector and c a scalar. Define $f(x) = 1/2 x^T A x + b^T x + c$.
Find $\partial f(x)/\partial x$ and $\partial^2 f(x)/\partial x^2$.

Solution: The first derivative is:

$$\frac{\partial f(x)}{\partial x} = Ax + b$$

and the second derivative is:

$$\frac{\partial^2 f(x)}{\partial x^2} = \frac{\partial (Ax + b)}{\partial x} = A$$

2.9.3 Derivative of a Vector Function with Respect to a Vector

Let $f(x) \in R^m$ be a vector function of vector $x \in R^n$, then by definition:

$$\frac{\partial f(x)}{\partial x} = \begin{bmatrix} \frac{\partial f_1(x)}{\partial x_1} & \frac{\partial f_2(x)}{\partial x_1} & \cdots & \frac{\partial f_m(x)}{\partial x_1} \\ \frac{\partial f_1(x)}{\partial x_2} & \frac{\partial f_2(x)}{\partial x_2} & \cdots & \frac{\partial f_m(x)}{\partial x_2} \\ \vdots & \vdots & \vdots & \vdots \\ \frac{\partial f_1(x)}{\partial x_n} & \frac{\partial f_2(x)}{\partial x_n} & \cdots & \frac{\partial f_m(x)}{\partial x_n} \end{bmatrix} \qquad (2.110)$$

The $m \times n$ matrix $\partial f(x)/\partial x$ is also known as the Jacobian of $f(x)$ with respect to x.

Example 2.32

Let $x \in R^3$ and $f(x) = \begin{bmatrix} x_1^2 x_2 + 3 x_2 + 4 x_3 - 5 \\ x_1 x_2 + x_2 - 2 x_1 x_2 x_3 \end{bmatrix}$, then Jacobian of $f(x)$ with respect to x is:

$$\frac{\partial f(x)}{\partial x} = \begin{bmatrix} \frac{\partial f_1(x)}{\partial x_1} & \frac{\partial f_2(x)}{\partial x_1} \\ \frac{\partial f_1(x)}{\partial x_2} & \frac{\partial f_2(x)}{\partial x_2} \\ \frac{\partial f_1(x)}{\partial x_3} & \frac{\partial f_2(x)}{\partial x_3} \end{bmatrix} = \begin{bmatrix} 2 x_1 x_2 & x_2 - 2 x_2 x_3 \\ x_1^2 + 3 & x_1 - 2 x_1 x_3 + 1 \\ 4 & -2 x_1 x_2 \end{bmatrix}$$

Problems

PROBLEM 2.1

a. Find determinant of the following matrices:

$$A = \begin{bmatrix} 1 & 2 & 3 \\ 4 & 5 & 6 \\ 7 & 8 & 9 \end{bmatrix}, B = \begin{bmatrix} 1 & 0 & -1 \\ 0 & 1 & 2 \\ 1 & 2 & -1 \end{bmatrix}$$

b. Show that $\det(AB) = \det(A)\det(B)$.

PROBLEM 2.2

Given matrices A and B, compute:

a. $\det(A)$, $\det(B)$, and $\det(A+B)$.
b. $\det(A+2I)$.
c. $\det(A \times B)$, where $\times$ stands for array multiplication.
d. $\det(A \otimes B)$, where $\otimes$ stands for Kronecker product.

$$A = \begin{bmatrix} 1 & 2 & 3 \\ 0 & 1 & 2 \\ 0 & 0 & 1 \end{bmatrix}, B = \begin{bmatrix} 1 & 0 & 0 \\ 2 & 1 & 0 \\ 3 & 2 & 1 \end{bmatrix}$$

PROBLEM 2.3

Show that the determinant of a permutation matrix is either 1 or –1.

PROBLEM 2.4

Let A and B be two $n \times n$ matrices. Show that:

a. $\det(AB) = \det(BA) = \det(A)\det(B)$.
b. $\det(A^T) = \det(A)$.
c. $\det(\alpha A) = \alpha^n \det(A)$.
d. If A has a row or column of zeros then, $\det(A) = 0$.
e. If A has two identical rows or columns then, $\det(A) = 0$.

PROBLEM 2.5

Find the inverse of the following matrices:

$$A = \begin{bmatrix} 2 & 3 \\ -1 & 4 \end{bmatrix}, B = \begin{bmatrix} \cos(\theta) & -\sin(\theta) \\ \sin(\theta) & \cos(\theta) \end{bmatrix}, C = \begin{bmatrix} 1 & 2 & 3 \\ 4 & 5 & 6 \\ 7 & 8 & 9 \end{bmatrix}, \text{ and } D = \begin{bmatrix} 1-j & j & 1+j & 2 \\ -j & 4 & 2-j & 3 \\ 1-j & 2+j & j & 3-j \\ 2 & 3 & 3+j & 1 \end{bmatrix}$$

PROBLEM 2.6

Show that:

$$A^{-1} = B^{-1} - A^{-1}(A-B)B^{-1}$$

PROBLEM 2.7
Show that if A and B are nonsingular $n \times n$ square matrices, then:
a. $AA^{-1} = A^{-1}A = I$.
b. $(AB)^{-1} = B^{-1}A^{-1}$.
c. $\det(A^{-1}) = 1/\det(A)$.

PROBLEM 2.8
Solve the following set of linear equations using matrix inversion

a. $\begin{cases} 2x_1 + 3x_2 + x_3 = 11 \\ -2x_1 + x_2 + x_3 = 3 \\ 4x_1 - 3x_2 + 4x_3 = 10 \end{cases}$; b. $\begin{cases} 2x_1 + 3x_2 + 4x_3 + x_4 = 24 \\ x_2 + x_3 - 3x_4 = 18 \\ 4x_3 + 5x_4 = 10 \\ x_1 - x_3 = 7 \end{cases}$

PROBLEM 2.9
Use MATLAB to solve the following set of equations

a. $\begin{bmatrix} 3 & 1.01 \\ 3 & 1 \end{bmatrix} \begin{bmatrix} x_1 \\ x_2 \end{bmatrix} = \begin{bmatrix} 5.02 \\ 5 \end{bmatrix}$

b. $\begin{bmatrix} 2 & -1 & 3 & 4 & -2 \\ 0 & -2 & 1 & 5 & 4 \\ 2 & 3 & 6 & -3 & -4 \\ 1 & 1 & -4 & 3 & 2 \\ 4 & 0 & -4 & 5 & -2 \end{bmatrix} \begin{bmatrix} x_1 \\ x_2 \\ x_3 \\ x_4 \\ x_5 \end{bmatrix} = \begin{bmatrix} 8 \\ -11 \\ 12 \\ 6 \\ 7 \end{bmatrix}$

c. $\begin{bmatrix} 1 & -1 & 2 & -2 & 3 & 2 & 4 \\ -1 & -5 & 4 & 3 & 2 & 0 & 1 \\ 0 & 2 & -5 & 4 & -5 & 2 & 3 \\ 1 & 2 & 1 & -3 & -2 & -5 & -1 \\ 3 & 4 & 5 & -2 & -3 & -1 & 0 \\ 2 & 0 & 5 & 4 & 6 & 7 & 1 \\ 1 & 1 & 1 & 2 & 2 & -2 & 3 \end{bmatrix} \begin{bmatrix} x_1 \\ x_2 \\ x_3 \\ x_4 \\ x_5 \\ x_6 \\ x_7 \end{bmatrix} = \begin{bmatrix} 8 \\ 26 \\ 5 \\ -1 \\ 5 \\ 13 \\ 21 \end{bmatrix}$

d. $\begin{bmatrix} 0 & 1 & -1 & 0 & -1 & 1 & 0 & -1 & 1 & 0 \\ 0 & 0 & 0 & -1 & -1 & 1 & 2 & -1 & -2 & 1 \\ -1 & 1 & -1 & 2 & 2 & -2 & -1 & -2 & 1 & 0 \\ 0 & 1 & 2 & 1 & 2 & -1 & 0 & 1 & 2 & -1 \\ 1 & 1 & -1 & -2 & 0 & 2 & 1 & -2 & -1 & 0 \\ -1 & 0 & -1 & 2 & 2 & 1 & 1 & 0 & 1 & 0 \\ 2 & 0 & -2 & -1 & -2 & -1 & 1 & -1 & 1 & -2 \\ -1 & -1 & 1 & 1 & 0 & 1 & -1 & 2 & 1 & 1 \\ -1 & -2 & -2 & 2 & -1 & 0 & -2 & 2 & 0 & 0 \\ 2 & -2 & -2 & -2 & -2 & 1 & -2 & -2 & 0 & 1 \end{bmatrix} \begin{bmatrix} x_1 \\ x_2 \\ x_3 \\ x_4 \\ x_5 \\ x_6 \\ x_7 \\ x_8 \\ x_9 \\ x_{10} \end{bmatrix} = \begin{bmatrix} 1 \\ -9 \\ 1 \\ 17 \\ -3 \\ 13 \\ -11 \\ 10 \\ -2 \\ -16 \end{bmatrix}$

Determinants, Matrix Inversion and Solutions

PROBLEM 2.10
Find the rank of the following matrices:

a. $A = \begin{bmatrix} 1 & 2 & 2 \\ 4 & 5 & 2 \\ 7 & 8 & 2 \end{bmatrix}$

b. $B = \begin{bmatrix} \cos(\theta) & -\sin(\theta) \\ \sin(\theta) & \cos(\theta) \end{bmatrix}$

c. $C = \begin{bmatrix} 2 & -2 & -3 & 4 \\ 0 & -2 & 3 & -9 \\ -1 & 0 & 7 & 3 \\ 0 & -2 & 0 & -8 \end{bmatrix}$

d. $D = \begin{bmatrix} 1 & -j & 3-j4 \\ 1-j & -1+j & -1+j \\ 4 & j2 & 12+j8 \end{bmatrix}$

PROBLEM 2.11
For the rotation matrix:

$$B(\theta) = \begin{bmatrix} \cos(\theta) & -\sin(\theta) \\ \sin(\theta) & \cos(\theta) \end{bmatrix}$$

Show that:

a. $B(\theta)B(-\theta) = I$.
b. $B(\theta_1 + \theta_2) = B(\theta_1)\, B(\theta_2) = B(\theta_2)\, B(\theta_1)$.
c. $B(n\theta) = [B(\theta)]^n$.

PROBLEM 2.12
Find the null and column space of the 2×3 matrix A

$$A = \begin{bmatrix} -2 & 3 & 1 \\ 3 & 5 & 2 \end{bmatrix}$$

PROBLEM 2.13
Let A be the complex $N \times N$ DFT matrix:

$$A = \begin{bmatrix} W_N^0 & W_N^0 & W_N^0 & \cdots & W_N^0 \\ W_N^0 & W_N^1 & W_N^2 & \cdots & W_N^{N-1} \\ \vdots & \vdots & \vdots & \cdots & \vdots \\ W_N^0 & W_N^{N-2} & W_N^{2(N-2)} & \cdots & W_N^{(N-2)(N-1)} \\ W_N^0 & W_N^{N-1} & W_N^{2(N-1)} & \cdots & W_N^{(N-1)(N-1)} \end{bmatrix}$$ where $W_N = \exp(-j\frac{2\pi}{N})$

Show that $A^{-1} = 1/NA^H$.

PROBLEM 2.14
Determine the null space of the following matrices:

$$A = \begin{bmatrix} -6 & 2 & -5 \\ 2 & 3 & -1 \end{bmatrix} \quad B = \begin{bmatrix} 2 & 3 \\ -3 & -4 \\ 2 & -6 \end{bmatrix}$$

PROBLEM 2.15
Show that if A, B, C and D are matrices of appropriate dimensions, then:

$$\begin{bmatrix} A & B \\ 0 & D \end{bmatrix}^{-1} = \begin{bmatrix} A^{-1} & -A^{-1}BD^{-1} \\ 0 & D^{-1} \end{bmatrix}, \quad \begin{bmatrix} A & 0 \\ C & B \end{bmatrix}^{-1} = \begin{bmatrix} A^{-1} & 0 \\ -B^{-1}CA^{-1} & B^{-1} \end{bmatrix}$$

PROBLEM 2.16
Let A and B be matrices of appropriate sizes. Show that:

a. rank(AB)$\leq$min[rank(A), rank(B)].
b. If matrix A is $m \times n$ and of full rank, then $B = A^T A$ is full rank or nonsingular.
c. If matrix A is $m \times n$ and B is $n \times l$, then:
rank(A)+rank(B)$-n \leq$ rank(AB)$\leq$min[rank(A), rank(B)].

PROBLEM 2.17
Assume that we represent complex numbers using real matrices, that is complex number $z = a+jb$ is represented by the real 2×2 matrix Z according to the following transformation:

$$z = a + jb \rightarrow Z = \begin{bmatrix} a & -b \\ b & a \end{bmatrix}$$

I. What is the matrix representation of complex number j? Show that $j^2 = -1$ and $j^3 = -j$.

II. Show that complex algebra can be performed using matrix algebra, that is:

a. $z_1 \pm z_2 \rightarrow Z_1 \pm Z_2$.
b. $z_1 z_2 \rightarrow Z_1 Z_2$.
c. $z_1 / z_2 \rightarrow Z_1 Z_2^{-1}$.
d. If $z \rightarrow Z$, then $z^* \rightarrow Z^T$.

PROBLEM 2.18
The bit-reverse operation used in FFT algorithm is a permutation operation. The following table shows the bit-reverse operation used in an eight-point FFT computation. Find a permutation matrix P such that:
y(binary)$= Px$(binary)

x Decimal	x Binary	y Bit-reverse	y Decimal
0	000	000	0
1	001	100	4
2	010	010	2
3	011	110	6
4	100	001	1
5	101	101	5
6	110	011	3
7	111	111	7

PROBLEM 2.19

a. Show that if $A(t)$ is a nonsingular $n \times n$ matrix with time dependent elements, then:

$$\frac{d}{dt}A^{-1}(t) = -A^{-1}(t)\frac{dA(t)}{dt}A^{-1}(t)$$

b. Find $\frac{d}{dt}A^{-1}(t)$ if $A(t) = \begin{bmatrix} t & 1 \\ e^{-t} & 1+t \end{bmatrix}$.

PROBLEM 2.20

Let $f(x)$ be defined as:

$$f(x) = (x^T A x)^2 + (b^T x)^2 + c^T x + d$$

where x, b, and c are $n \times 1$ vectors, A is a symmetric $n \times n$ matrix and d is scalar.

a. Find $\partial f(x)/\partial x$.
b. Find $\partial f^2(x)/\partial x^2$.

PROBLEM 2.21 (MATLAB)

a. Use MATLAB to generate $N=128$ samples of the signal given as:

$$x(n) = 2\cos\left(\frac{15\pi}{64}n\right) + 3\sin\left(\frac{17\pi}{64}n\right) \quad n = 0, 1, 2, \ldots, N-$$

b. Compute the DFT of this sequence using the MATLAB FFT routine. Plot the magnitude of the FFT and locate the locations of the peaks of the magnitude.
c. Repeat part using matrix multiplication. Plot the DFT magnitude, locate the peaks and compare the results with results obtained in part.

PROBLEM 2.22 (MATLAB)

a. Use MATLAB to generate $N=256$ samples of the signal given as:

$$x(n) = 2\cos\left(\frac{15.8\pi}{128}n\right) + 3\sin\left(\frac{27.3\pi}{128}n\right) \quad n = 0, 1, 2, \ldots, N-1$$

b. Compute the DFT of this sequence using the MATLAB FFT routine. Plot the magnitude of the FFT and locate the locations of the peaks of the magnitude.
c. Repeat part (b) using matrix multiplication. Plot the DFT magnitude, locate the peaks and compare the results with results obtained in part (b).

PROBLEM 2.23 (MATLAB)

Use MATLAB command $f=$**imread('cameraman.tif');** to read the cameraman image into MATLAB workspace.

a. Scale the image by 0.5 in both direction and display the result.
b. Rotate the image by 4° clockwise and display the result.
c. Rotate the image by 4° counterclockwise and display the result.

PROBLEM 2.24 (MATLAB)

a. Use MATLAB to generate a Nilpotent matrix of size 16×16. Call it A.
b. Find $\det(A)$.
c. Show that it is a Nilpotent matrix by computing A^{16}.
d. Compute $B = (A+I)^{-1}$. Is B a Nilpotent matrix?

PROBLEM 2.25 (MATLAB)

a. Use MATLAB to generate a Hadamard matrix of size 32×32. Call it A.
b. Find $\det(A)$.
c. Find $\text{Trace}(A)$.
d. Find $\det[(A+I)^{-2}(I-A)]$.

PROBLEM 2.26 (MATLAB)

The probability transition matrix of a binary symmetric communication channel with two inputs $x = \begin{bmatrix} x_1 \\ x_2 \end{bmatrix}$ and two outputs $y = \begin{bmatrix} y_1 \\ y_2 \end{bmatrix}$ is given by

$$P = \begin{bmatrix} p_{11} & p_{12} \\ p_{21} & p_{22} \end{bmatrix} = \begin{bmatrix} 1-\rho & \rho \\ \rho & 1-\rho \end{bmatrix}$$

where $p_{ij} = \Pr(y=y_j | x=x_i)$ and $0 \le \rho \le 0.5$ is probability of incorrect transmission.

a. Find the probability transition matrix of cascade of two such channels.
b. Repeat part (a) for cascade of three channels.
c. What is the limiting case as number of channels goes to infinity?

PROBLEM 2.27 (MATLAB)

Consider the linear regression model

$$y(k) = a\cos(0.56\pi k) + b\sin(0.675\pi k) + v(k) \quad k=0,1,2,\ldots,99$$

where $a=2b=2$ and $v(k)$ is zero mean Gaussian white noise with $\sigma_v^2 = 0.04$.

a. Find the signal to noise ratio in dB defined as. $SNR = 10\log(a^2+b^2/2\sigma_v^2)$
b. Use MATLAB to generate data $y(0), y(1),\ldots,y(99)$.
c. Use RLS algorithm to estimate a and b. Denote the estimated values by $\hat{a}$ and $\hat{b}$.
d. Find the resulting normalized MSE defined by $MSE = ((a-\hat{a})^2 + (b-\hat{b})^2)/a^2+b^2$.
e. Repeat parts (b)–(d) for $SNR=10$ dB. (Hint: use same values for a and b, but change σ_v^2).

3

Linear Vector Spaces

3.1 Introduction

Linear vector space is an important concept in linear algebra. Linear vector spaces have applications in different fields of engineering. In the discipline of electrical engineering, vector spaces are extensively used in detection and estimation theory, color engineering, signal and image processing and statistical communication theory. The concept of normed and inner product vector spaces is the foundation of optimization theory and is widely used in solving engineering problems such as optimal control and operations research type problems including inventory control. The rich theory of vector spaces is the foundation of modern engineering and computational algorithms used to solve various engineering problems. This chapter addresses the concept of linear vector spaces and normed and inner product spaces with applications in signal and image processing, communication and color engineering.

3.2 Linear Vector Space

The theory of linear vector spaces is the main topic of this section. Numerous practical examples are used to clarify the concept of a vector space.

3.2.1 Definition of Linear Vector Space

A linear vector space is a set V, together with two operations:

i. **Vector addition** $+: V \times V \rightarrow V$ that takes two vectors and produces the sum
ii. **Scalar multiplication** $\cdot : R^1(C^1) \times V \rightarrow V$

such that for every x, y and $z \in V$ and every scalar $\alpha \in R^1(C^1)$ (field of real or complex numbers) the following conditions hold:

a. $x+y=y+x$ (commutative).
b. $x+(y+z)=(x+y)+z$ (associative).
c. There is a unique element, denoted by 0, such that $x+0=x$ for every $x \in V$.
d. There is a unique element, denoted by $-x$, such that $x+(-x)=0$, where 0 is zero element of V.
e. $\alpha(x+y)=\alpha x+\alpha y$.
f. $\alpha(\beta x)=(\alpha\beta)x$.
g. $(\alpha+\beta)x=\alpha x+\beta x$.
h. $1.x=x$.
i. $0.x=0$.

Elements of V are called vectors. If the field associated with the vector space is complex, it is called a complex vector space. The real field is denoted by R^1 and the complex field by C^1. In summary, a vector space is a set of vectors that is closed under vector addition and scalar multiplication satisfying the above axioms.

3.2.2 Examples of Linear Vector Spaces

Example 3.1

The set R^n consisting of all n dimensional vectors is a linear vector space, with addition and scalar multiplication defined as:

$$x+y = [x_1+y_1 \quad x_2+y_2 \quad \cdots \quad x_n+y_n]^T \tag{3.1}$$

and

$$\alpha x = [\alpha x_1 \quad \alpha x_2 \quad \cdots \quad \alpha x_n]^T \tag{3.2}$$

where $x=[x_1 \ x_2 \ \cdots \ x_n]^T$ and $y=[y_1 \ y_2 \ \cdots \ y_n]^T$ are elements of V and α is a scalar. Note that the set of real numbers is by itself a linear vector space. The complex linear vector space C^n consists of all n dimensional vectors and is defined similarly to R^n.

Example 3.2

The set S of all real-valued continuous functions defined over the open interval of $(a\ b)$, with addition and scalar multiplication defined by $(f+g)(x)=f(x)+g(x)$ and $(\alpha\ f)(x)=\alpha\ f(x)$ (in an ordinary way) is a linear vector space. It is easy to show that all the linear vector space axioms hold; therefore, it is a vector space. We refer to this vector space as $L(a, b)$.

Linear Vector Spaces

Example 3.3

Let P_n be the set of all polynomials with real coefficients of degree less than or equal to n. A typical element of this set is:

$$p(x) = \sum_{i=0}^{n} p_i x^i \tag{3.3}$$

Let us define polynomial addition and scalar multiplication as:

$$(p+g)(x) = p(x) + g(x) = \sum_{i=0}^{n} p_i x^i + \sum_{i=0}^{n} g_i x^i = \sum_{i=0}^{n} (p_i + g_i) x^i \tag{3.4}$$

and

$$(\alpha\, p)(x) = \alpha\, p(x) = \sum_{i=0}^{n} \alpha p_i x^i \tag{3.5}$$

P_n is a vector space since all the vector space axioms hold. The zero vector of this space is $p(x) = 0x^n + 0x^{n-1} + \cdots + 0 = 0$.

3.2.3 Additional Properties of Linear Vector Spaces

If V is a vector space and $x \in V$, then:

i. $x + y = 0$ implies that $x = -y$.
ii. $(-1)x = -x$.

These properties can be easily proven as follows:

i. Since $x + y = 0$, then $-y + (x + y) = -y$, therefore $x = -y$.
ii. $(-1)x + (1)x = (-1 + 1)x = 0x = 0$, therefore $(-1)x + x = 0$, hence $(-1)x = -x$.

3.2.4 Subspace of a Linear Vector Space

A subset M of a linear vector space V is a subspace of V if:

a. $x + y \in M$, whenever $x \in M$ and $y \in M$.
b. $\alpha x \in M$, whenever $x \in M$ and $\alpha \in R^1(C^1)$.

Notice that M is also a linear vector space. As an example, R^n is a subspace of C^n and R^2 is a subspace of R^3.

Example 3.4

Consider the vector space R^2. Let S be the set of all vectors in R^2 whose components are nonnegative that is:

$$S = \{v = [x \quad y]^T \in R^2 \mid x \geq 0 \text{ and } y \geq 0\} \qquad (3.6)$$

S is not a subspace of R^2 since if $v \in S$, then αv is not in S for negative values of α.

Example 3.5

Let V be a vector space consisting of all 3×3 matrices. The set of all lower triangular 3×3 matrices is a subspace of vector space V. This can be easily verified since:

a. The sum of two lower triangular matrices is a lower triangular matrix, and
b. multiplying a lower triangular matrix by a scale factor results in another lower triangular matrix.

Example 3.6

Let V be a vector space of $n \times n$ matrices. The set of all symmetric $n \times n$ matrices is a subspace of the vector space V. The proof is left as an exercise (see Problem 3.2).

Example 3.7

Consider set S, a subset of vector space R^3, defined by:

$$S = \{v = [x \quad y \quad z]^T \in R^3 \mid x = y\} \qquad (3.7)$$

S is nonempty since $x = [1 \ 1 \ 0]^T \in S$. Furthermore:

i. If $v = [x \quad x \quad z] \in S$, then $\alpha v = [\alpha x \quad \alpha x \quad \alpha z] \in S$
ii. If $v_1 = [x_1 \quad x_1 \quad z_1] \in S$ and $v_2 = [x_2 \quad x_2 \quad z_2] \in S$ then the sum

$$v_1 + v_2 = [x_1 + x_2 \quad x_1 + x_2 \quad z_1 + z_2] \in S.$$

Since S is nonempty and satisfies the two closure conditions, S is a subspace of R^3.

3.3 Span of a Set of Vectors

A set of vectors can be combined to form a new space, called the span of a set of vectors. The span of a set of vectors can be defined in term of linear combinations.

Linear Vector Spaces

Definition of linear combination: Vector V is a linear combination of vectors $V_1, V_2, \ldots, V_n$ if there are scalar constants $\alpha_1, \alpha_2, \ldots, \alpha_n$ such that:

$$V = \alpha_1 V_1 + \alpha_2 V_2 + \cdots + \alpha_n V_n \tag{3.8}$$

Example 3.8

Show that vector $V = [1 \quad 8 \quad 11]^T$ is a linear combination of $V_1 = [1 \quad 2 \quad 1]^T$ and $V_2 = [-1 \quad 1 \quad 4]^T$

Solution:

$$V = \alpha_1 V_1 + \alpha_2 V_2 \rightarrow \begin{bmatrix} 1 \\ 8 \\ 11 \end{bmatrix} = \alpha_1 \begin{bmatrix} 1 \\ 2 \\ 1 \end{bmatrix} + \alpha_2 \begin{bmatrix} -1 \\ 1 \\ 4 \end{bmatrix} \rightarrow \begin{cases} \alpha_1 - \alpha_2 = 1 \\ 2\alpha_1 + \alpha_2 = 8 \\ \alpha_1 + 4\alpha_2 = 11 \end{cases}$$

The above set of three equations with two unknowns has a unique solution $\alpha_1 = 3$, $\alpha_2 = 2$. Therefore, vector V is a linear combination of the two vectors V_1 and V_2.

Definition: The set of all linear combinations of $V_1, V_2, \ldots, V_n$ is called the span of $V_1, V_2, \ldots, V_n$ and is denoted by span($V_1, V_2, \ldots, V_n$).

Example 3.9

Let e_1, e_2 and e_3 be three unit vectors in R^3 as shown in Figure 3.1. Then, the span of e_1 and e_2 is:

$$\text{span}(e_1, e_2) = \alpha_1 \begin{bmatrix} 1 \\ 0 \\ 0 \end{bmatrix} + \alpha_2 \begin{bmatrix} 0 \\ 1 \\ 0 \end{bmatrix} = \begin{bmatrix} \alpha_1 \\ \alpha_2 \\ 0 \end{bmatrix} = \text{set of all vectors in } x\text{--}y \text{ plane. It is a subspace}$$

of vector space R^3. Also, the span of e_1, e_2 and e_3 is:

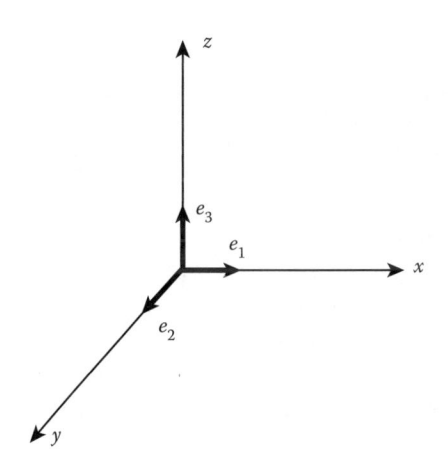

FIGURE 3.1
Vector space R^3.

$$\text{span}(e_1, e_2, e_3) = \alpha_1 \begin{bmatrix} 1 \\ 0 \\ 0 \end{bmatrix} + \alpha_2 \begin{bmatrix} 0 \\ 1 \\ 0 \end{bmatrix} + \alpha_3 \begin{bmatrix} 0 \\ 0 \\ 1 \end{bmatrix} = \begin{bmatrix} \alpha_1 \\ \alpha_2 \\ \alpha_3 \end{bmatrix} = R^3 \qquad (3.9)$$

The following theorem relates the span of a set of vectors to the subspace of the same vector space.

Theorem 3.1: If $V_1, V_2, \ldots, V_n$ are elements of a vector space V, then: span($V_1, V_2, \ldots, V_n$) is a subspace of V.

Proof: Let $V = \alpha_1 V_1 + \alpha_2 V_2 + \cdots + \alpha_n V_n$ be an arbitrary element of span($V_1, V_2, \ldots, V_n$), then for any scalar β.

$$\beta V = \beta(\alpha_1 V_1 + \alpha_2 V_2 + \cdots + \alpha_n V_n) = (\beta \alpha_1)V_1 + (\beta \alpha_2)V_2 + \cdots + (\beta \alpha_n)V_n$$
$$= \gamma_1 V_1 + \gamma_2 V_2 + \cdots + \lambda_n V_n \qquad (3.10)$$

Therefore, $\beta V \in \text{span}(V_1, V_2, \ldots, V_n)$. Next, let $V = \alpha_1 V_1 + \alpha_2 V_2 + \cdots + \alpha_n V_n$ and $W = \beta_1 V_1 + \beta_2 V_2 + \ldots + \beta_n V_n$, then

$$V + W = \alpha_1 V_1 + \alpha_2 V_2 + \cdots + \alpha_n V_n + \beta_1 V_1 + \beta_2 V_2 + \cdots + \beta_n V_n$$
$$= (\alpha_1 + \beta_1)V_1 + (\alpha_2 + \beta_2)V_2 + \cdots + (\alpha_n + \beta_n)V_n \qquad (3.11)$$
$$= \gamma_1 V_1 + \gamma_2 V_2 + \cdots + \gamma_n V_n \in \text{span}(V_1, V_2, \ldots, V_n)$$

Therefore, span($V_1, V_2, \ldots, V_n$) is a subspace of V.

3.3.1 Spanning Set of a Vector Space

Definition: The set $\{V_1, V_2, \ldots, V_n\}$ is a spanning set of vector space V if and only if every vector in vector space V can be written as a linear combination of $V_1, V_2, \ldots, V_n$.

It can be easily proved (see Problem 3.3) that if $V_1, V_2, \ldots, V_n$ spans the vector space V and one of these vectors can be written as a linear combination of the other $n-1$ vectors, then these $n-1$ vectors span the vector space V.

3.3.2 Linear Dependence

Definition: A set of vectors $V_1, V_2, \ldots, V_n$ is linearly dependent if there are scalars $\alpha_1, \alpha_2, \ldots, \alpha_n$ at least one of which is not zero, such that $\alpha_1 V_1 + \alpha_2 V_2 + \cdots + \alpha_n V_n = 0$. This means that a set of vectors is linearly dependent if one of them can be written as a linear combination of the others. A set of vectors that is not linearly dependent is linearly independent.

Example 3.10

Consider the following two vectors $V_1 = \begin{bmatrix} 1 \\ 1 \end{bmatrix}$ and $V_2 = \begin{bmatrix} 1 \\ 2 \end{bmatrix}$ in R^2. Are V_1 and V_2 linearly independent?

Linear Vector Spaces

Solution:

$$\alpha_1 V_1 + \alpha_2 V_2 = 0 \to$$

$$\alpha_1 \begin{bmatrix} 1 \\ 1 \end{bmatrix} + \alpha_2 \begin{bmatrix} 1 \\ 2 \end{bmatrix} = 0 \to \begin{cases} \alpha_1 + \alpha_2 = 0 \\ \alpha_1 + 2\alpha_2 = 0 \end{cases} \to \begin{cases} \alpha_1 = 0 \\ \alpha_2 = 0 \end{cases}$$

Therefore, V_1 and V_2 are linearly independent.

Example 3.11

Are the following three vectors linearly independent?

$$V_1 = \begin{bmatrix} 1 \\ 2 \\ 0 \end{bmatrix}, V_2 = \begin{bmatrix} 1 \\ -1 \\ 1 \end{bmatrix}, V_3 = \begin{bmatrix} 2 \\ 1 \\ 1 \end{bmatrix}$$

Solution:

$$\alpha_1 \begin{bmatrix} 1 \\ 2 \\ 0 \end{bmatrix} + \alpha_2 \begin{bmatrix} 1 \\ -1 \\ 1 \end{bmatrix} + \alpha_3 \begin{bmatrix} 2 \\ 1 \\ 1 \end{bmatrix} = 0 \to \begin{cases} \alpha_1 + \alpha_2 + 2\alpha_3 = 0 \\ 2\alpha_1 - \alpha_2 + \alpha_3 = 0 \\ 0\alpha_1 + \alpha_2 + \alpha_3 = 0 \end{cases} \to \begin{cases} \alpha_1 = 1 \\ \alpha_2 = 1 \\ \alpha_3 = -1 \end{cases}$$

Since α_1, α_2, and $\alpha_3 \neq 0$, therefore V_1, V_2 and V_3 are not linearly independent. Theorem 3.2 provides the necessary and sufficient condition for linear independence of a set of vectors.

Theorem 3.2: Let $X_1, X_2, \ldots, X_n$ be n vectors in R^n. Define the $n \times n$ matrix X as:

$$X = [X_1 \mid X_2 \mid \cdots \mid X_n] \qquad (3.12)$$

The vectors $X_1, X_2, \ldots, X_n$ are linearly dependent if and only if X is singular, that is:

$$\det(X) = 0 \qquad (3.13)$$

Proof: The proof is left as an exercise (see Problem 4).

Example 3.12

Are the following three vectors linearly independent?

$$X_1 = \begin{bmatrix} 4 \\ 2 \\ 3 \end{bmatrix}, X_2 = \begin{bmatrix} 2 \\ 3 \\ 1 \end{bmatrix}, X_3 = \begin{bmatrix} 2 \\ -5 \\ 3 \end{bmatrix}$$

Solution: Form the 3×3 matrix

$$X = [X_1 \quad X_2 \quad X_3] = \begin{bmatrix} 4 & 2 & 2 \\ 2 & 3 & -5 \\ 3 & 1 & 3 \end{bmatrix}$$

The determinant of matrix X is:

$$\det(X) = 4\begin{vmatrix} 3 & -5 \\ 1 & 3 \end{vmatrix} - 2\begin{vmatrix} 2 & -5 \\ 3 & 3 \end{vmatrix} + 2\begin{vmatrix} 2 & 3 \\ 3 & 1 \end{vmatrix} = 56 - 42 - 14 = 0$$

$\det(X) = 0$, therefore X_1, X_2 and X_3 are linearly dependent.

Example 3.13

Let $C^{n-1}[a, b]$ represent the set of all continuous real-valued differentiable functions defined on the closed interval $[a, b]$. It is assumed that these functions are $n-1$ times differentiable. Let vectors $f_1, f_2, f_3, \ldots, f_n \in C^{n-1}[a, b]$. If $f_1, f_2, f_3, \ldots, f_n$ are linearly independent, then:

$$c_1 f_1(x) + c_2 f_2(x) + \cdots + c_n f_n(x) = 0 \tag{3.14}$$

has a trivial solution $c_1 = c_2 = \cdots = c_n = 0$. Differentiation of the above equation $n-1$ times yields:

$$c_1 f_1'(x) + c_2 f_2'(x) + \cdots + c_n f_n'(x) = 0 \tag{3.15}$$

$$c_1 f_1''(x) + c_2 f_2''(x) + \cdots + c_n f_n''(x) = 0 \tag{3.16}$$

$$\vdots$$

$$c_1 f_1^{n-1}(x) + c_2 f_2^{n-1}(x) + \cdots + c_n f_n^{n-1}(x) = 0 \tag{3.17}$$

Writing these equations in matrix form, we have:

$$\begin{bmatrix} f_1(x) & f_2(x) & \cdots & f_n(x) \\ f_1'(x) & f_2'(x) & \cdots & f_n'(x) \\ \vdots & \vdots & \ddots & \vdots \\ f_1^{n-1}(x) & f_2^{n-1}(x) & \cdots & f_n^{n-1}(x) \end{bmatrix} \begin{bmatrix} c_1 \\ c_2 \\ \vdots \\ c_n \end{bmatrix} = \begin{bmatrix} 0 \\ 0 \\ \vdots \\ 0 \end{bmatrix} \tag{3.18}$$

Let us define:

$$W(f_1, f_2, \ldots, f_n) = \begin{bmatrix} f_1(x) & f_2(x) & \cdots & f_n(x) \\ f_1'(x) & f_2'(x) & \cdots & f_n'(x) \\ \vdots & \vdots & \ddots & \vdots \\ f_1^{n-1}(x) & f_2^{n-1}(x) & \cdots & f_n^{n-1}(x) \end{bmatrix} \text{ and } c = \begin{bmatrix} c_1 \\ c_2 \\ \vdots \\ c_n \end{bmatrix} \tag{3.19}$$

Then, $Wc = 0$ has a unique trivial solution $c = 0$ if the $n \times n$ matrix $W(f_1, f_2, \ldots, f_n)$ is nonsingular, that is:

$$\det[W(f_1, f_2, \ldots, f_n)] \neq 0 \tag{3.20}$$

Theorem 3.3 is the result of the above observation. The proof is left as an exercise to the reader.

Linear Vector Spaces

Theorem 3.3: Let $f_1, f_2, \ldots, f_n \in C^{n-1}[a, b]$. If there exists a point x_0 in $[a, b]$ such that $|W[f_1(x_0), f_2(x_0), \ldots, f_n(x_0)]| \neq 0$, then $f_1, f_2, \ldots, f_n$ are linearly independent.

Example 3.14

Let $f_1(x) = e^x$ and $f_2(x) = e^{-x}$. $f_1(x), f_2(x) \in C(-\infty, +\infty)$. Since:

$$\det(W[f_1, f_2]) = \begin{vmatrix} e^x & e^{-x} \\ e^x & -e^{-x} \end{vmatrix} = -e^x e^{-x} - e^x e^{-x} = -2 \neq 0$$

Hence, $f_1(x), f_2(x)$ are linearly independent.

Example 3.15

Let $f_1(x) = 1$, $f_2(x) = x$, $f_3(x) = x^2$ and $f_4(x) = x^3$ be elements of a linear vector space consisting of polynomials of degree less than or equal to 3. Then:

$$W[f_1, f_2, f_3, f_4] = \begin{vmatrix} 1 & x & x^2 & x^3 \\ 0 & 1 & 2x & 3x^2 \\ 0 & 0 & 2 & 6x \\ 0 & 0 & 0 & 6 \end{vmatrix} = 12 \neq 0$$

Therefore, the set f_1, f_2, f_3, f_4 is linearly independent.

If $V_1, V_2, \ldots, V_n$ is the smallest set of vectors spanning the vector space V, then $V_1, V_2, \ldots, V_n$ are linearly independent and are called the minimal spanning set of V. On the other hand, if $V_1, V_2, \ldots, V_n$ are linearly independent and capable of spanning V, then they are a minimal spanning set for the vector space V. This minimal spanning set forms a set of basis vectors capable of generating (spanning) vector space V. In the next section, we study the basis vectors.

3.3.3 Basis Vectors

The minimal spanning set of a vector space forms the basis vector set for that space. A vector space V is called finite dimensional if it has a basis consisting of a finite number of vectors. The dimension of V, denoted by $\dim(V)$, is the number of vectors in a basis for V. A vector space that has no finite basis is called infinite dimensional. From the above, it is obvious that the vectors $V_1, V_2, \ldots, V_n$ form a basis for a finite dimensional vector space V if and only if:

i. $V_1, V_2, \ldots, V_n$ are linearly independent.
ii. $V_1, V_2, \ldots, V_n$ spans V.

Example 3.16

Given R^3, the standard basis set is $e_1 = \begin{bmatrix} 1 \\ 0 \\ 0 \end{bmatrix}$, $e_2 = \begin{bmatrix} 0 \\ 1 \\ 0 \end{bmatrix}$, $e_3 = \begin{bmatrix} 0 \\ 0 \\ 1 \end{bmatrix}$

Since:

$$A = [e_1 \mid e_2 \mid e_3] = \begin{bmatrix} 1 & 0 & 0 \\ 0 & 1 & 0 \\ 0 & 0 & 1 \end{bmatrix} \text{ and}$$

$$\det(A) = 1$$

Therefore, the set of vectors e_1, e_2, e_3 are linearly independent.

Example 3.17

Consider the vector space V consisting of all 2×2 matrices. Show that the set $\{E_{11}, E_{12}, E_{21}, E_{22}\}$ given by:

$$E_{11} = \begin{bmatrix} 1 & 0 \\ 0 & 0 \end{bmatrix}, E_{12} = \begin{bmatrix} 0 & 1 \\ 0 & 0 \end{bmatrix}, E_{21} = \begin{bmatrix} 0 & 0 \\ 1 & 0 \end{bmatrix}, E_{22} = \begin{bmatrix} 0 & 0 \\ 0 & 1 \end{bmatrix}$$

forms basis vectors for the vector space V.

Solution: E_{11}, E_{12}, E_{21}, and E_{22} are linearly independent since $c_1 E_{11} + c_2 E_{12} + c_3 E_{21} + c_4 E_{22} = 0$ implies that:

$$c_1 \begin{bmatrix} 1 & 0 \\ 0 & 0 \end{bmatrix} + c_2 \begin{bmatrix} 0 & 1 \\ 0 & 0 \end{bmatrix} + c_3 \begin{bmatrix} 0 & 0 \\ 1 & 0 \end{bmatrix} + c_4 \begin{bmatrix} 0 & 0 \\ 0 & 1 \end{bmatrix} = \begin{bmatrix} 0 & 0 \\ 0 & 0 \end{bmatrix}$$

$$\begin{bmatrix} c_1 & 0 \\ 0 & 0 \end{bmatrix} + \begin{bmatrix} 0 & c_2 \\ 0 & 0 \end{bmatrix} + \begin{bmatrix} 0 & 0 \\ c_3 & 0 \end{bmatrix} + \begin{bmatrix} 0 & 0 \\ 0 & c_4 \end{bmatrix} = \begin{bmatrix} 0 & 0 \\ 0 & 0 \end{bmatrix}$$

$$\begin{bmatrix} c_1 & c_2 \\ c_3 & c_4 \end{bmatrix} = \begin{bmatrix} 0 & 0 \\ 0 & 0 \end{bmatrix}$$

$E_{11}, E_{12}, E_{21}, E_{22}$ are linearly independent and form a basis for vector space V, i.e. if $A \in V$, A can be written as a linear combination of $E_{11}, E_{12}, E_{21}, E_{22}$. That is:

$$A = \begin{bmatrix} a_{11} & a_{12} \\ a_{21} & a_{22} \end{bmatrix} = a_{11} \begin{bmatrix} 1 & 0 \\ 0 & 0 \end{bmatrix} + a_{12} \begin{bmatrix} 0 & 1 \\ 0 & 0 \end{bmatrix} + a_{21} \begin{bmatrix} 0 & 0 \\ 1 & 0 \end{bmatrix} + a_{22} \begin{bmatrix} 0 & 0 \\ 0 & 1 \end{bmatrix}$$

$$= a_{11} E_{11} + a_{12} E_{12} + a_{21} E_{21} + a_{22} E_{22}$$

3.3.4 Change of Basis Vectors

Let V be an n dimensional linear vector space with basis $E = [v_1 \mid v_2 \mid \cdots \mid v_n]$. Assume that $v \in V$ is an element of vector space V, then v can be written as linear combination of $v_1, v_2, \ldots, v_n$. This implies that:

$$v = c_1 v_1 + c_2 v_2 + \cdots + c_n v_n = Ec \tag{3.21}$$

where $c = [c_1 \quad c_2 \quad \cdots \quad c_n]^T$. The set of scalars $c_1, c_2, \ldots, c_n$ are called coordinates of vector v with respect to the basis $v_1, v_2, \ldots, v_n$. Now assume that we would like to change the basis vectors to $F = [u_1 \mid u_2 \mid \cdots \mid u_n]$, then:

Linear Vector Spaces

$$v = d_1 u_1 + d_2 u_2 + \cdots + d_n u_n = Fd \qquad (3.22)$$

where $d = [d_1 \quad d_2 \quad \cdots \quad d_n]^T$. Comparing Equations 3.21 and 3.22, we have:

$$Fd = Ec \qquad (3.23)$$

Equation 3.23 can be solved for d, the coordinates with respect to the new basis $F = [u_1 \mid u_2 \mid \cdots \mid u_n]$

$$d = F^{-1} Ec = Sc \qquad (3.24)$$

The matrix $S = F^{-1}E$ is called a transition matrix.

Example 3.18

Let $X = 3v_1 + 2v_2 - v_3$ with respect to basis vectors

$$E = [v_1 \quad v_2 \quad v_3] = \begin{bmatrix} 1 & 2 & 1 \\ 1 & 3 & 5 \\ 1 & 2 & 4 \end{bmatrix}$$

Find the coordinates of vector X with respect to a new basis defined by:

$$F = [u_1 \quad u_2 \quad u_3] = \begin{bmatrix} 1 & 1 & 1 \\ 1 & 2 & 2 \\ 0 & 0 & 1 \end{bmatrix}$$

Solution:

$$d = F^{-1}Ec = \begin{bmatrix} 1 & 1 & 1 \\ 1 & 2 & 2 \\ 0 & 0 & 1 \end{bmatrix}^{-1} \begin{bmatrix} 1 & 2 & 1 \\ 1 & 3 & 5 \\ 1 & 2 & 4 \end{bmatrix} \begin{bmatrix} 3 \\ 2 \\ -1 \end{bmatrix} = \begin{bmatrix} 1 & 1 & -3 \\ -1 & -1 & 0 \\ 1 & 2 & 4 \end{bmatrix} \begin{bmatrix} 3 \\ 2 \\ -1 \end{bmatrix} = \begin{bmatrix} 8 \\ -5 \\ 3 \end{bmatrix}$$

Therefore, vector X can be written as:

$$X = 8u_1 - 5u_2 + 3u_3.$$

Example 3.19

Let:

$$u_1 = \begin{bmatrix} 1 \\ 1 \\ 1 \end{bmatrix}; u_2 = \begin{bmatrix} 1 \\ 2 \\ 2 \end{bmatrix}; u_3 = \begin{bmatrix} 2 \\ 3 \\ 4 \end{bmatrix}$$

a. Find the transition matrix corresponding to the change of basis from (e_1, e_2, e_3) to (u_1, u_2, u_3).
b. Find the coordinate of $[3 \quad 2 \quad 5]^T$ with respect to the basis (u_1, u_2, u_3).

Solution:

a. $S = F^{-1}E = \begin{bmatrix} 1 & 1 & 2 \\ 1 & 2 & 3 \\ 1 & 2 & 4 \end{bmatrix}^{-1} \begin{bmatrix} 1 & 0 & 0 \\ 0 & 1 & 0 \\ 0 & 0 & 1 \end{bmatrix} = \begin{bmatrix} 2 & 0 & -1 \\ -1 & 2 & -1 \\ 0 & -1 & 1 \end{bmatrix}$

b. The coordinates of vector $[3 \quad 2 \quad 5]^T$ with respect to the new basis are:

$$d = Sc = \begin{bmatrix} 2 & 0 & -1 \\ -1 & 2 & -1 \\ 0 & -1 & 1 \end{bmatrix} \begin{bmatrix} 3 \\ 2 \\ 5 \end{bmatrix} = \begin{bmatrix} 1 \\ -4 \\ 3 \end{bmatrix}$$

3.4 Normed Vector Spaces

3.4.1 Definition of Normed Vector Space

A normed linear vector space is a vector space V together with a real-valued function on V (called the norm), such that:

a. $\|x\| \geq 0$, for every $x \in V$. $\|x\| = 0$ if and only if $x = 0$.
b. $\|\alpha x\| = |\alpha| \|x\|$, for every $x \in V$ and every scalar α.
c. $\|x + y\| \leq \|x\| + \|y\|$, for every $x, y \in V$. This is called triangular inequality.

The norm is a generalization of the concept of the length of a vector in 2-d or 3-d space. Given a vector x in V, the nonnegative number $\|x\|$ can be thought of as the length of vector x or the distance of x from the origin of the vector space.

3.4.2 Examples of Normed Vector Spaces

Example 3.20

Consider the linear vector space R^n or C^n. Let us define the norm as:

$$\|x\|_p = [|x_1|^p + |x_2|^p + \cdots + |x_n|^p]^{\frac{1}{p}} \tag{3.25}$$

where p is an integer $1 \leq p < \infty$. Different values of p result in different normed vector spaces. If $p = 1$, it is called L_1 norm, and when $p = 2$, it is called L_2 norm and they are given by:

$$\|x\|_1 = |x_1| + |x_2| + \cdots + |x_n| \tag{3.26}$$

and

$$\|x\|_2 = [|x_1|^2 + |x_2|^2 + \cdots + |x_n|^2]^{\frac{1}{2}} \tag{3.27}$$

Linear Vector Spaces

As $p \to \infty$, the term having maximum absolute value becomes very dominant compared to the remaining terms. Therefore:

$$\|x\|_\infty = \lim_{p \to \infty}[|x_1|^p + |x_2|^p + \cdots + |x_n|^p]^{\frac{1}{p}} = \lim_{p \to \infty}[\max_i |x_i|^p]^{\frac{1}{p}} = \max_i |x_i| \qquad (3.28)$$

This is called L_∞ norm. The MATLAB® command to compute the p norm of an n dimensional real or complex vector is **norm(x,p)**. For $p=\infty$, the command is **norm(x,inf)**.

Example 3.21

Compute L_1, L_2 and L_∞ norms of the following two vectors in R^3.

$$x = [1 \quad -2 \quad 3 \quad -4]^T,\ y = [3 \quad 0 \quad 5 \quad -2]^T.$$

Solution:

$$\|x\|_1 = |x_1| + |x_2| + |x_3| + |x_4| = 10$$

$$\|x\|_2 = [|x_1|^2 + |x_2|^2 + |x_3|^2 + |x_4|^2]^{\frac{1}{2}} = 5.4772$$

$$\|x\|_\infty = \max_i |x_i| = 4$$

$$\|y\|_1 = |y_1| + |y_2| + |y_3| + |y_4| = 10$$

$$\|y\|_2 = [|y_1|^2 + |y_2|^2 + |y_3|^2 + |y_4|^2]^{\frac{1}{2}} = 6.1644$$

$$\|y\|_\infty = \max_i |y_i| = 5$$

Example 3.22

Compute L_1, L_2 and L_∞ norms of complex vector $x \in C^3$.

$$x = [1-j \quad j \quad 3+j4]^T$$

Solution:

$$\|x\|_1 = |x_1| + |x_2| + |x_3| = \sqrt{2} + 1 + 5 = 6 + \sqrt{2} = 7.4142$$

$$\|x\|_2 = [|x_1|^2 + |x_2|^2 + |x_3|^2]^{\frac{1}{2}} = \sqrt{28} = 5.2915$$

$$\|x\|_\infty = \max_i |x_i| = 5$$

3.4.3 Distance Function

Given two vectors x and y in a normed vector space, the metric defined by:

$$d(x, y) = \|x - y\|_p \qquad (3.29)$$

is called the *p*-norm distance between vector *x* and vector *y*. The distance function defined by the above equation satisfies the following properties:

a. $d(x, y) \geq 0$, for every x and $y \in V$. $d(x, y) = 0$ if and only if $x = y$.
b. $d(x, y) = d(y, x)$.
c. $d(x, z) \leq d(x, y) + d(y, z)$, for every x, y, and $z \in V$.

The proof is left as an exercise (see Problem 3.6).

Example 3.23

Compute the 2-norm distance $d(x, y)$ between the following two vectors:

$$x = [1-j \quad j \quad 3+j4]^T \text{ and } y = [1+j \quad 2 \quad 1+j3]^T$$

Solution:

$$x - y = [-j2 \quad -2+j \quad 2+j]^T$$

$$d = \|x - y\|_2 = [|-j2|^2 + |-2+j|^2 + |2+j|^2]^{\frac{1}{2}} = \sqrt{4+5+5} = 3.7416$$

3.4.4 Equivalence of Norms

The L_1, L_2 and L_∞ norms on R^n (or C^n) are equivalent in the sense that if a converging sequence x_m is converging to $x_\infty = \lim_{x \to \infty} x_m$, as determined by one of the norms, then it converges to x_∞ as determined in all three norms. It can also be proven that the L_1, L_2 and L_∞ norms on R^n (or C^n) satisfy the following inequalities (see Problem 3.10):

$$\frac{\|x\|_2}{\sqrt{n}} \leq \|x\|_\infty \leq \|x\|_2 \qquad (3.30)$$

$$\|x\|_2 \leq \|x\|_1 \leq \sqrt{n} \|x\|_2 \qquad (3.31)$$

$$\frac{\|x\|_1}{n} \leq \|x\|_\infty \leq \|x\|_1 \qquad (3.32)$$

Example 3.24

Consider the linear vector space consisting of the set of all continuous functions defined over the closed interval [*a b*]. This vector spaced is a normed vector space with the L_p norm defined as:

$$\|f(x)\|_p = \left[\int_a^b |f(x)|^p dx \right]^{\frac{1}{p}}, \quad 1 \leq p < \infty \qquad (3.33)$$

Linear Vector Spaces

Similar to R^n, we have:

$$\|f(x)\|_1 = \int_a^b |f(x)|\, dx \tag{3.34}$$

$$\|f(x)\|_2 = \left[\int_a^b |f(x)|^2\, dx\right]^{\frac{1}{2}} \tag{3.35}$$

and

$$\|f(x)\|_\infty = \sup_{a \le x \le b} |f(x)| \tag{3.36}$$

where "sup" stands for supremum, which is the generalization of the maximum and is defined below.

Definition of supremum: Given set S, a subset of real numbers, the smallest number a such that $a \ge x$ for every $x \in S$ is called the supremum (sup) of set S. Therefore:

$$a = \sup(S) \quad \text{if } a \ge x \text{ for every } x \in S \tag{3.37}$$

Example 3.25

Compute L_1, L_2 and L_∞ norm of the function $f(x) = x - 0.5$ defined over the closed interval $[0\ 1]$. The function $f(x)$ is shown in Figure 3.2.

Solution:

$$\|f(x)\|_1 = \int_a^b |f(x)|\, dx = \int_0^1 |x - 0.5|\, dx = \int_0^{0.5} (-x + 0.5)\, dx + \int_{0.5}^1 (x - 0.5)\, dx = \frac{1}{8} + \frac{1}{8} = \frac{1}{4}$$

$$\|f(x)\|_2 = \left[\int_a^b |f(x)|^2\, dx\right]^{\frac{1}{2}} = \left[\int_0^1 (x - 0.5)^2\, dx\right]^{\frac{1}{2}} = \frac{1}{\sqrt{12}}$$

$$\|f(x)\|_\infty = \sup_{a \le x \le b} |f(x)| = 0.5$$

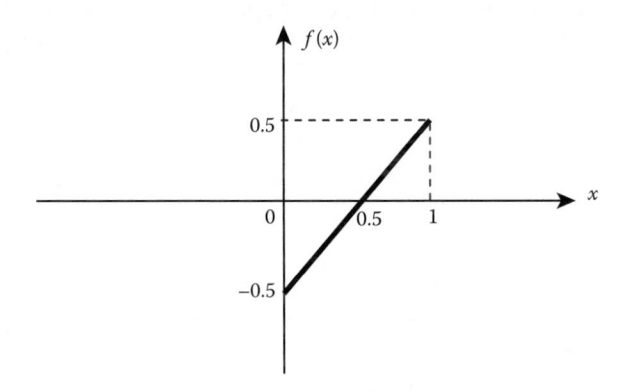

FIGURE 3.2
Function $f(x)$.

Example 3.26

Consider the linear vector space consisting of the set of all continuous complex valued functions defined over the closed interval [0 T]. Compute L_1, L_2 and L_∞ norm of an element of this vector space given by:

$$f(t) = \sqrt{P} \exp\left(jk\frac{2\pi}{T}t\right)$$

where k is an integer.

Solution:

$$\|f(t)\|_1 = \int_a^b |f(t)|\, dt = \int_0^T \sqrt{P}\, dt = \sqrt{P}\,T$$

$$\|f(t)\|_2 = \left[\int_a^b |f(t)|^2\, dt\right]^{\frac{1}{2}} = \left[\int_0^T P\, dt\right]^{\frac{1}{2}} = \sqrt{PT}$$

$$\|f(t)\|_\infty = \sup_{a \le t \le b} |f(t)| = \sqrt{P}$$

3.5 Inner Product Spaces

The norm defines a metric for a vector space that is the generalization of Euclidean vector length in R^n. To define orthogonality in a vector space, we need another metric that is similar to the dot product between two vectors in R^2 or R^3. This metric is known as the inner product and is defined below.

3.5.1 Definition of Inner Product

An inner product space is a linear vector space V with associated field R^1 or C^1, together with a function on $V \times V \to R^1$ (C^1) called the inner product (or dot product):

$$< x, y > : V \times V \to R^1\ (C^1) \tag{3.38}$$

Such that:

a. $< x+y, z > = < x, z > + < y, z >$
b. $< x, y > = \overline{< x, y >}$ (bar stands for complex conjugate)
c. $< \alpha x, y > = \alpha < x, y >$
d. $< x, x > \ge 0$ ($=0$ if and only if $x=0$)

Linear Vector Spaces

Notice that, as a result of the above properties, $<x, y>$ has further properties such as:

$$<x, y+z> = <x, y> + <x, z> \tag{3.39}$$

$$<x, \alpha y> = \bar{\alpha} <x, y> \tag{3.40}$$

3.5.2 Examples of Inner Product Spaces

Example 3.27

Consider the complex linear vector space C^n. The vector space C^n is an inner product space with inner product:

$$<x, y> = \sum_{i=1}^{n} x_i \bar{y}_i \tag{3.41}$$

The MATLAB® command to compute the dot product between two vectors x and y in $C^n(R^n)$ is **dot(x,y)**.

Example 3.28

Find the dot product between $x = [1-j \quad 2-j \quad 3]^T$ and $y = [1 \quad 2j \quad 4]^T$.

Solution:

$$<x, y> = \sum_{i=1}^{n} x_i \bar{y}_i = (1-j) \times 1 + (2-j) \times (-j2) + 3 \times 4 = 11 - j5$$

Example 3.29

The linear vector space of square integrable continuous functions defined over the interval $[a, b]$ is an inner product space with inner product:

$$<x(t), y(t)> = \int_a^b x(t)\overline{y(t)}\,dt \tag{3.42}$$

An interesting property of an inner product space is that we can use the inner product to define the norm. The resulting norm is called the induced norm. To show this, we first prove a theorem known as Schwarz's inequality.

3.5.3 Schwarz's Inequality

Schwarz's inequality is an extremely useful inequality used in different engineering optimization problems. As an example, it is used to derive the matched filter as well as the maximum ratio combining technique found in digital communication systems. We will first prove the theorem and then apply it to two communication examples.

Theorem 3.4: Schwarz's inequality: For any inner product space V, we have

$$|<x,y>| \leq <x,x>^{\frac{1}{2}} <y,y>^{\frac{1}{2}} \quad (3.43)$$

The equality holds if and only if $x = \alpha y$ (linearly dependent).

Proof: For any scalar β and t, we have:

$$<tx + \beta y, tx + \beta y> \geq 0 \quad (3.44)$$

Expanding the above inner product yields:

$$t\bar{t} <x,x> + t\bar{\beta} <x,y> + \beta \bar{t} <y,x> + \beta \bar{\beta} <y,y> \geq 0 \quad (3.45)$$

If we assume that t is real, then:

$$t^2 <x,x> + t\bar{\beta} <x,y> + t\beta <y,x> + \beta \bar{\beta} <y,y> \geq 0 \quad (3.46)$$

Let us choose β such that:

$$\beta = \begin{cases} \dfrac{<x,y>}{|<x,y>|} & \text{if } <x,y> \neq 0 \\ 1 & \text{if } <x,y> = 0 \end{cases} \quad (3.47)$$

Then $\bar{\beta}<x,y> = <y,x>\beta = |<x,y>|$ and $\beta\bar{\beta} = 1$. Therefore, Equation 3.46 is reduced to:

$$t^2 <x,x> + 2t|<x,y>| + <y,y> \geq 0 \text{ for any real number } t \quad (3.48)$$

Now if we let $t = -\dfrac{|<x,y>|}{<x,x>}$, then:

$$\dfrac{|<x,y>|^2}{<x,x>} - 2\dfrac{|<x,y>|^2}{<x,x>} + <y,y> \geq 0 \quad (3.49)$$

Therefore:

$$-|<x,y>|^2 + <x,x><y,y> \geq 0 \quad (3.50)$$

Hence:

$$|<x,y>| \leq <x,x>^{\frac{1}{2}} <y,y>^{\frac{1}{2}} \quad (3.51)$$

Linear Vector Spaces

The following inequalities are the results of applying Swartz's inequality to C^n and $L(a, b)$:

$$\left| \sum_{i=1}^{n} x_i \bar{y}_i \right| \le \left[\sum_{i=1}^{n} |x_i|^2 \right]^{\frac{1}{2}} \left[\sum_{i=1}^{n} |y_i|^2 \right]^{\frac{1}{2}} \quad (3.52)$$

and

$$\left| \int_a^b x(t)\bar{y}(t)\, dt \right| \le \left[\int_a^b |x(t)|^2\, dt \right]^{\frac{1}{2}} \left[\int_a^b |y(t)|^2\, dt \right]^{\frac{1}{2}} \quad (3.53)$$

3.5.4 Norm Derived from Inner Product

Theorem 3.5: If V is an inner product space with inner product $<x, y>$, then it is also a normed vector space with induced norm defined by:

$$\|x\| = \sqrt{<x, x>} \quad (3.54)$$

The proof is left as an exercise. In the following section, we consider the application of Schwarz's inequality in communication systems.

3.5.5 Applications of Schwarz Inequality in Communication Systems

As an important application of Schwarz inequality, we consider the matched filter detector used in communication and radar systems. We first consider the discrete and then the continuous matched filter detector.

3.5.5.1 Detection of a Discrete Signal "Buried" in White Noise

Assume that the received signal plus noise is given by:

$$r_i = s_i + n_i \quad i = 1, 2, \ldots, m \quad (3.55)$$

The values of the signal samples s_i, $i=1, 2, \ldots, m$, are assumed to be known. The noise samples n_i are assumed to be independent and have zero mean. The variance of each noise sample is assumed to be σ_n^2. The goal is to process the samples r_i to determine the presence or absence of the signal s_i. A common approach is to compute the weighted average of the received signal and compare that with a threshold. The weighted average is given by:

$$y = \sum_{i=1}^{m} h_i r_i = \underbrace{\sum_{i=1}^{m} h_i s_i}_{\text{Signal}} + \underbrace{\sum_{i=1}^{m} h_i n_i}_{\text{Noise}} \quad (3.56)$$

This is equivalent to filtering the received signal by a filter with coefficients h_i. This is shown in Figure 3.3.

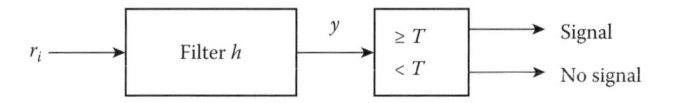

FIGURE 3.3
The matched filter detector.

The filter coefficients are computed by maximizing the signal-to-noise ratio (SNR) at the output of the filter defined by:

$$\text{SNR} = \frac{\text{Signal power}}{\text{Noise power}} = \frac{\left(\sum_{i=1}^{m} h_i s_i\right)^2}{\sigma_n^2 \sum_{i=1}^{m} h_i^2} \tag{3.57}$$

The SNR can be written in vector form as:

$$\text{SNR} = \frac{\text{Signal power}}{\text{Noise power}} = \frac{(<h, s>)^2}{\sigma_n^2 \|h\|^2} \tag{3.58}$$

where $h = [h_1 \ h_2 \ \ldots \ h_m]^T$ and $s = [s_1 \ s_2 \ \ldots \ s_m]^T$. The filter coefficients represented by vector h that maximize SNR are found by using Schwarz inequality:

$$\text{SNR} = \frac{(<h, s>)^2}{\sigma_n^2 \|h\|^2} \le \frac{\|s\|^2 \|h\|^2}{\sigma_n^2 \|h\|^2} = \frac{\|s\|^2}{\sigma_n^2} \tag{3.59}$$

Equality occurs when:

$$h = \alpha\, s \tag{3.60}$$

Therefore SNR is maximized when the filter coefficients are proportional to the signal samples. The proportionality constant α is chosen to be one, since it is does not affect the SNR. Hence:

$$h_i = s_i \quad i = 1, 2, \ldots, m \quad \text{and} \quad \text{SNR}_{\max} = \frac{1}{\sigma_n^2} \sum_{i=1}^{m} s_i^2 \tag{3.61}$$

The threshold T is chosen to minimize the probability of error. It is given by:

$$T = \frac{\|s\|^2}{2} \tag{3.62}$$

The above detector is known as the "matched filter" detector.

Linear Vector Spaces

3.5.5.2 Detection of Continuous Signal "Buried" in Noise

Consider the binary digital data communication system shown in Figure 3.4, in which the transmitted signals that represent binary bits "1" and "0" are s(t) and the absence of s(t), respectively. It is assumed that the signal duration is T_b seconds.

The communication channel is modeled as an ideal additive zero mean white Gaussian noise channel with power spectral density:

$$S_n(f) = \frac{N_0}{2} \qquad (3.63)$$

The receiver consists of a linear filter called the matched filter with impulse response $h(t)$ of duration T_b, an ideal sampler that samples the output of the matched filter at the bit rate $f_s = R_b = 1/T_b$, and a one-bit quantizer, as shown in Figure 3.5. As the result, the output of the quantizer is an estimate of the kth transmitted binary bit.

The one-bit quantizer $Q[.]$ is given by:

$$Q[x] = \begin{cases} 1 & x \geq T \\ 1 & x < T \end{cases} \qquad (3.64)$$

where T is a constant threshold. The goal of filtering the received signal is to reduce the effect of noise thus increasing SNR and the signal detectability. Hence, the optimum filter, also known as the matched filter, is the one that maximizes SNR. The threshold T is chosen to minimize the bit error rate (BER).

We would like to determine the filter $h(t)$ that maximizes the SNR at the input to the quantizer. The filter output $y(t)$ is the convolution of the input with the filter impulse response and is given by:

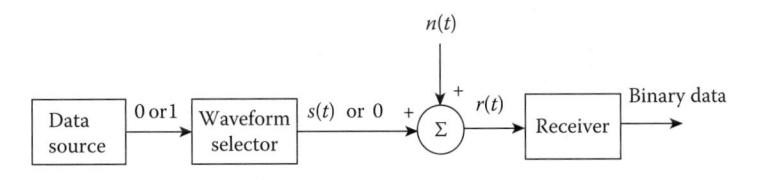

FIGURE 3.4
A digital data communication system.

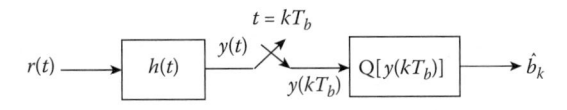

FIGURE 3.5
Matched filter receiver.

$$y(t) = r(t) * h(t) = \int_0^t h(\tau)r(t-\tau)d\tau \qquad (3.65)$$

Given that the binary bit "1" is transmitted, the sampler output is:

$$y(T_b) = \int_0^{T_b} h(T_b - \tau)r(\tau)d\tau = \int_0^{T_b} h(T_b - \tau)s(\tau)d\tau + N_1 \qquad (3.66)$$

If the binary bit "0" is transmitted, the sampler output is:

$$y(T_b) = \int_0^{T_b} h(T_b - \tau)r(\tau)d\tau = \int_0^{T_b} h(T_b - \tau)n(\tau)d\tau + N_0 \qquad (3.67)$$

where N_0 and N_1 are identically distributed zero mean Gaussian random variables given by:

$$N = \int_0^{T_b} h(T_b - \tau)n(\tau)d\tau \qquad (3.68)$$

The variance of the random variable N_0 is the same as N_1 and can be computed as:

$$\sigma^2 = E\left[\left(\int_0^{T_b} h(T_b - \tau)n(\tau)d\tau\right)^2\right] = E\left[\int_0^{T_b}\int_0^{T_b} h(T_b - \tau_1)h(T_b - \tau_2)n(\tau_1)n(\tau_2)d\tau_1 d\tau_2\right] \qquad (3.69)$$

which can be reduced to:

$$\sigma^2 = \int_0^{T_b}\int_0^{T_b} h(T_b - \tau_1)h(T_b - \tau_2)E(n(\tau_1)n(\tau_2))d\tau_1 d\tau_2$$

$$= \int_0^{T_b}\int_0^{T_b} h(T_b - \tau_1)h(T_b - \tau_2)\frac{N_0}{2}\delta(\tau_1 - \tau_2)d\tau_1 d\tau_2$$

$$= \frac{N_0}{2}\int_0^{T_b} h^2(T_b - \tau_2)\int_0^{T_b}\delta(\tau_1 - \tau_2)d\tau_1 d\tau_2$$

$$= \frac{N_0}{2}\int_0^{T_b} h^2(T_b - \tau_2)(u(T_b - \tau_2) - u(-\tau_2))d\tau_2$$

$$= \frac{N_0}{2}\int_0^{T_b} h^2(T_b - \tau_2)d\tau_2 = \frac{N_0}{2}\int_0^{T_b} h^2(t)dt \qquad (3.70)$$

Linear Vector Spaces

The SNR at the input to the quantizer is given by:

$$\text{SNR} = \frac{\left[\int_0^{T_b} h(T_b - \tau)s(\tau)d\tau\right]^2}{\sigma^2} = \frac{2\left[\int_0^{T_b} h(T_b - \tau)s(\tau)d\tau\right]^2}{N_0 \int_0^{T_b} h^2(\tau)d\tau} \quad (3.71)$$

Applying Schwarz inequality to the Equation 3.71 we have:

$$\text{SNR} \leq \frac{2\int_0^{T_b} s^2(\tau)d\tau \int_0^{T_b} h^2(T_b - \tau)d\tau}{N_0 \int_0^{T_b} h^2(\tau)d\tau} = \frac{2\int_0^{T_b} s^2(\tau)d\tau \int_0^{T_b} h^2(\tau)d\tau}{N_0 \int_0^{T_b} h^2(\tau)d\tau} = \frac{2\int_0^{T_b} s^2(\tau)d\tau}{N_0}$$

The equality is achieved if:

$$s(\tau) = h(T_b - \tau) \quad (3.72)$$

or

$$h(t) = s(T_b - t) \quad (3.73)$$

The variance of the noise is given by:

$$\sigma^2 = \frac{N_0}{2}\int_0^{T_b} h^2(t)dt = \frac{N_0}{2}\int_0^{T_b} s^2(t)dt = \frac{N_0}{2}E \quad (3.74)$$

and the maximum SNR is: $\text{SNR}_{max} = 2E/N_0$, where $E = \int_0^{T_b} s^2(t)dt$ is the energy in the signal $s(t)$. As the result, the input to the quantizer is:

$$y(kT_b) = \begin{cases} E + N_1 & \text{if } b_k = 1 \\ N_0 & \text{if } b_k = 0 \end{cases} \quad (3.75)$$

The constant threshold T is chosen to minimize the BER. For a symmetric binary data source where $p(0) = p(1) = 0.5$, the optimum choice for the threshold T is $E/2$ and the resulting BER is:

$$\text{BER} = p(e|0)p(0) + p(e|1)p(1) = 0.5p(N_0 > T) + 0.5p(N_1 + E < T) \quad (3.76)$$

where:

$$p(N_0 > T) = \frac{1}{\sqrt{2\pi}\sigma}\int_T^{\infty} e^{-\frac{x^2}{2\sigma^2}}dx = Q\left(\frac{\alpha}{\sigma}\right) \quad (3.77)$$

and

$$p(N_1 + E < T) = \frac{1}{\sqrt{2\pi}\sigma} \int_{-\infty}^{T} e^{-\frac{x^2}{2\sigma^2}} dx = Q\left(\frac{\alpha}{\sigma}\right) \quad (3.78)$$

The function $Q(x)$ is defined as: $Q(x) = (1/\sqrt{2\pi})\int_x^\infty e^{-u^2/2} du$. By substituting Equations 3.77 and 3.78 into Equation 3.76, the result simplifies to:

$$\text{BER} = 0.5 Q\left(\frac{\alpha}{\sigma}\right) + 0.5 Q\left(\frac{\alpha}{\sigma}\right) = Q\left(\frac{\alpha}{\sigma}\right) = Q\left(\frac{\frac{E}{2}}{\sqrt{\frac{N_0 E}{2}}}\right) = Q\left(\sqrt{\frac{2E}{N_0}}\right) \quad (3.79)$$

As an example, assume that a rectangular pulse of amplitude one and duration T_b is used for transmitting binary bit one and the absence of this pulse is used for transmitting binary bit zero, as shown in Figure 3.6. Assume that the input bit stream is $b(n) = [1\ 1\ 0\ 1]$ and SNR = 10 dB.

Then the received signal $r(t)$ is:

$$r(t) = 1 \times s(t) + 1 \times s(t - T_b) + 0 \times s(t - 2T_b) + 1 \times s(t - 3T_b) + n(t) \quad (3.80)$$

where $n(t)$ is zero mean white Gaussian noise. The impulse response of the matched filter is given by:

$$h(t) = s(T_b - t) = s(t) \quad (3.81)$$

The threshold T is:

$$T = 0.5E = \int_0^{T_b} s^2(t) dt = 0.5 T_b \quad (3.82)$$

The received signal $r(t)$ and the output of the matched filter $y(t)$ are shown in Figure 3.7.

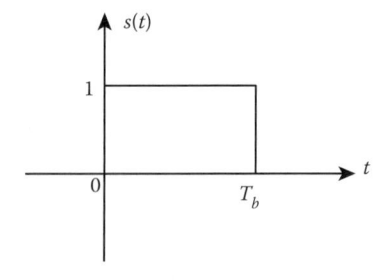

FIGURE 3.6
Signal used to transmitt binary bit "1".

Linear Vector Spaces

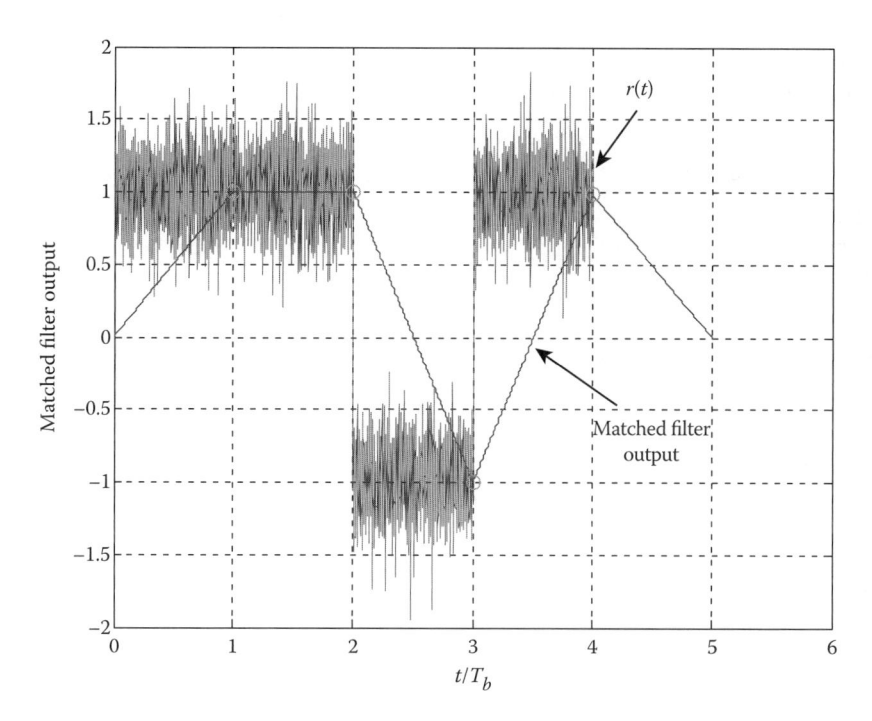

FIGURE 3.7
(**See color insert following page 174.**) The received signal and the output of the matched filter.

It is clear from Figure 3.7 that the matched filter has successfully filtered the high frequency noise component of the received signal. The bit rate matched filter output samples are:

$$y(kT_b) = [1.0018 \quad 1.0031 \quad -0.999 \quad 0.9881] \quad (3.83)$$

The corresponding output bits are:

$$\hat{b}_k = [1 \quad 1 \quad 0 \quad 1] \quad (3.84)$$

The MATLAB® code implementation of the matched filter receiver is shown in Table 3.1.

3.5.6 Hilbert Space

A complete inner product space is called a Hilbert space. A vector space is complete if every Cauchy sequence of vectors in that space is convergent to a vector in that space. For example, if $\{x_n\}$, $n=0, 1, \ldots$ is a Cauchy sequence in a complete vector space V, then there exists a vector x^* in V such that:

$$\lim_{n \to \infty} x_n \to x^* \quad (3.85)$$

TABLE 3.1

MATLAB® Code for the Matched Filter Receiver

```
% Matched Filter Receiver
% Generate Signal s(t)
Tb=1;
t1=linspace(0,Tb,1000);
s=ones(size(t1));
h=s(end:-1:1);
% Generate bits
M=4;
b=sign(rand(4,1)-0.5);
bin=0.5*(b+1);
% Set SNR in dB and compute the noise power
EbN0=10;
EbN0=10^(EbN0/10);
Eb=Tb;
N0=Eb/EbN0;
sigma=sqrt(Eb*N0/2);
% Received signal
r=[];
tr=[];
for i=1:size(b);
  r=[r b(i)*s];
  tr=[tr t1+(i-1)*Tb];
end
r=r+sigma*randn(size(r));
% Filter the received signal using matched filter
DT=t1(2)-t1(1);
y=DT*conv(r,h)-DT*r(1)*h(1);
t=0:DT:(length(y)-1)*DT;
k=length(t1);
% Sample the matched filter output at bit rate
z=y(k:k:size(b)*k);
plot(tr,r)
hold on
plot(t,y,'r')
k1=1:1:length(z);
plot(k1,z,'Og')
hold off
xlabel('t/Tb')
ylabel('Matched Filter Output')
% Quantize the output of the matched filter sampled at the bit rate
bh=0.5*(sign(z)+1);
```

A Cauchy sequence is a convergent sequence. This means that if x_n is a Cauchy sequence in vector space V, then given $\varepsilon > 0$ there exists a positive integer n_0 such that for any $m, n > n_0$, $||x_m - x_n|| < \varepsilon$, where $||x||$ is the norm defined for vector space V. An example of a Hilbert space is the space of all square integrable continuous functions. Not that there are inner product spaces which are not complete. For example, the vector space V of all real

Linear Vector Spaces 131

and rational numbers is not complete since the Cauchy sequence $x_n = 1 + 1/1! + 1/2! + 1/3! + \cdots + 1/n!$ converges to the irrational number e, which is not in the vector space V.

3.6 Orthogonality

The orthogonality concept is very useful in engineering problems related to estimation, prediction, signal modeling, and transform calculus (Fourier transform, Z-transform, etc.). In this section, we discuss orthogonal vectors in a given vector space with applications such as Fourier series and discrete Fourier transform (DFT).

3.6.1 Orthonormal Set

Definition: Two vectors x and y in an inner product space are said to be orthogonal if the inner product between x and y is zero. This means that:

$$<x, y> = 0 \qquad (3.86)$$

A set S in an inner product vector space is called an orthonormal set if:

$$<x, y> = \begin{cases} 1 & \text{if } x = y \\ 0 & \text{if } x \neq y \end{cases} \text{ for every pair } x, y \in S \qquad (3.87)$$

Example 3.30

The set $S = \{e_1\ e_2\ e_3\} \in R^3$ where $e_1 = [1\ \ 0\ \ 0]^T$, $e_2 = [0\ \ 1\ \ 0]^T$, and $e_3 = [0\ \ 0\ \ 1]^T$ is an orthonormal set since:

$$<e_i, e_j> = \begin{cases} 1 & \text{if } i = j \\ 0 & \text{if } i \neq j \end{cases}$$

3.6.2 Gram–Schmidt Orthogonalization Process

The Gram–Schmidt orthogonalization process is used to convert a set of independent or dependent vectors in a given vector space to an orthonormal set. Given a set of n nonzero vectors $S = \{x_1\ x_2\ \ldots\ x_n\}$ in vector space V, we would like to find an orthonormal set of vectors $\hat{S} = \{u_1\ u_2\ \ldots\ u_m\}$ with the same span as S. It is obvious that $m \leq n$. If the vectors forming set S are linearly independent, then $m = n$. This is accomplished by using the

Gram–Schmidt orthogonalization process. The steps of the process as explained below:

1. Determine u_1 by normalizing the first vector x_1 that is:

$$u_1 = \frac{x_1}{\|x_1\|} \qquad (3.88)$$

2. Form the error vector e_2 by finding the difference between the projection of x_2 onto u_1 and x_2. If $e_2 = 0$, discard x_2; otherwise, obtain u_2 by normalizing e_2 (See Figure 3.8).

$$e_2 = x_2 - <x_2, u_1> u_1 \qquad (3.89)$$

$$u_2 = \frac{e_2}{\|e_2\|} \qquad (3.90)$$

It is easy to see that u_1 and u_2 are orthogonal and each have unit norm.

3. Now, we need to find a unit norm vector u_3 that is orthogonal to both u_1 and u_2. This is done by finding the difference between x_3 and the projection of x_3 onto the subspace formed by the span of u_1 and u_2. This difference vector is normalized to produce u_3.

$$e_3 = x_3 - <x_3, u_1> u_1 - <x_3, u_2> u_2 \qquad (3.91)$$

Thus:

$$u_3 = \frac{e_3}{\|e_3\|} \qquad (3.92)$$

4. This is continued until the vector x_n has been processed, i.e. $k = 4, 5, \ldots, n$. At the kth stage, vector e_k is formed as:

$$e_k = x_k - \sum_{i=1}^{k-1} <x_k, u_i> u_i \qquad (3.93)$$

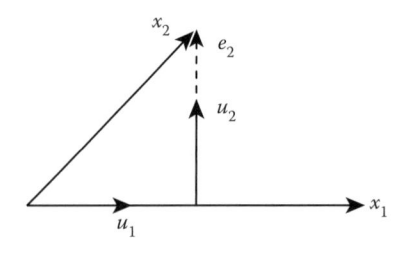

FIGURE 3.8
Error vector.

Linear Vector Spaces 133

$$u_k = \frac{e_k}{\|e_k\|} \qquad (3.94)$$

Example 3.31

Consider the set of polynomials $S=\{1\ t\ t^2\}$ defined over the interval of $-1\le t\le 1$. Using the Gram–Schmidt orthogonalization process, obtain an orthonormal set.

Solution:

$$u_1(t) = \frac{x_1(t)}{\|x_1(t)\|} = \frac{1}{\sqrt{\int_{-1}^{1} 1^2\,dt}} = \frac{1}{\sqrt{2}}$$

$$e_2(t) = x_2(t) - <x_2(t), u_1(t)> u_1(t) = t - \frac{1}{\sqrt{2}}\int_{-1}^{1} t\frac{1}{\sqrt{2}}\,dt = t - \frac{1}{2}\int_{-1}^{1} t\,dt = t - 0 = t$$

$$u_2(t) = \frac{e_2(t)}{\|e_2(t)\|} = \frac{t}{\sqrt{\int_{-1}^{1} t^2\,dt}} = \frac{t}{\sqrt{\frac{2}{3}}} = \sqrt{\frac{3}{2}}\,t$$

$$e_3(t) = x_3(t) - <x_3(t), u_1(t)> u_1(t) - <x_3(t), u_2(t)> u_2(t)$$

$$= t^2 - \frac{1}{\sqrt{2}}\int_{-1}^{1} t^2 \frac{1}{\sqrt{2}}\,dt - \sqrt{\frac{3}{2}}t\int_{-1}^{1} t^2\sqrt{\frac{3}{2}}t\,dt = t^2 - \frac{1}{2}\int_{-1}^{1} t^2\,dt - \frac{3}{2}t\int_{-1}^{1} t^3\,dt = t^2 - \frac{1}{3}$$

$$u_3(t) = \frac{e_3(t)}{\|e_3(t)\|} = \frac{t^2 - \frac{1}{3}}{\sqrt{\int_{-1}^{1}(t^2 - \frac{1}{3})^2\,dt}} = \frac{t^2 - \frac{1}{3}}{\sqrt{\frac{8}{45}}} = \frac{1}{2}\sqrt{\frac{5}{2}}(3t^2 - 1)$$

Example 3.32

Consider the set $S=\{x_1\ x_2\ x_3\ x_4\}\in R^3$ where:

$$x_1 = \begin{bmatrix} 1 \\ 2 \\ -2 \end{bmatrix},\ x_2 = \begin{bmatrix} -1 \\ 3 \\ 1 \end{bmatrix},\ x_3 = \begin{bmatrix} 1 \\ -2 \\ 5 \end{bmatrix},\ x_4 = \begin{bmatrix} 4 \\ 3 \\ 0 \end{bmatrix}$$

Use the Gram–Schmidt orthogonalization process to obtain an orthonormal set.

Solution:

$$u_1 = \frac{x_1}{\|x_1\|} = \frac{x_1}{3} = \begin{bmatrix} \frac{1}{3} & \frac{2}{3} & -\frac{2}{3} \end{bmatrix}^T = [0.3333\quad 0.6667\quad -0.6667]^T$$

$$e_2 = x_2 - <x_2, u_1> u_1 = x_2 - u_1 = \begin{bmatrix} -\frac{4}{3} & \frac{7}{3} & \frac{5}{3} \end{bmatrix}^T$$

$$u_2 = \frac{e_2}{\|e_2\|} = \frac{e_2}{\sqrt{10}} = \left[-\frac{4}{3\sqrt{10}} \quad \frac{7}{3\sqrt{10}} \quad \frac{5}{3\sqrt{10}}\right]^T = [-0.4216 \quad 0.7379 \quad 0.527]^T$$

$$e_3 = x_3 - <x_3, u_1> u_1 - <x_3, u_2> u_2$$

$$= x_3 + \frac{13}{3} u_1 - \frac{7}{3\sqrt{10}} u_2 = [2.7556 \quad 0.3444 \quad 1.7222]^T$$

$$u_3 = \frac{e_3}{\|e_3\|} = \frac{e_3}{3.2677} = [0.8433 \quad 0.1054 \quad 0.527]^T$$

$$e_4 = x_4 - <x_4, u_1> u_1 - <x_4, u_2> u_2 - <x_4, u_3> u_3$$

$$= x_4 - 3.333 u_1 - 0.527 u_2 - 3.6893 u_3 = 0$$

Since $e_4 = 0$, we discard x_4. This means that the dimension of the space spanned by S is three as expected, since S is a subspace of R^3 and its dimension cannot exceed three. The MATLAB® function file shown in Table 3.2 can be used to extract an orthonormal set from a given set of vectors.

3.6.3 Orthogonal Matrices

A real $n \times n$ matrix Q is said to be an orthogonal matrix if:

$$QQ^T = Q^T Q = I \qquad (3.95)$$

For complex matrices, the above definition holds except for the fact that the transpose is replaced by the Hermitian transpose. From the above definition,

TABLE 3.2

MATLAB Code Implementation of Gram–Schmidt Orthogonalization Algorithm

```
function [V] = gramschmidt(A)
% A = a matrix of size mxn containing n vectors.
% The dimension of each vector is m.
% V = output matrix: Columns of V form an orthonormal set.
[m,n] = size(A);
V = [A(:,1)/norm(A(:,1))];
for j = 2:1:n
  v = A(:,j);
  for i = 1:size(V,2)
    a = v'*V(:,i);
    v = v - a*V(:,i);
  end
  if (norm(v))^4 >= eps
    V = [V v/norm(v)];
  else
  end
end
```

Linear Vector Spaces

it is obvious that the inverse of an orthogonal matrix is equal to its transpose. Also, the determinant of an orthogonal matrix is one.

Orthogonal matrices are also known as unitary matrices. To see the relationship to inner product vector spaces, consider an orthogonal matrix Q and an n dimensional vector x. Then:

$$\|Qx\|^2 = <Qx, Qx> = (Qx)^T Qx = x^T Q^T Qx = x^T x = \|x\|^2 \qquad (3.96)$$

This means that orthogonal matrices are norm preserving matrices. The set of $n \times n$ orthogonal matrices form an algebraic group that is used in many engineering applications. They form the foundation to numerical algorithms such as QR decomposition. As another example, the discrete cosine transform used in JPEG still image compression and MPEG video compression can be represented by an orthogonal matrix. The following matrices are examples of orthogonal matrices:

$$A = \begin{bmatrix} 1 & 0 \\ 0 & 1 \end{bmatrix}, B = \begin{bmatrix} \dfrac{\sqrt{2}}{2} & \dfrac{\sqrt{2}}{2} \\ -\dfrac{\sqrt{2}}{2} & \dfrac{\sqrt{2}}{2} \end{bmatrix}, C = \begin{bmatrix} 0.4243 & 0.7702 & -0.4762 \\ 0.5657 & 0.1852 & 0.8036 \\ 0.7071 & -0.6103 & -0.3571 \end{bmatrix} \qquad (3.97)$$

3.6.3.1 Complete Orthonormal Set

Definition: An orthonormal set S in vector space V is a complete orthonormal set if there is no nonzero vector $x \in V$ for which $\{S\ x\}$ forms an orthonormal set. In other words, if S is a complete orthonormal set, there is no other vector in V being orthogonal to every element of vector space V.

3.6.4 Generalized Fourier Series (GFS)

Dimension of a vector space: Let V be a vector space and let $S = \{u_i\}_{i=1}^{n}$ be a finite orthonormal set in the space V. The set S is a complete orthonormal set if and only if there is no other vector in vector space V being orthogonal to all elements of set S. In this case, the dimension of vector space V is n and set S can span the whole space in the sense that every vector in V is a linear combination of $u_1, u_2, \ldots, u_n$. That is, any vector $x \in V$ can be written as:

$$x = \alpha_1 u_1 + \alpha_2 u_2 + \cdots + \alpha_n u_n = \sum_{i=1}^{n} \alpha_i u_i \qquad (3.98)$$

where

$$\alpha_i = <x, u_i> \qquad (3.99)$$

This is obvious by the orthogonality relations:

$$<x, u_i> = \alpha_1 <u_1, u_i> + \alpha_2 <u_2, u_i> + \cdots + \alpha_n <u_n, u_i> = \alpha_i \quad (3.100)$$

If $x=0$, then $\alpha_1 = \alpha_2 = \cdots = \alpha_n = 0$. Hence, the $\{u_i\}_{i=1}^{n}$ form a linearly independent set and hence a basis for the space V which they generate. This concept can be generalized to an infinite dimension space through Theorem 3.6.

Theorem 3.6: Fourier series

Let V be a complete vector space and let $S = \{u_i\}_{i=-\infty}^{\infty}$ be a countably infinite orthonormal set in V. Then every element of vector space V can be written as a linear combination of the elements of S as follows:

$$x = \sum_{i=-\infty}^{\infty} \alpha_i u_i \quad (3.101)$$

where

$$\alpha_i = <x, u_i> \quad (3.102)$$

This is known as GFS.

Proof: Let $y = \sum_{i=-N}^{N} \alpha_i u_i$, then:

$$\|y - x\|^2 = <y-x, y-x> = <y, y> - <y, x> - <x, y> + <x, x> \quad (3.103)$$

Now, we find each term in the above expansion.

$$<y, y> = <\sum_{i=-N}^{N} \alpha_i u_i, \sum_{i=-N}^{N} \alpha_i u_i> = \sum_{i=-N}^{N}\sum_{j=-N}^{N} <\alpha_i u_i, \alpha_j u_j>$$

$$= \sum_{i=-N}^{N}\sum_{j=-N}^{N} \alpha_i \overline{\alpha_i} <u_i, u_j> = \sum_{i=-N}^{N} \alpha_i \overline{\alpha_i} = \sum_{i=-N}^{N} |\alpha_i|^2 \quad (3.104)$$

$$<y, x> = <\sum_{i=-N}^{N} \alpha_i u_i, x> = \sum_{j=-N}^{N} <\alpha_i u_i, x> = \sum_{j=-N}^{N} \alpha_i <u_i, x> = \sum_{i=-N}^{N} \alpha_i \overline{\alpha_i} = \sum_{i=-N}^{N} |\alpha_i|^2 \quad (3.105)$$

$$<x, y> = \overline{<y, x>} = \sum_{i=-N}^{N} |\alpha_i|^2 \quad (3.106)$$

$$<x, x> = <\sum_{i=-\infty}^{\infty} \alpha_i u_i, \sum_{i=-\infty}^{\infty} \alpha_i u_i> = \sum_{i=-\infty}^{\infty}\sum_{j=-\infty}^{\infty} <\alpha_i u_i, \alpha_j u_j>$$

$$= \sum_{i=-\infty}^{\infty}\sum_{j=-\infty}^{\infty} \alpha_i \overline{\alpha_i} <u_i, u_j> = \sum_{i=-\infty}^{\infty} \alpha_i \overline{\alpha_i} = \sum_{i=-\infty}^{\infty} |\alpha_i|^2 \quad (3.107)$$

Linear Vector Spaces 137

Therefore:

$$\|y - x\|^2 = \sum_{i=-\infty}^{\infty} |\alpha_i|^2 - \sum_{i=-N}^{N} |\alpha_i|^2 \qquad (3.108)$$

Taking the limit as $N \to \infty$, we have:

$$\lim_{N \to \infty} \|y - x\|^2 = \lim_{N \to \infty} \left[\sum_{i=-\infty}^{\infty} |\alpha_i|^2 - \sum_{i=-N}^{N} |\alpha_i|^2 \right] = 0 \qquad (3.109)$$

Since $y \to x$ as $N \to \infty$, we have $x = \sum_{i=-\infty}^{\infty} \alpha_i u_i$.

We now consider special cases of GFS. We first consider the continuous Fourier transform (CFT).

3.6.5 Applications of GFS

The GFS is used to derive the continuous Fourier series as well as the discrete Fourier series.

3.6.5.1 Continuous Fourier Series

Consider the set $S = \{u_n(t)\}_{n=-\infty}^{\infty}$, where:

$$u_n(t) = \frac{1}{\sqrt{T}} \exp\left(j \frac{2\pi}{T} nt \right), \quad 0 < t < T \quad \text{and} \quad n = 0, \pm 1, \pm 2, \ldots \qquad (3.110)$$

This is an orthonormal set since:

$$< u_n(t), u_m(t) > = \int_0^T u_n(t) \bar{u}_m(t) dt = \frac{1}{T} \int_0^T \exp\left(j \frac{2\pi}{T} (n-m)t \right) dt$$

$$= \frac{\exp(j 2\pi(n-m)) - 1}{j 2\pi(n-m)} = \begin{cases} 1 & \text{if } n = m \\ 0 & \text{if } n \neq m \end{cases} \qquad (3.111)$$

It is also a complete set. Therefore, any function $x(t)$ defined over the interval of $0 < t < T$ can be written as:

$$x(t) = \sum_{n=-\infty}^{\infty} \alpha_n u_n(t) = \sum_{n=-\infty}^{\infty} \alpha_n \frac{1}{\sqrt{T}} \exp\left(j \frac{2\pi}{T} nt \right) \qquad (3.112)$$

where

$$\alpha_n = < x(t), u_n(t) > = \int_0^T x(t) \bar{u}_n(t) dt = \int_0^T \frac{1}{\sqrt{T}} x(t) \exp\left(-j \frac{2\pi}{T} nt \right) dt \qquad (3.113)$$

Let us define:

$$X_n = \frac{\alpha_n}{\sqrt{T}} \quad (3.114)$$

Then, Equations 3.112 and 3.113 can be written as:

$$x(t) = \sum_{n=-\infty}^{\infty} X_n \exp\left(j\frac{2\pi}{T}nt\right) \quad (3.115)$$

$$X_n = \frac{1}{T}\int_0^T x(t)\exp\left(-j\frac{2\pi}{T}nt\right)dt \quad (3.116)$$

This is called the exponential form of the continuous Fourier series. Fourier series can be applied to signals with finite duration T, or signals which are periodic with period T. If $x(t)$ is a real periodic signal, the expansion given in Equation 3.115 can be written in terms of the trigonometric sine and cosine functions as given by the following equations:

$$x(t) = \frac{a_0}{2} + \sum_{n=1}^{\infty} a_n \cos(n\omega_0 t) + \sum_{n=1}^{\infty} b_n \sin(n\omega_0 t) \quad (3.117)$$

where $\omega_0 = 2\pi/T$ is the fundamental frequency of signal $x(t)$ and a_n, b_n are trigonometric Fourier series coefficients given by:

$$a_n = \frac{2}{T}\int_0^T x(t)\cos(n\omega_0 t)dt = 2\text{real}(X_n), \quad n = 0, 1, 2, \ldots \quad (3.118)$$

$$b_n = \frac{2}{T}\int_0^T x(t)\sin(n\omega_0 t)dt = -2\text{imag}(X_n), \quad n = 1, 2, \ldots \quad (3.119)$$

Note that as Equations 3.118 and 3.119 indicate, the trigonometric Fourier series coefficients a_n, b_n are computed directly or by using the complex Fourier series coefficients X_n. The Fourier series coefficients can be computed numerically using MATLAB®. The MATLAB function file shown in Table 3.3 is used to compute a_n and b_n.

Example 3.33

Compute the Fourier series coefficients of the periodic square wave signal shown in Figure 3.9.

Linear Vector Spaces

TABLE 3.3

MATLAB Code to Compute Fourier Series Coefficients

```
% M-file to Compute Fourier Series Coefficients
% function [a,b] = fsc(t,x,M)
% a(i) = (2/T)*integral of x(t)cos((i-1)w0t) from 0 to T.
% b(i) = (2/T)*integral of x(t)sin(iw0t) from 0 to T.
% w0 = 2pi/T;
% M = Number of Fourier Series Coefficients
% t = Nx1 = time axis (0 to T) uniformly sampled
% x = Nx1 = x(t) at the points in vector t.
% The DC coefficient is a(1)/2
%
function [a,b] = fsc(t,x,M)
T = max(t) - min(t);
c = 2.0*pi/T;
a = zeros(M,1);
b = zeros(M-1,1);
a(1) = (2/T)*trapz(t,x);
for i = 2:M
 gc = x.*cos(c*(i-1)*t);
 gs = x.*sin(c*(i-1)*t);
 a(i) = (2/T)*trapz(t,gc);
 b(i-1) = (2/T)*trapz(t,gs);
end
```

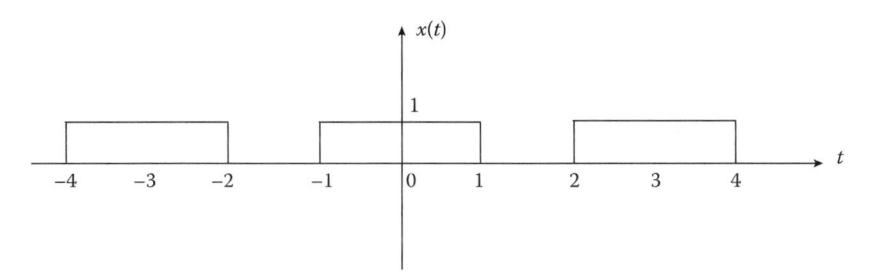

FIGURE 3.9
Periodic square wave signal.

The period of this signal is $T=3$. The Fourier series coefficients are:

$$X_n = \frac{1}{T}\int_T x(t)e^{-j\frac{2\pi}{T}nt}dt = \frac{1}{3}\int_{-1}^{1} e^{-j\frac{2\pi}{3}nt}dt = \frac{1}{-j2\pi n}e^{-j\frac{2\pi}{3}nt}\bigg|_{-1}^{1}$$

$$= \frac{1}{-j2\pi n}\left(e^{-j\frac{2\pi}{3}n} - e^{j\frac{2\pi}{3}n}\right) = \frac{-j2\sin\left(\frac{2\pi}{3}n\right)}{-j2n\pi}$$

or

$$X_n = \begin{cases} \dfrac{2}{3} & n=0 \\ \dfrac{\sin\left(\dfrac{2\pi}{3}n\right)}{n\pi} & n \neq 0 \end{cases}$$

Therefore:

$$x(t) = \frac{2}{3} + \frac{1}{\pi}\sum_{\substack{n=-\infty \\ n\neq 0}}^{\infty} \frac{\sin\left(\dfrac{2\pi}{3}n\right)}{n} e^{j\frac{2n}{3}\pi t} = \frac{2}{3} + \frac{2}{\pi}\sum_{n=1}^{\infty} \frac{\sin\left(\dfrac{2\pi}{3}n\right)}{n}\cos\left(\frac{2\pi}{3}nt\right)$$

The plots of x(t) and the truncated Fourier series expansions of x(t) using M terms for different values of M are shown in Figures 3.10 through 3.12.

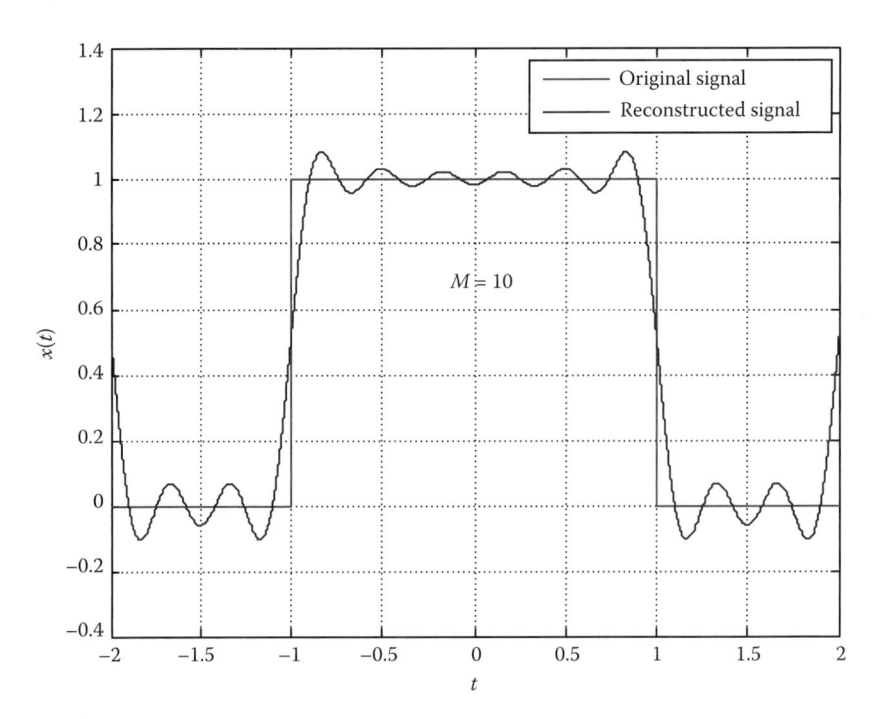

FIGURE 3.10
(See color insert following page 174.) Fourier series expansion of x(t) with M=10.

Linear Vector Spaces 141

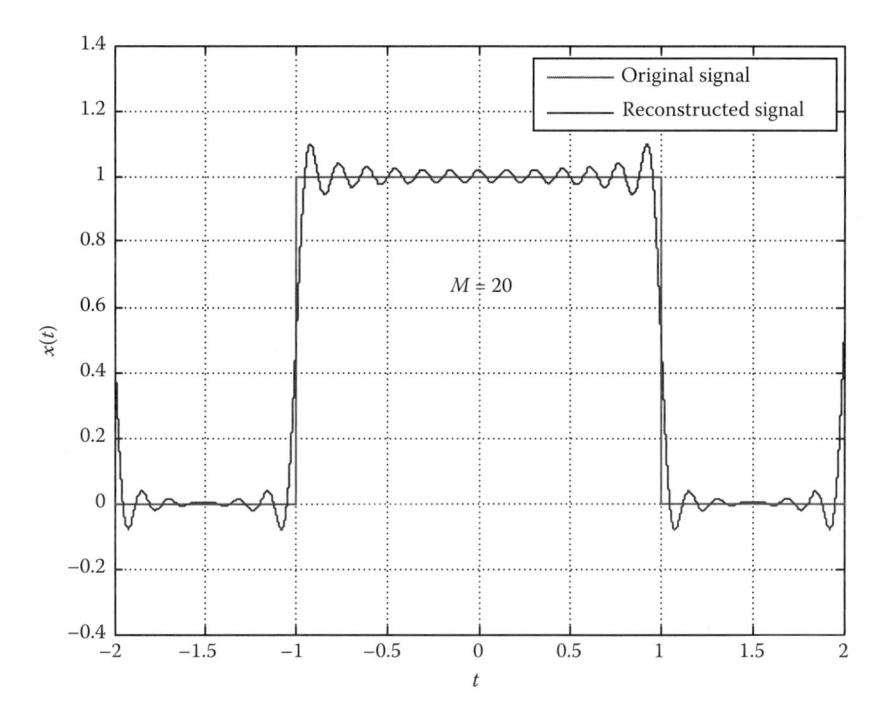

FIGURE 3.11
(See color insert following page 174.) Fourier series expansion of $x(t)$ with $M=20$.

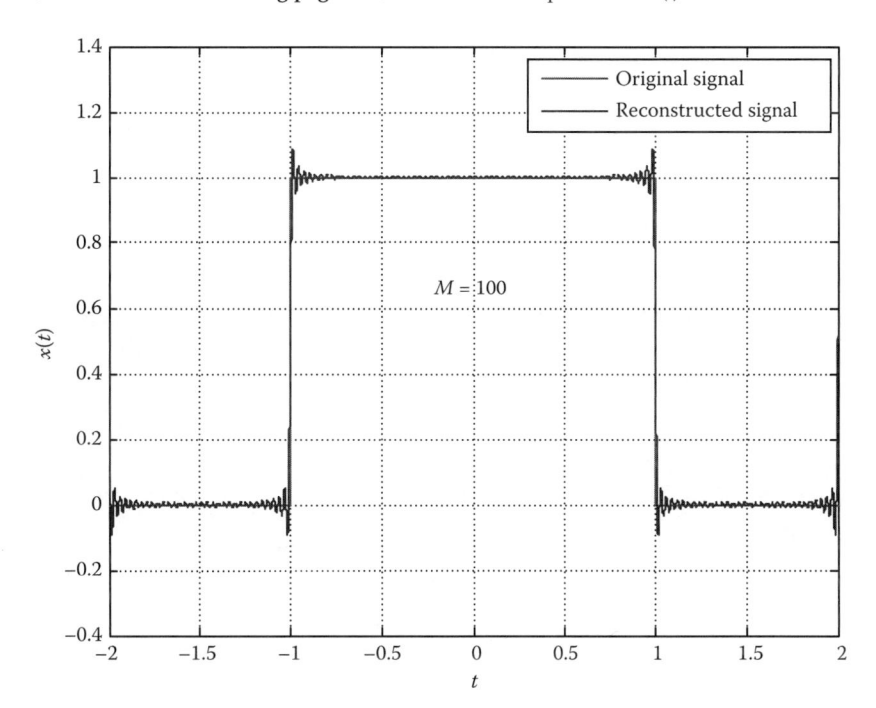

FIGURE 3.12
(See color insert following page 174.) Fourier series expansion of $x(t)$ with $M=100$.

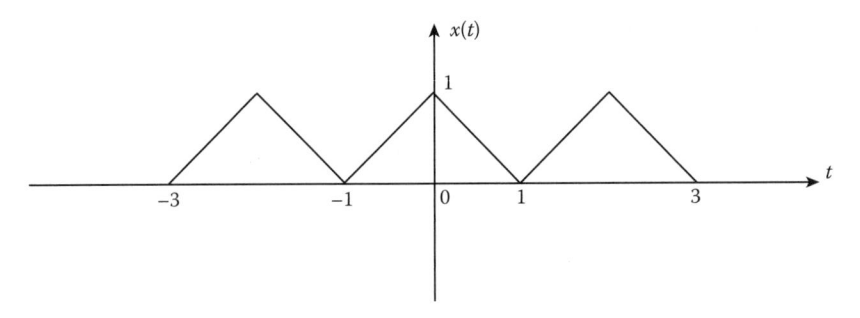

FIGURE 3.13
Triangular periodic signal.

Example 3.34

Compute the Fourier series coefficients for the periodic signal shown in Figure 3.13. The period of the signal is $T=2$. The Fourier series coefficients are:

$$X_n = \frac{1}{T}\int_0^T x(t)e^{-j\frac{2\pi}{T}nt}dt = \frac{1}{2}\int_0^1 (1-t)e^{-jn\pi t}dt + \frac{1}{2}\int_1^2 (t-1)e^{-jn\pi t}dt$$

$$= \frac{1}{2}\left(\frac{1}{-jn\pi}e^{-jn\pi t} - \frac{t}{-jn\pi}e^{-jn\pi t} + \frac{1}{(-jn\pi)^2}e^{-jn\pi t}\right)\bigg|_0^1$$

$$+ \frac{1}{2}\left(-\frac{1}{-jn\pi}e^{-jn\pi t} + \frac{t}{-jn\pi}e^{-jn\pi t} - \frac{1}{(-jn\pi)^2}e^{-jn\pi t}\right)\bigg|_1^2$$

$$= \frac{1}{2}\left(\frac{1}{-jn\pi}e^{-jn\pi} - \frac{1}{-jn\pi} - \frac{1}{-jn\pi}e^{-jn\pi} + \frac{1}{(-jn\pi)^2}e^{-jn\pi} - \frac{1}{(-jn\pi)^2}\right)$$

$$+ \frac{1}{2}\left(-\frac{1}{-jn\pi}e^{-j2n\pi} + \frac{1}{-jn\pi}e^{-jn\pi} + \frac{2}{-jn\pi}e^{-j2n\pi} - \frac{1}{(-jn\pi)^2}e^{-j2n\pi}\right.$$

$$\left. -\frac{1}{-jn\pi}e^{-jn\pi} + \frac{1}{(-jn\pi)^2}e^{-jn\pi}\right)$$

$$= \frac{e^{-jn\pi}-1}{(-jn\pi)^2} = \begin{cases} 0.5 & n=0 \\ \dfrac{2}{n^2\pi^2} & n=\text{odd} \\ 0 & n=\text{even} \end{cases}$$

Therefore:

$$x(t) = 0.5 + \frac{2}{\pi^2}\sum_{\substack{n=-\infty \\ n=\text{odd}}}^{\infty} \frac{1}{n^2}e^{jn\pi t} = 0.5 + \frac{4}{\pi^2}\sum_{\substack{n=1 \\ n=\text{odd}}}^{\infty} \frac{1}{n^2}\cos(n\pi t)$$

The plots of $x(t)$ and the truncated Fourier series expansion of $x(t)$ using M terms are shown in Figures 3.14 and 3.15.

Linear Vector Spaces 143

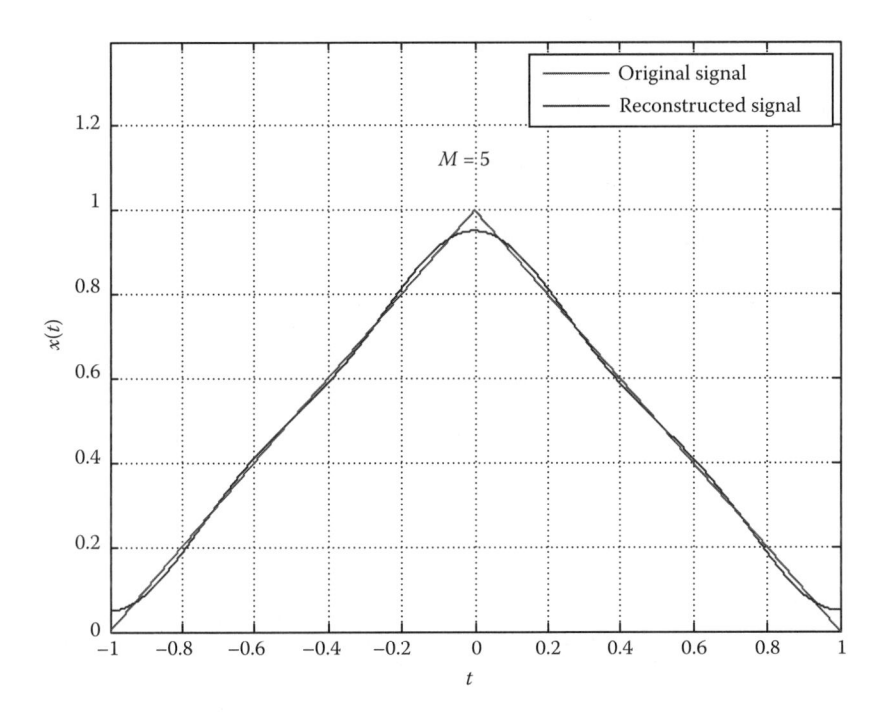

FIGURE 3.14
(See color insert following page 174.) Fourier series expansion of $x(t)$ with $M=5$.

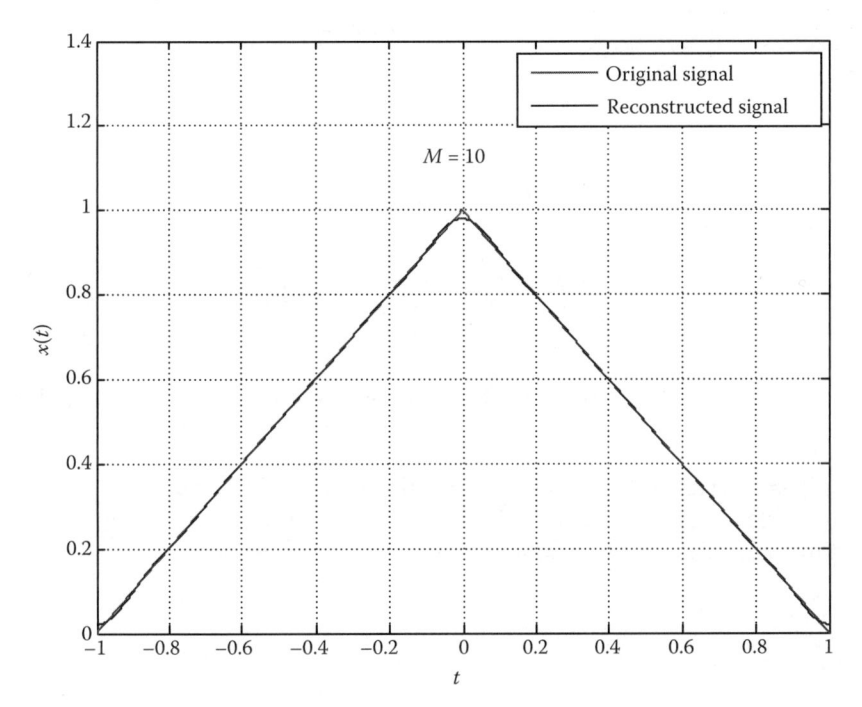

FIGURE 3.15
(See color insert following page 174.) Fourier series expansion of $x(t)$ with $M=10$.

3.6.5.2 Discrete Fourier Transform (DFT)

Consider the set $S = \{u_m(n)\}_{m=0}^{N-1}$, where:

$$u_m(n) = \frac{1}{\sqrt{N}} \exp\left(j\frac{2\pi}{N}mn\right), \quad n,m = 0, 1, \ldots, N-1 \qquad (3.120)$$

This is a orthonormal set because:

$$<u_m(n), u_l(n)> = \sum_{n=0}^{N-1} u_m(n)\bar{u}_l(n) = \frac{1}{N}\sum_{n=0}^{N-1} \exp\left(j\frac{2\pi}{N}(m-l)n\right) \qquad (3.121)$$

$$= \frac{1}{N}\frac{1-\exp(j2\pi(m-l))}{1-\exp\left(j\frac{2\pi}{N}(m-l)\right)} = \begin{cases} 1 & \text{if } m = l \\ 0 & \text{if } m \neq l \end{cases}$$

It is also complete. Therefore, any discrete function $x(n)$ defined over the interval $0 \leq n \leq N-1$ can be written as:

$$x(n) = \sum_{k=0}^{N-1} \alpha_k u_k(n) = \sum_{k=0}^{N-1} \alpha_k \frac{1}{\sqrt{N}} \exp\left(j\frac{2\pi}{N}nk\right) \qquad (3.122)$$

where

$$\alpha_k = <x(n), u_k(n)> = \sum_{n=0}^{N-1} x(n)\bar{u}_k(n) = \frac{1}{\sqrt{N}}\sum_{n=0}^{N-1} x(n)\exp\left(-j\frac{2\pi}{N}nk\right) \qquad (3.123)$$

Let us define:

$$X(k) = \sqrt{N}\alpha_k \qquad (3.124)$$

Then, Equations 3.124 and 3.122 can be written as:

$$X(k) = \sum_{n=0}^{N-1} x(n)\exp\left(-j\frac{2\pi}{N}nk\right) \qquad (3.125)$$

$$x(n) = \frac{1}{N}\sum_{k=0}^{N-1} X(k)\exp\left(j\frac{2\pi}{N}nk\right) \qquad (3.126)$$

This is known as DFT. Equation 3.125 is the forward DFT and Equation 3.126 is the backward or inverse DFT. The MATLAB® commands to compute DFT and inverse DFT are $X=$**fft**(x) and $x=$**ifft**(X), respectively.

Example 3.35

Compute the DFT of the four point sequence below:

$$x(n) = [2 -3 \quad 4 \quad 6]$$

Linear Vector Spaces

Solution:

$$X(k) = \sum_{n=0}^{N-1} x(n)\exp\left(-j\frac{2\pi}{N}nk\right) = \sum_{n=0}^{3} x(n)\exp\left(-j\frac{\pi}{2}nk\right)$$

$$= 2 - 3e^{-j\frac{\pi}{2}k} + 4e^{-j\pi k} + 6e^{-j3\frac{\pi}{2}k}$$

$X(0) = 2 - 3 + 4 + 6 = 9$

$X(1) = 2 - 3e^{-j\frac{\pi}{2}} + 4e^{-j\pi} + 6e^{-j3\frac{\pi}{2}} = 2 + j3 - 4 + j6 = -2 + j9$

$X(2) = 2 - 3e^{-j\pi} + 4e^{-j2\pi} + 6e^{-j3\pi} = 2 + 3 + 4 - 6 = 3$

$X(3) = 2 - 3e^{-j\frac{3\pi}{2}} + 4e^{-j3\pi} + 6e^{-j9\frac{\pi}{2}} = 2 - j3 - 4 - j6 = -2 - j9$

Therefore:

$$X(k) = \begin{bmatrix} 9 & -2+j9 & 3 & -2-j9 \end{bmatrix}$$

The MATLAB code to compute the DFT of this sequence **x=[2 −3 4 6]** is:
X=fft(x).

3.6.5.3 Legendre Polynomial

Another example of a complete orthonormal set is Legendre polynomials. These polynomials are defined over the closed interval [−1 1]. The first three Legendre polynomials are:

$$l_0(x) = 1, \; l_1(x) = x \; \text{ and } \; l_2(x) = \frac{3}{2}x^2 - \frac{1}{2} \tag{3.127}$$

Legendre polynomials for $n \geq 3$ can be obtained using the following recursive equation:

$$l_{n+1}(x) = \frac{2n+1}{n+1}xl_n(x) - \frac{n}{n+1}l_{n-1}(x) \tag{3.128}$$

For example, $l_3(x)$ is computed as:

$$l_3(x) = \frac{4+1}{2+1}xl_2(x) - \frac{2}{2+1}l_1(x) = \frac{5}{3}x\left(\frac{3}{2}x^2 - \frac{1}{2}\right) - \frac{2}{3}x = \frac{5}{2}x^3 - \frac{3}{2}x \tag{3.129}$$

Note that this set of polynomials are orthogonal and not orthonormal. To make the set orthonormal, each polynomial should be normalized, that is:

$$\hat{l}_n(x) = \frac{l_n(x)}{\|l_n(x)\|} = \frac{l_n(x)}{\sqrt{\int_{-1}^{1} l_n^2(x)dx}} \tag{3.130}$$

3.6.5.4 Sinc Functions

The sinc function can be used to form a complete orthonormal set. Let:

$$\phi_n(t) = \sqrt{2B}\operatorname{sinc}\left(2B\left(t - \frac{n}{2B}\right)\right) \quad n = 0, \pm 1, \pm 2, \ldots \quad (3.131)$$

where the sinc function is a bandlimited signal and is given by:

$$\operatorname{sinc}(t) = \frac{\sin(\pi t)}{\pi t} \quad (3.132)$$

The Fourier transform of $\varphi_n(t)$ is given by:

$$\Phi_n(\omega) = \int_{-\infty}^{\infty} \phi_n(t) e^{-j\omega t} dt = \begin{cases} \dfrac{1}{\sqrt{2B}} e^{-j\frac{n}{2B}\omega} & -2\pi B < \omega < 2\pi B \\ 0 & \text{otherwise} \end{cases} \quad (3.133)$$

Therefore, the set is a bandlimited set since the Fourier transform of $\varphi_n(t)$, an element of the set, satisfies:

$$\Phi_n(\omega) = 0 \quad \text{for } \omega \notin (-2\pi B, 2\pi B) \quad (3.134)$$

It is also an orthonormal set since, by Parseval's theorem, we have:

$$\int_{-\infty}^{\infty} \varphi_n(t)\overline{\varphi}_m(t)dt = \frac{1}{2\pi} \int_{-2\pi B}^{2\pi B} \Phi_n(\omega)\overline{\Phi}_m(\omega)d\omega$$

$$= \frac{1}{2\pi} \int_{-2\pi B}^{2\pi B} |\Phi_n(\omega)| e^{-j\omega\frac{n}{2B}} |\overline{\Phi}_m(\omega)| e^{j\omega\frac{m}{2B}} d\omega$$

$$= \frac{1}{2\pi} \int_{-2\pi B}^{2\pi B} \frac{1}{\sqrt{2B}}\frac{1}{\sqrt{2B}} e^{-j\omega\frac{n-m}{2B}} d\omega = \frac{1}{4B\pi} \left. \frac{e^{-j\omega\frac{n-m}{2B}}}{-j\frac{n-m}{2B}} \right|_{-2\pi B}^{2\pi B}$$

$$= \frac{e^{-j\omega(n-m)\pi} - e^{j\omega(n-m)\pi}}{-j2\pi(n-m)} = \frac{\sin(n-m)\pi}{(n-m)\pi} = \begin{cases} 1 & n = m \\ 0 & n \neq m \end{cases} \quad (3.135)$$

Therefore, any bandlimited signal $f(t)$ can be expanded in terms of these basis functions as:

$$f(t) = \sum_{n=-\infty}^{\infty} \alpha_n \varphi_n(t) \quad (3.136)$$

where the coefficients α_n are given by the dot product between $f(t)$ and $\varphi_n(t)$ as follows:

$$\alpha_n = <f(t), \varphi_n(t)> = \int_{-\infty}^{\infty} f(t)\sqrt{2B}\,\text{sinc}\left(2B\left(t - \frac{n}{2B}\right)\right) dt \qquad (3.137)$$

It can be shown that $\alpha_n = 1/\sqrt{2B}\,f(n/2B)$; that is, the coefficients α_n are samples of the continuous signal $f(t)$ taken every $1/2B$ seconds apart (see Problem 3.13).

3.7 Matrix Factorization

Matrix factorization deals with the general problem of decomposing a matrix into the sum of a product of matrices. The general factorization problem is to factor matrix A as:

$$A = A_1 B_1 + A_2 B_2 + \cdots + A_k B_k \qquad (3.138)$$

There are different techniques to factorize a matrix. Each factorization methodology has its own properties and applications. In the next section, QR factorization with its application in solving a set of simultaneous linear equations will be discussed.

3.7.1 QR Factorization

QR factorization factors a rectangular matrix into the product of a unitary and an upper triangular matrix. An important application of QR factorization is solving linear systems of equations. This is due to the fact that the inverse of unitary matrices is obtained from the conjugate transpose (Hermitian operation). It is also a useful tool in solving least squares problems. Different techniques, such as Householder transformation and Gram–Schmidt decomposition, are used for QR factorization.

Theorem 3.7: Let A be an $n \times m$ matrix ($n \geq m$) of rank $r = m$. Then A can be factored into a product of two matrices Q and R, where Q is $n \times m$ with pairwise orthogonal column vectors and R is an $m \times m$ upper triangular matrix with positive diagonal elements.

Proof: The proof is by construction. Partition matrix A as:

$$A = [a_1 \quad A_2] = [q_1 \quad Q_2]\begin{bmatrix} r_{11} & R_{12} \\ 0 & R_{22} \end{bmatrix} \qquad (3.139)$$

where $a_1 \in R^n$ is the first column of matrix A, A_2 is $n\times(m-1)$, q_1 is $n\times 1$, and Q_2 is $n\times(m-1)$. The matrix $[q_1 \quad Q_2]$ is an orthogonal matrix and is formed by applying the Gram–Schmitt orthogonalization process to the columns of matrix A. Therefore,

$$q_1 = \frac{a_1}{\|a_1\|} \tag{3.140}$$

Since Q is an orthogonal matrix, we have:

$$\begin{bmatrix} q_1^H \\ Q_2^H \end{bmatrix} [q_1 \quad Q_2] = \begin{bmatrix} q_1^H q_1 & q_1^H Q_2 \\ Q_2^H q_1 & Q_2^H Q_2 \end{bmatrix} = \begin{bmatrix} 1 & 0 \\ 0 & I \end{bmatrix} \tag{3.141}$$

Therefore, $q_1^H q_1 = 1$, $Q_2^H Q_2 = I$ and $q_1^H Q_2 = 0$. From Equation 3.139 we have:

$$A_2 = q_1 R_{12} + Q_2 R_{22} \tag{3.142}$$

Premultiplying both sides of Equation 3.142 by q_1^H and using the fact that $q_1^H q_1 = 1$ and $q_1^H Q_2 = 0$ yields:

$$R_{12} = q_1^H A_2 \tag{3.143}$$

Next, premultiplying both sides of Equation 3.142 by Q_2^H and using the fact that $q_1^H Q_2 = 0$ and $Q_2^H Q_2 = I$ results in:

$$R_{22} = Q_2^H A_2 \tag{3.144}$$

Note that $A_2 - q_1 R_{12} = Q_1 R_{22}$ is the QR factorization of the $n\times(m-1)$ matrix $A_2 - q_1 R_{12}$. In summary, to compute the QR factorization of the $n\times m$ matrix A, we first compute the Q matrix using the Gram–Schmidt orthogonalization process and then obtain R by premultiplying A by Q^H, that is:

$$R = Q^H A \tag{3.145}$$

The MATLAB® command to perform QR factorization of matrix A is $[Q,R] = qr(A)$.

Example 3.36

Find the QR factorization of the matrix below:

$$A = \begin{bmatrix} 1 & 2 & 3 \\ -1 & -2 & 4 \\ 2 & 0 & 3 \\ -5 & -3 & 1 \end{bmatrix}$$

Solution:
Applying the Gram–Schmidt orthogonalization to the columns of A yields:

$$Q = [q_1 \quad q_2 \quad q_3] = \begin{bmatrix} 0.1796 & 0.5994 & 0.7799 \\ -0.1796 & -0.5994 & 0.4911 \\ 0.3592 & -0.5297 & 0.3369 \\ -0.898 & 0.0279 & 0.1925 \end{bmatrix}$$

The R matrix is obtained as:

$$R = Q^H A = \begin{bmatrix} 0.1796 & 0.5994 & 0.7799 \\ -0.1796 & -0.5994 & 0.4911 \\ 0.3592 & -0.5297 & 0.3369 \\ -0.898 & 0.0279 & 0.1925 \end{bmatrix}^T \begin{bmatrix} 1 & 2 & 3 \\ -1 & -2 & 4 \\ 2 & 0 & 3 \\ -5 & -3 & 1 \end{bmatrix} = \begin{bmatrix} 5.5678 & 3.4125 & 0 \\ 0 & 2.3141 & -2.1607 \\ 0 & 0 & 5.5074 \end{bmatrix}$$

Therefore:

$$A = QR = \begin{bmatrix} 0.1796 & 0.5994 & 0.7799 \\ -0.1796 & -0.5994 & 0.4911 \\ 0.3592 & -0.5297 & 0.3369 \\ -0.898 & 0.0279 & 0.1925 \end{bmatrix} \begin{bmatrix} 5.5678 & 3.4125 & 0 \\ 0 & 2.3141 & -2.1607 \\ 0 & 0 & 5.5074 \end{bmatrix}$$

3.7.2 Solution of Linear Equations Using QR Factorization

Consider the set of simultaneous linear equations given by:

$$Ax = b \tag{3.146}$$

where A is an $n \times n$ matrix, and b, x are $n \times 1$ vectors. Replacing A by its QR factors transforms Equation 3.146 into:

$$QRx = b \tag{3.147}$$

Premultiplying the above equation by Q^T yields:

$$Q^T QRx = Q^T b \tag{3.148}$$

Since Q is an orthogonal matrix, $Q^T Q = 1$. Hence:

$$Rx = Q^T b \tag{3.149}$$

Equation 3.149 can be solved recursively for x_i as follows:

$$x_i = \frac{1}{r_{ii}} \left(b'_i - \sum_{k=i+1}^{n} r_{ik} x_k \right) \quad i = n, n-1, \ldots, 1 \tag{3.150}$$

where r_{ik} is the ik^{th} element of matrix R and b'_i is the ith element of vector Q^Tb.

Example 3.37

Solve the following set of linear equations using QR factorization.

$$\begin{bmatrix} 1 & 2 & 1 \\ -2 & 3 & 4 \\ 1 & 0 & 5 \end{bmatrix} \begin{bmatrix} x_1 \\ x_2 \\ x_3 \end{bmatrix} = \begin{bmatrix} 6 \\ 20 \\ 14 \end{bmatrix}$$

Solution:
Using MATLAB® command [Q, R]=qr(A) to perform QR factorization, we have

$$Q = \begin{bmatrix} -0.4082 & -0.8296 & -0.381 \\ 0.8165 & -0.51851 & 0.254 \\ -0.4082 & -0.2074 & 0.889 \end{bmatrix}, R = \begin{bmatrix} -2.4495 & 1.633 & 0.8165 \\ 0 & -3.2146 & -3.9404 \\ 0 & 0 & 5.08 \end{bmatrix}, \text{ and}$$

$$Q^Tb = \begin{bmatrix} -0.4082 & -0.8296 & -0.381 \\ 0.8165 & -0.51851 & 0.254 \\ -0.4082 & -0.2074 & 0.889 \end{bmatrix} \begin{bmatrix} 6 \\ 20 \\ 14 \end{bmatrix} = \begin{bmatrix} 8.165 \\ -18.2508 \\ 15.24 \end{bmatrix}$$

Hence,

$$Rx = Q^Tb \rightarrow \begin{bmatrix} -2.4495 & 1.633 & 0.8165 \\ 0 & -3.2146 & -3.9404 \\ 0 & 0 & 5.08 \end{bmatrix} \begin{bmatrix} x_1 \\ x_2 \\ x_3 \end{bmatrix} = \begin{bmatrix} 8.165 \\ -18.2508 \\ 15.24 \end{bmatrix}$$

We can use back-substitution to solve the above system.

$$5.08 x_3 = 15.24 \Rightarrow x_3 = 3$$

$$-3.2146 x_2 - 3.9404 x_3 = -18.2508$$

$$\Rightarrow -3.2146 x_2 = -18.2508 + 3.9404 x_3 = -18.2508 + 3.9404 \times 3 = -6.4291$$

$$\Rightarrow x_2 = \frac{-6.4291}{-3.2146} = 2$$

$$-2.4495 x_1 + 1.633 x_2 + 0.8165 x_3 = 8.165$$

$$\Rightarrow -2.4495 x_1 = 8.165 - 1.633 x_2 - 0.8165 x_3 = 8.165 - 1.633 \times 2 - 0.8165 \times 3$$

$$\Rightarrow -2.4495 x_1 = 2.4495 \Rightarrow x_1 = \frac{2.4495}{-2.4495} = -1$$

The solution to the above set of equations is: $[x_1 \ x_2 \ x_3] = [-1 \ 2 \ 3]$.

Problems

PROBLEM 3.1
Let V denote the set of vectors in R^2. If $x=[x_1 \quad x_2]^T$ and $y=[y_1 \quad y_2]^T$ are elements of V and $\alpha \in R^1$, define:

$$x+y=[x_1+y_1 \quad x_2 y_2]^T \quad \text{and} \quad \alpha[x_1 \quad x_2]^T=[\alpha x_1 \quad x_2]^T$$

Is V a vector space with these operations?

PROBLEM 3.2
Let V denote the set of all $n \times n$ matrices and let S denote the set of all symmetric $n \times n$ matrices. Show that V is a vector space and S is a subspace of V.

PROBLEM 3.3
Show that if $V_1, V_2, \ldots, V_n$ spans the vector space V and one of these vectors can be written as a linear combination of the other $n-1$ vectors, then these $n-1$ vectors span the vector space V.

PROBLEM 3.4
Let $X_1, X_2, \ldots, X_n$ be n vectors in R^n. Define the $n \times n$ matrix X as $X=[X_1| X_2| \ldots | X_n]$. Show that the vectors $X_1, X_2, \ldots, X_n$ are linearly dependent if and only if X is singular, that is $\det(X)=0$.

PROBLEM 3.5
Consider the sequence of vectors in R^3

$$x_m = \begin{bmatrix} \dfrac{2}{m} & \dfrac{2m}{m+3} & \dfrac{m+1}{m} \end{bmatrix}^T$$

Show that x_m converges to x_∞ by showing that $\lim_{m \to \infty} \|x_m - x_\infty\|_p \to 0$ for any $p \geq 1$.

PROBLEM 3.6
Let the distance between two vectors x and y in a linear normed vector space be defined as:

$$d(x,y) = \|x-y\|_p$$

Show that this distance function satisfies the distance properties, that is

a. $d(x,y) \geq 0$, for every x and $y \in V$. $d(x,y)=0$ if and only if $x=y$.
b. $d(x,y)=d(y,x)$
c. $d(x,z) \leq d(x,y)+d(y,z)$, for every x, y, and $z \in V$

PROBLEM 3.7
Let $s=[2 \quad -3 \quad 1]^T$. Find vector x to maximize the function y defined by:

$$y = \frac{|<x,s>|^2}{\|x\|^2}$$

Is the solution unique? If not, find another vector.

PROBLEM 3.8
Suppose that V is a real vector space with an inner product. Prove that:

$$\|x+y\|^2 = \|x\|^2 + \|y\|^2$$

if and only if x and y are orthogonal.

PROBLEM 3.9
Show that the following two vectors are orthogonal:

$$x(n) = 3\exp\left(j\frac{2\pi}{N}n\right) \quad \text{and} \quad y(n) = 4\exp\left(j\frac{6\pi}{N}n\right), \quad n = 0, 1, \ldots, N-1$$

PROBLEM 3.10
Show that the L_1, L_2 and L_∞ norms on R^n (or C^n) satisfy the following inequalities:

$$\frac{\|x\|_2}{\sqrt{n}} \leq \|x\|_\infty \leq \|x\|_2$$

$$\|x\|_2 \leq \|x\|_1 \leq \sqrt{n}\|x\|_2$$

$$\frac{\|x\|_1}{n} \leq \|x\|_\infty \leq \|x\|_1$$

PROBLEM 3.11
If V is a real inner product space, show that:

$$<x, y> = \frac{1}{4}[\|x+y\|^2 - \|x-y\|^2]$$

PROBLEM 3.12
If V is a complex inner product space, show that:

a. Real $<x, y> = \frac{1}{4}[\|x+y\|^2 - \|x-y\|^2]$

b. Im $<x, y> = -\frac{1}{4}[\|jx+y\|^2 - \|jx-y\|^2]$

PROBLEM 3.13
Show that if $f(t)$ is bandlimited to B Hz, then:

$$\alpha_n = \int_{-\infty}^{+\infty} f(t)\sqrt{2B}\,\text{sinc}\left(2B\left(t - \frac{n}{2B}\right)\right)dt = \frac{1}{\sqrt{2B}}f\left(\frac{n}{2B}\right)$$

where sinc(t) is the bandlimited sinc function defined as:

$$\text{sinc}(t) = \frac{\sin(\pi t)}{\pi t}$$

Linear Vector Spaces

PROBLEM 3.14

Let $S = \{\phi_n(t)\}_{n=0}^{M-1}$ be a set of functions defined over the time interval $[0 \quad T]$ as:

$$\varphi_n(t) = \frac{\sqrt{2}}{\sqrt{T}} \cos(2\pi f_c t + 2\pi n \Delta f t)$$

where $f_c = k/T$ represent the carrier frequency for some integer k, and Δf is the frequency separation. This is a set of signals used to transmit digital data using M-ary FSK. Find the minimum frequency separation Δf that makes the set S an orthonormal set.

PROBLEM 3.15

For the inner product $<x(t), y(t)> = \int_0^\infty x(t)\bar{y}(t)dt$, prove the Schwarz inequality if:

a. $x(t) = 2e^{-t} u(t)$ and $y(t) = 3e^{-2t} u(t)$
b. $x(t) = te^{-t} u(t)$ and $y(t) = 2e^{-t} u(t)$
c. $x(t) = e^{-(1-j)t} u(t)$ and $y(t) = 2e^{-(1+j2)t} u(t)$

PROBLEM 3.16

Show that in a normed vector space the following inequality holds:

$$|\,\|x\| - \|y\|\,| \leq \|x - y\|$$

PROBLEM 3.17

Solve the following set of linear equations using QR factorization:

$$x_1 + 2x_2 + 3x_3 = 14$$
$$4x_1 + 5x_2 + 6x_3 = 32$$
$$7x_1 - 3x_2 - 2x_3 = -5$$

PROBLEM 3.18

Find the QR factorization of the 4×3 matrix A.

$$A = \begin{bmatrix} 1 & -2 & 3 \\ -1 & 1 & 0 \\ 6 & 3 & 4 \\ 1 & 2 & 4 \end{bmatrix}$$

PROBLEM 3.19 (MATLAB)

The following three functions are defined over the interval of $(0, \infty)$. Use the Gram–Schmidt orthogonalization process to obtain an orthonormal set over the interval of $(0, \infty)$.

a. $f_1(x) = e^{-x} u(x)$
b. $f_2(x) = 2e^{-x} u(x) - 3e^{-2x} u(x)$
c. $f_3(x) = xe^{-x} u(x)$

PROBLEM 3.20

Show that for any two real sequences $\{a_i\}_{i=1}^n$ and $\{b_i\}_{i=1}^n$

$$\left|\sum_{i=1}^n a_i b_i\right|^2 \leq \sum_{i=1}^n |a_i|^2 \sum_{i=1}^n |b_i|^2$$

with equality if and only if $b_i = k a_i$ for $i = 1, 2, \ldots, n$.

PROBLEM 3.21 (MATLAB)

Consider the set of vectors $S = \{x_1\ x_2\ x_3\ x_4\} \in R^4$

$$x_1 = \begin{bmatrix} 1 \\ -1 \\ 2 \\ 4 \end{bmatrix}, \quad x_2 = \begin{bmatrix} -2 \\ 5 \\ 6 \\ 9 \end{bmatrix}, \quad x_3 = \begin{bmatrix} 1 \\ 2 \\ 0 \\ -3 \end{bmatrix}, \quad x_4 = \begin{bmatrix} 4 \\ -2 \\ -4 \\ -7 \end{bmatrix}$$

Use the Gram–Schmidt orthogonalization process to obtain an orthonormal set of vectors.

PROBLEM 3.22 (MATLAB)

a. Write a MATLAB code to compute L_p norm of a continuous function $f(x)$ defined over the interval of $a \leq x \leq b$. Assume that the function f is given by a look-up table with vector x and y containing N samples of the function uniformly spaced between a and b
b. Test your code with $a = 0$, $b = 1$, $N = 2000$, and $f(x) = 2x^2 - x + 1$
c. Use the code to find L_p norm of the following functions for $p = 1, 2,$ and ∞

$$f(x) = 1 - x\ln(x) - (1-x)\ln(1-x)$$

$$g(x) = 1 + x + x\cos(2\pi x) + 4\sin(\pi x)$$

where $0 \leq x \leq 1$

PROBLEM 3.23 (MATLAB)

The periodic signal $x(t)$ is the input to a lowpass filter with frequency response

$$H(j\omega) = \frac{200\pi}{j\omega + 200\pi}$$

a. Find the Fourier series coefficient of $x(t)$ if one period of $x(t)$ is

$$x(t) = e^{-200t} \quad 0 < t < T \quad T = 0.01 \ (\text{sec})$$

b. Find the Fourier series coefficient of $y(t)$, the output of the filter.
c. Find the average input and output power.

Linear Vector Spaces

PROBLEM 3.24 (MATLAB)

In a binary data transmission system, the signal representing binary bit "1" is given by:

$$s(t) = \frac{t}{T_b}, \quad 0 < t < T_b$$

where T_b is the bit rate. The signal representing bit "0" is the absence of this pulse.

a. Generate 400,000 bits randomly and simulate the system for a bit rate of $R_b = 2$ Kbps. Assume SNR to be between 0 and 15 dB. Plot the theoretical and simulated BER as a function of SNR.

b. Replace signal $s(t)$ by:

$$s(t) = \sin\left(\frac{2\pi t}{T_b}\right), \quad 0 < t < T_b$$

and repeat part (a). Assume all the other parameters are the same.

PROBLEM 3.25 (MATLAB)

The 95% bandwidth of gray scale image $f(x, y)$ of size $N \times M$ is defined to be ρ if:

$$\frac{1}{4\pi^2}\iint_D |F(\omega_1, \omega_2)|^2 d\omega_1 d\omega_2 = 0.95 \sum_{y=0}^{M-1}\sum_{x=0}^{N-1} |f(x, y)|^2$$

where D is the circle:.

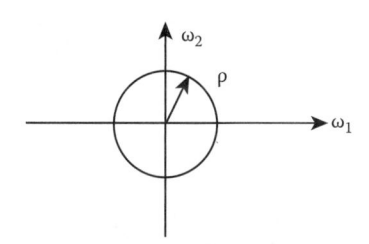

a. Develop a FFT based algorithm to compute the image bandwidth.
b. Implement the algorithm in MATLAB and compute the numerical value of ρ for the cameraman image (MATLAB command to load the cameraman image is:

$$f = \text{imread}('cameraman.tif').$$

c. Repeat part (b) if:

$$f(x, y) = \begin{bmatrix} 10 & 20 & 30 & 20 \\ 18 & 20 & 24 & 26 \\ 16 & 30 & 25 & 15 \\ 26 & 27 & 29 & 18 \end{bmatrix}$$

d. Can you extend the definition of bandwidth given above to an RGB image?

4

Eigenvalues and Eigenvectors

4.1 Introduction

The words "eigenvalues" and "eigenvector" are derived from the German word "eigen" which means "proper" or "characteristic". Eigenvalues and eigenvectors have many applications in several engineering disciplines ranging from the analysis of structures in civil engineering to circuit analysis, signal and image processing, and control in electrical engineering. In this chapter, we first introduce the mathematical concepts of eigenvalues and eigenvectors and then discuss their usage in engineering related applications.

4.2 Matrices as Linear Transformations

4.2.1 Definition: Linear Transformation

Let V and $\tilde{V}$ be two vector spaces over the field of real (complex) numbers. The transformation $T: V \to \tilde{V}$ that maps a vector from V to $\tilde{V}$ is a linear transformation from V to $\tilde{V}$ if, for any $x, y \in V$ and any scalar α, we have:

$$\text{(a) } T(x+y) = T(x) + T(y) \tag{4.1}$$

and

$$\text{(b) } T(\alpha x) = \alpha T(x) \tag{4.2}$$

Example 4.1 Convolution integral

Let V be the real vector space of square integrable continuous functions defined over the interval $(-\infty, \infty)$ and let T be a transformation from V to V defined by:

$$T[x(t)] = \int_{-\infty}^{\infty} x(\tau) h(t-\tau) d\tau \tag{4.3}$$

This is a linear transformation. To show that the transformation is linear, let $\alpha \in R^1$ and $x(t), y(t) \in V$. Then:

$$T[x(t)+y(t)] = \int_{-\infty}^{\infty} (x(\tau)+y(\tau))h(t-\tau)d\tau$$

$$= \int_{-\infty}^{\infty} x(\tau)h(t-\tau)d\tau + \int_{-\infty}^{\infty} y(\tau)h(t-\tau)d\tau = T[x(t)] + T[y(t)] \quad (4.4)$$

Also,

$$T[\alpha x(t)] = \int_{-\infty}^{\infty} \alpha x(\tau)h(t-\tau)d\tau = \alpha \int_{-\infty}^{\infty} x(\tau)h(t-\tau)d\tau = \alpha T[x(t)] \quad (4.5)$$

Therefore, T is linear. This integral is known as the "convolution integral". If $x(t)$ is the input to a linear time invariant system with impulse response $h(t)$, then the output of the system $y(t)$ is given by the convolution integral. That is:

$$y(t) = \int_{-\infty}^{\infty} x(\tau)h(t-\tau)d\tau \quad (4.6)$$

Example 4.2

Consider the transformation from R^2 to R^2 given by:

$$T(x) = T(x_1, x_2) = \begin{bmatrix} a_{11}x_1 + a_{12}x_2 \\ a_{21}x_1 + a_{22}x_2 \end{bmatrix} = \begin{bmatrix} a_{11} & a_{12} \\ a_{21} & a_{22} \end{bmatrix} \begin{bmatrix} x_1 \\ x_2 \end{bmatrix} = Ax \quad (4.7)$$

It is easy to show that this is a linear transformation and, depending on the choice of matrix A, it can be a rotation, reflection, projection, etc. Each of these cases is discussed below.

Case I: Counterclockwise rotation by an angle θ
In this case the matrix A is defined as:

$$A = \begin{bmatrix} \cos\theta & -\sin\theta \\ \sin\theta & \cos\theta \end{bmatrix} \quad (4.8)$$

This transformation is shown in Figure 4.1.

Case II: Clockwise rotation by an angle θ
In this case the matrix A is defined as:

$$A = \begin{bmatrix} \cos\theta & \sin\theta \\ -\sin\theta & \cos\theta \end{bmatrix} \quad (4.9)$$

Case III: Reflection around the x_1 axis
In this case the matrix A is defined as:

$$A = \begin{bmatrix} 1 & 0 \\ 0 & -1 \end{bmatrix} \quad (4.10)$$

This is shown in Figure 4.2.

Eigenvalues and Eigenvectors

Case IV: Projection on the x_1 axis

In this case the matrix A is defined as:

$$A = \begin{bmatrix} 1 & 0 \\ 0 & 0 \end{bmatrix} \qquad (4.11)$$

This is shown in Figure 4.3.

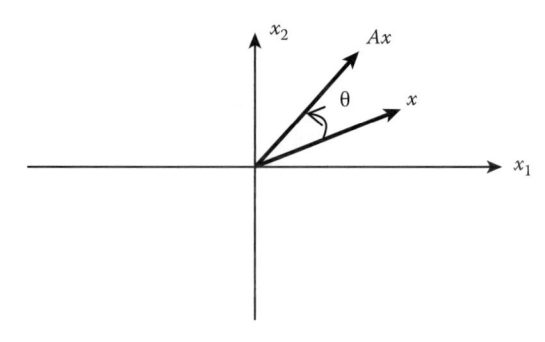

FIGURE 4.1
Rotation by angle θ counterclockwise.

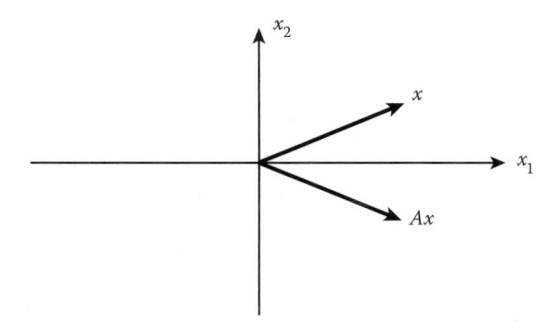

FIGURE 4.2
Reflection along first axis.

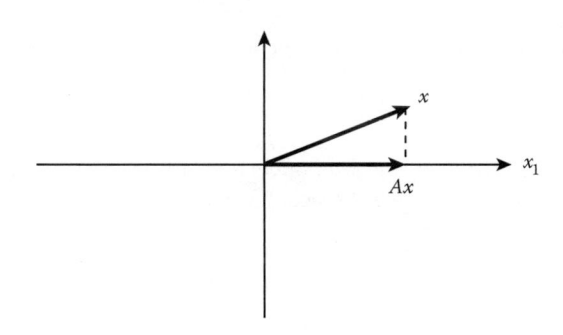

FIGURE 4.3
Projection on the first axis.

4.2.2 Matrices as Linear Operators

An $m \times n$ matrix can be considered a mapping from R^n to R^m defined by:

$$y = Ax \qquad (4.12)$$

where $x \in R^n$ and $y \in R^m$. This operator maps an n-dimensional vector in R^n to an m-dimensional vector in R^m.

4.2.3 Null Space of a Matrix

Let A be a $m \times n$ matrix. The set of nonzero vectors in R^n for which $Ax = 0$ is called the null space of matrix A. It is denoted by $N(A)$ and is defined as:

$$N(A) = \{x \in R^n \text{ such that } Ax = 0\} \qquad (4.13)$$

The null space of A is always a subspace of R^n.

Example 4.3

Find the null space of the 2×3 matrix A

$$A = \begin{bmatrix} 2 & -3 & 4 \\ 1 & 3 & -2 \end{bmatrix}$$

Solution: The null space of A is obtained by solving the equation $Ax = 0$.

$$Ax = \begin{bmatrix} 2 & -3 & 4 \\ 1 & 3 & -2 \end{bmatrix} \begin{bmatrix} x_1 \\ x_2 \\ x_3 \end{bmatrix} = \begin{bmatrix} 2x_1 - 3x_2 + 4x_3 \\ x_1 + 3x_2 - 2x_3 \end{bmatrix} = \begin{bmatrix} 0 \\ 0 \end{bmatrix}$$

Therefore,

$$\begin{cases} 2x_1 - 3x_2 + 4x_3 = 0 \\ x_1 + 3x_2 - 2x_3 = 0 \end{cases}$$

The above two equations are linearly dependent. The resulting equation is:

$$2x_1 + x_2 = 0$$

Therefore, the null space of A is a set of vectors in R^3 given by $x = [a \ -2a \ b]^T$. This defines a 2-d plane in the 3-d space. It can be written as:

$$x = a \begin{bmatrix} 1 \\ -2 \\ 0 \end{bmatrix} + b \begin{bmatrix} 0 \\ 0 \\ 1 \end{bmatrix}$$

Notice that the null space of A is also a vector space. It is a subspace of R^3.

Eigenvalues and Eigenvectors

Example 4.4

Find the null space of the 2×2 operator A given below:

$$A = \begin{bmatrix} 1 & -1 \\ -1 & 1 \end{bmatrix}$$

Solution: The null space of A is obtained by solving the equation $Ax=0$.

$$Ax = \begin{bmatrix} 1 & -1 \\ -1 & 1 \end{bmatrix}\begin{bmatrix} x_1 \\ x_2 \end{bmatrix} = \begin{bmatrix} x_1 - x_2 \\ -x_1 + x_2 \end{bmatrix} = \begin{bmatrix} 0 \\ 0 \end{bmatrix}$$

This yields:

$$x_1 = x_2$$

Therefore, the null space of A is a set of vectors in R^2 given by $x = [a\ \ a]^T$. This defines the 45 degree line in R^2. Hence:

$$N(A) = \left\{ x \in R^2 : x = a\begin{bmatrix} 1 \\ 1 \end{bmatrix} \right\}$$

4.2.4 Projection Operator

Definition: Let P be a linear transformation from the inner product space V onto itself. P is a projection operator if $P^2 = P$. A few examples of projection operators are given below.

Bandlimited operator: Let $f(t)$ be a signal with Fourier transform $F(j\omega)$. Define P as:

$$P[F(j\omega)] = \begin{cases} F(j\omega) & \text{if } -\omega_M \leq \omega \leq \omega_M \\ 0 & \text{otherwise} \end{cases} \quad (4.14)$$

This is a projection operator since $P\{P[F(j\omega)]\} = P[F(j\omega)]$ This operation involves filtering a signal using an ideal low-pass filter and produces a bandlimited signal at the output of the filter. This operator is called the bandlimited operator.

Example 4.5

Let $x \in R^3$ and let P be a mapping from vector space R^3 to R^3 defined by:

$$P = \begin{bmatrix} 1 & 0 & 0 \\ 0 & 1 & 0 \\ 0 & 0 & 0 \end{bmatrix}$$

P is a projection operator since:

$$P^2 = \begin{bmatrix} 1 & 0 & 0 \\ 0 & 1 & 0 \\ 0 & 0 & 0 \end{bmatrix}\begin{bmatrix} 1 & 0 & 0 \\ 0 & 1 & 0 \\ 0 & 0 & 0 \end{bmatrix} = \begin{bmatrix} 1 & 0 & 0 \\ 0 & 1 & 0 \\ 0 & 0 & 0 \end{bmatrix} = P$$

Theorem 4.1: Let P be a projection operator defined over a linear vector space V. Then the range (denoted by $R(P)$) and null space of P are subspaces of V such that $V = R(P) + N(P)$.

Proof: Let the range of P be defined as $R(P) = P(x)$, where P represents the projector operator and x denotes any element in V. Then, the null space of P is $N(P) = x - P(x)$ because:

$$P(x - P(x)) = P(x) - P^2(x) = P(x) - P(x) = 0 \qquad (4.15)$$

Therefore:

$$R(P) + V(P) = P(x) + x - P(x) = x = V \qquad (4.16)$$

4.2.5 Orthogonal Projection

Definition: P is an orthogonal projection if its range and null space are orthogonal. That is:

$$R(P) \perp N(P) \qquad (4.17)$$

Example 4.6

Let P be a mapping from R^3 to R^2 defined by:

$$P = \begin{bmatrix} 1 & 0 & 0 \\ 0 & 1 & 0 \\ 0 & 0 & 0 \end{bmatrix}$$

Is operator P an orthogonal operator?

Solution: The range space of P is:

$$R(P) \quad \rightarrow \quad y = Px = \begin{bmatrix} 1 & 0 & 0 \\ 0 & 1 & 0 \\ 0 & 0 & 0 \end{bmatrix} \begin{bmatrix} x_1 \\ x_2 \\ x_3 \end{bmatrix} = \begin{bmatrix} x_1 \\ x_2 \\ 0 \end{bmatrix},$$

and the null space of P is:

$$N(P) \quad \rightarrow \quad z = (I - P)x = \begin{bmatrix} 0 & 0 & 0 \\ 0 & 0 & 0 \\ 0 & 0 & 1 \end{bmatrix} \begin{bmatrix} x_1 \\ x_2 \\ x_3 \end{bmatrix} = \begin{bmatrix} 0 \\ 0 \\ x_3 \end{bmatrix}$$

The dot product between an arbitrary vector y in the range space and an arbitrary vector z in the null space of the operator is:

$$<y, z> = y^T z = 0$$

Thus, it is an orthogonal projection.

Eigenvalues and Eigenvectors 163

Theorem 4.2: Any Hermitian (symmetric) projection matrix P is an orthogonal projection matrix.

Proof: Let $x \in R(P)$, that is $x = Pu$, and let $y \in N(P)$, that is $y = (I-P)v$. Then:

$$<x, y> = x^H y = u^H P^H (I-P)v = u^H P(I-P)v = u^H(P-P^2)v = u^H(P-P)v = 0 \quad (4.18)$$

Therefore, operator P is an orthogonal projection.

4.2.5.1 Projection Theorem

Theorem 4.3: Let V be a complete normed vector space and let S be a subspace of V. Then, for every vector $x \in V$, there is a unique vector $x_0 \in S$ such that:

$$\|x - x_0\| \leq \|x - y\| \quad (4.19)$$

for all $y \in S$. Furthermore, $e = x - x_0$ is orthogonal to any vector in S. x_0 is called the projection of x onto subspace S and is written as:

$$x_0 = P(x) \quad (4.20)$$

This is shown in Figure 4.4.

4.2.5.2 Matrix Representation of Projection Operator

Let P be a projection operator from vector space $V = R^n$ to S, a subspace of V. Since S is a subspace of $V = R^n$, we can form a basis $B = \{b_1 \ b_2 \ \dots \ b_m\}$ (an $n \times m$ matrix) for S where $m \leq n$ is the dimension of S. Therefore, $x_0 = P(x)$ is a linear combination of $b_1, b_2, \dots, b_m$ and can be written as:

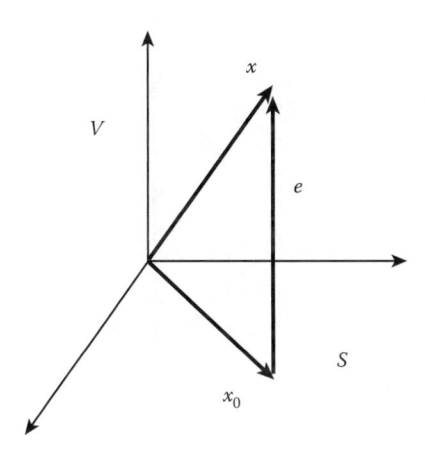

FIGURE 4.4
Projection theorem.

$$x_0 = \sum_{i=1}^{m} \alpha_i b_i = B\alpha \qquad (4.21)$$

where $\alpha = [\alpha_1 \ \alpha_2 \ \ldots \ \alpha_m]^T$. Then, the projection of x onto subspace S is the solution to the following minimization problem:

$$x_0 = \arg\min_{\alpha} \| x - B\alpha \|^2 = \arg\min_{\alpha} (x - B\alpha)^T (x - B\alpha) \qquad (4.22)$$

The optimum solution is obtained by setting the gradient of $(x-B\alpha)^T(x-B\alpha)$ with respect to α equal to zero as shown below:

$$\frac{\partial (x - B\alpha)^T (x - B\alpha)}{\partial \alpha} = 2B^T (x - B\alpha) = 0 \qquad (4.23)$$

Solving for α yields:

$$\alpha = (B^T B)^{-1} B^T x \qquad (4.24)$$

Therefore:

$$x_0 = B(B^T B)^{-1} B^T x \qquad (4.25)$$

and

$$P = B(B^T B)^{-1} B^T \qquad (4.26)$$

Example 4.7

Let $V = R^3$ and $x_0 = P(x)$ be the projection onto the 2-d subspace S as shown in Figure 4.5. Find the corresponding projection operator.

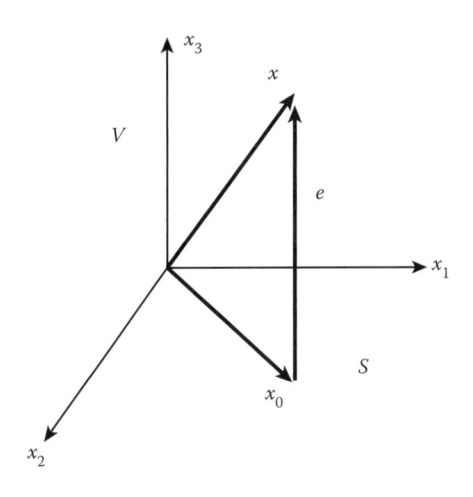

FIGURE 4.5
Projection onto the 2-d subspace S.

Eigenvalues and Eigenvectors

Solution: The B matrix consists of the basis of S and is given by $B = \begin{bmatrix} 1 & 0 \\ 0 & 1 \\ 0 & 0 \end{bmatrix}$. Hence:

$$P = \begin{bmatrix} 1 & 0 \\ 0 & 1 \\ 0 & 0 \end{bmatrix} \left(\begin{bmatrix} 1 & 0 & 0 \\ 0 & 1 & 0 \end{bmatrix} \begin{bmatrix} 1 & 0 \\ 0 & 1 \\ 0 & 0 \end{bmatrix} \right)^{-1} \begin{bmatrix} 1 & 0 & 0 \\ 0 & 1 & 0 \end{bmatrix}$$

$$= \begin{bmatrix} 1 & 0 \\ 0 & 1 \\ 0 & 0 \end{bmatrix} \begin{bmatrix} 1 & 0 \\ 0 & 1 \end{bmatrix} \begin{bmatrix} 1 & 0 & 0 \\ 0 & 1 & 0 \end{bmatrix} = \begin{bmatrix} 1 & 0 & 0 \\ 0 & 1 & 0 \\ 0 & 0 & 0 \end{bmatrix}$$

Notice that:

$$x_0 = Px = \begin{bmatrix} 1 & 0 & 0 \\ 0 & 1 & 0 \\ 0 & 0 & 0 \end{bmatrix} \begin{bmatrix} x_1 \\ x_2 \\ x_3 \end{bmatrix} = \begin{bmatrix} x_1 \\ x_2 \\ 0 \end{bmatrix}$$

4.3 Eigenvalues and Eigenvectors

4.3.1 Definition of Eigenvalues and Eigenvectors

The nonzero vector x is an eigenvector of the square $n \times n$ matrix A if there is a scale factor λ such that:

$$Ax = \lambda x \tag{4.27}$$

The scale factor λ is called the eigenvalue corresponding to the eigenvector x. The above equation can be considered as an operator operating on x. The eigenvectors of A are vectors that are not changed by the operator A except for the scale factor λ. This means that eigenvectors are invariant with respect to the operator A. This is similar to the concept of eigenfunctions for linear time invariant (LTI) systems. For example, the steady state response of LTI systems to a sinusoidal signal is a sinusoidal signal having the same frequency as the input signal, but a different magnitude and phase. Therefore, sinusoidal signals are eigenfunctions of LTI systems. For instance, Gaussian noise can be considered as an eigennoise of LTI systems since the probability density function (PDF) of the output of an LTI linear system driven by Gaussian noise is Gaussian.

The equation $Ax = \lambda x$ can be written as:

$$(\lambda I - A)x = 0 \tag{4.28}$$

This equation has a nontrivial solution if and only if matrix $\lambda I - A$ is singular. That is:

$$\det(\lambda I - A) = 0 \quad (4.29)$$

The above determinant is a polynomial of degree n and is denoted by $P(\lambda)$. This polynomial is called the characteristic polynomial of matrix A. The characteristic polynomial has n roots that represent the eigenvalues of matrix A. There is an eigenvector corresponding to each eigenvalue. The eigenvalues can be distinct or repeat and they may also be complex.

The MATLAB® command to find the characteristic polynomial of matrix A is **p=poly(A)**. The vector p contains the coefficients of the characteristic polynomial of A. Once the eigenvalues of A are computed, Equation 4.28 is used to compute the corresponding eigenvectors. The MATLAB command to compute eigenvalues and eigenvectors of matrix A is **[U,S]=eig(A)**, where A is a $n \times n$ input matrix, U is the matrix of eigenvectors (columns of U are eigenvectors) and S is a $n \times n$ diagonal matrix of eigenvalues. The following simple examples show the process of computing eigenvalues and eigenvectors.

Example 4.8

Find the eigenvalues and eigenvectors of the 2×2 matrix $A = \begin{bmatrix} -1 & 1 \\ -12 & 6 \end{bmatrix}$.

Solution: First, we compute the two eigenvalues λ_1 and λ_2. The characteristic polynomial of A is:

$$P(\lambda) = |\lambda I - A| = \begin{vmatrix} \lambda + 1 & -1 \\ 12 & \lambda - 6 \end{vmatrix} = \lambda^2 - 5\lambda + 6 = (\lambda - 2)(\lambda - 3) = 0$$

Therefore, $\lambda_1 = 2$, and $\lambda_2 = 3$. Now we compute the eigenvectors corresponding to the two eigenvalues. The eigenvector corresponding to $\lambda_1 = 2$ is computed as:

$$Ax_1 = \lambda_1 x_1$$

$$\begin{bmatrix} -1 & 1 \\ -12 & 6 \end{bmatrix} \begin{bmatrix} a \\ b \end{bmatrix} = 2 \begin{bmatrix} a \\ b \end{bmatrix} \rightarrow \begin{cases} -a + b = 2a \\ -12a + 6b = 2b \end{cases} \rightarrow b = 3a$$

Let $a = 1$, then $b = 3$ and

$$x_1 = \begin{bmatrix} a \\ b \end{bmatrix} = \begin{bmatrix} 1 \\ 3 \end{bmatrix}$$

The eigenvector corresponding to $\lambda_2 = 3$ is:

$$Ax_2 = \lambda_2 x_2$$

$$\begin{bmatrix} -1 & 1 \\ -12 & 6 \end{bmatrix} \begin{bmatrix} a \\ b \end{bmatrix} = 3 \begin{bmatrix} a \\ b \end{bmatrix} \rightarrow \begin{cases} -a + b = 3a \\ -12a + 6b = 3b \end{cases} \rightarrow b = 4a$$

Eigenvalues and Eigenvectors

Let $a=1$, then $b=4$ and

$$x_2 = \begin{bmatrix} a \\ b \end{bmatrix} = \begin{bmatrix} 1 \\ 4 \end{bmatrix}$$

Example 4.9

Find the eigenvalues and eigenvectors of the matrix A given below:

$$A = \begin{bmatrix} 8.25 & -9.75 & -7.5 \\ 2.25 & 2.25 & -1.5 \\ 2.25 & -12.75 & -4.5 \end{bmatrix}$$

Solution: The characteristic polynomial is:

$$P(\lambda) = |\lambda I - A| = \lambda^3 - 6\lambda^2 - 9\lambda + 54 = (\lambda+3)(\lambda-3)(\lambda-6) = 0$$

The eigenvalues are $\lambda_1 = -3$, $\lambda_2 = 3$, and $\lambda_3 = 6$. The corresponding eigenvectors are computed as follows:

$$Ax_1 = \lambda_1 x_1 \rightarrow \begin{bmatrix} 8.25 & -9.75 & -7.5 \\ 2.25 & 2.25 & -1.5 \\ 2.25 & -12.75 & -4.5 \end{bmatrix} \begin{bmatrix} a \\ b \\ c \end{bmatrix}$$

$$= -3 \begin{bmatrix} a \\ b \\ c \end{bmatrix} \rightarrow \begin{cases} 8.25a - 9.75b - 7.5c = -3a \\ 2.25a + 2.25b - 1.5c = -3b \\ 2.25a - 12.75b - 4.5c = -3c \end{cases}$$

These three equations are not linearly independent. Only two of them are linearly independent. Using the first two equations with $a=1$, we have:

$$\begin{cases} -9.75b - 7.5c = -11.25 \\ 5.25b - 1.5c = -2.25 \end{cases} \rightarrow \begin{cases} b = 0 \\ c = 1.5 \end{cases} \rightarrow x_1 = \begin{bmatrix} 1 \\ 0 \\ 1.5 \end{bmatrix}$$

Similarly,

$$Ax_2 = \lambda_2 x_2 \rightarrow \begin{bmatrix} 8.25 & -9.75 & -7.5 \\ 2.25 & 2.25 & -1.5 \\ 2.25 & -12.75 & -4.5 \end{bmatrix} \begin{bmatrix} a \\ b \\ c \end{bmatrix}$$

$$= 3 \begin{bmatrix} a \\ b \\ c \end{bmatrix} \rightarrow \begin{cases} 8.25a - 9.75b - 7.5c = 3a \\ 2.25a + 2.25b - 1.5c = 3b \\ 2.25a - 12.75b - 4.5c = 3c \end{cases}$$

$$\begin{cases} -9.75b - 7.5c = -5.25 \\ -0.75b - 1.5c = -2.25 \end{cases} \rightarrow \begin{cases} b = -1 \\ c = 2 \end{cases} \rightarrow x_2 = \begin{bmatrix} 1 \\ -1 \\ 2 \end{bmatrix}$$

And finally, the third eigenvector is computed as:

$$Ax_3 = \lambda_3 x_3 \rightarrow \begin{bmatrix} 8.25 & -9.75 & -7.5 \\ 2.25 & 2.25 & -1.5 \\ 2.25 & -12.75 & -4.5 \end{bmatrix} \begin{bmatrix} a \\ b \\ c \end{bmatrix}$$

$$= 6 \begin{bmatrix} a \\ b \\ c \end{bmatrix} \rightarrow \begin{cases} 8.25a - 9.75b - 7.5c = 6a \\ 2.25a + 2.25b - 1.5c = 6b \\ 2.25a - 12.75b - 4.5c = 6c \end{cases}$$

$$\begin{cases} -9.75b - 7.5c = -2.25 \\ -3.75b - 1.5c = -2.25 \end{cases} \rightarrow \begin{cases} b = 1 \\ c = -1 \end{cases} \rightarrow x_3 = \begin{bmatrix} 1 \\ 1 \\ -1 \end{bmatrix}$$

4.3.2 Properties of Eigenvalues and Eigenvectors

Eigenvalues and eigenvectors have several useful properties. This section will cover a few of these properties.

4.3.2.1 Independent Property

The first property is related to the independence of eigenvectors. The eigenvectors of matrix A may not be independent. The following theorem states the necessary condition for eigenvectors to be linearly independent.

Theorem 4.4: If the n eigenvalues of $n \times n$ matrix A are distinct, that is $\lambda_1 \neq \lambda_2 \neq \cdots \neq \lambda_n$, then the corresponding eigenvectors are linearly independent.

Proof: Without loss of generality, assume that matrix A is 2×2 with two distinct eigenvalues, and let the two corresponding eigenvectors x_1 and x_2 be linearly dependent. That is, there are two nonzero scale factors α and β such that:

$$\alpha x_1 + \beta x_2 = 0 \tag{4.30}$$

Multiply both sides of the above equation by matrix A. This yields:

$$\alpha A x_1 + \beta A x_2 = 0 \tag{4.31}$$

Since $Ax_1 = \lambda_1 x_1$ and $Ax_2 = \lambda_2 x_2$, the above equations simplifies to:

$$\alpha \lambda_1 x_1 + \beta \lambda_2 x_2 = 0 \tag{4.32}$$

Given that $\alpha x_1 + \beta x_2 = 0 \Rightarrow \beta x_2 = -\alpha x_1$ (see Equation 4.30), x_2 can be eliminated from Equation 4.32 yielding:

$$\alpha \lambda_1 x_1 - \alpha \lambda_2 x_2 = \alpha(\lambda_1 - \lambda_2) x_1 = 0 \tag{4.33}$$

Since $x_1 \neq 0$, we must have:

$$\lambda_1 - \lambda_2 = 0 \Rightarrow \lambda_1 = \lambda_2 \qquad (4.34)$$

and the two eigenvalues cannot be distinct. On the other hand, if the eigenvalues are not distinct, there is no guarantee that the eigenvectors will be independent. For example, consider the 2×2 matrix:

$$A = \begin{bmatrix} 2 & -3 \\ 0 & 2 \end{bmatrix}$$

The characteristic polynomial of this matrix is:

$$P(\lambda) = |\lambda I - A| = \begin{vmatrix} \lambda - 2 & 3 \\ 0 & \lambda - 2 \end{vmatrix} = (\lambda - 2)^2$$

Therefore, $\lambda_1 = \lambda_2 = 2$. The eigenvectors are the solution to:

$$Ax = \lambda x$$

or

$$\begin{bmatrix} 2 & -3 \\ 0 & 2 \end{bmatrix} \begin{bmatrix} a \\ b \end{bmatrix} = 2 \begin{bmatrix} a \\ b \end{bmatrix} \rightarrow \begin{cases} 2a - 3b = 2a \\ 2b = 2b \end{cases} \rightarrow b = 0 \rightarrow x = \begin{bmatrix} a \\ 0 \end{bmatrix}$$

Therefore, there is only one independent eigenvector and that is $x = \begin{bmatrix} 1 \\ 0 \end{bmatrix}$.

Example 4.10

Consider the 2×2 matrix:

$$A = \begin{bmatrix} 1 & 1 \\ -1 & 3 \end{bmatrix}$$

The characteristic polynomial of this matrix is:

$$P(\lambda) = |\lambda I - A| = \begin{vmatrix} \lambda - 1 & -1 \\ 1 & \lambda - 3 \end{vmatrix} = (\lambda - 2)^2 = 0$$

Hence, $\lambda_1 = \lambda_2 = 2$. The corresponding eigenvectors are found by solving:

$$Ax = \lambda x$$

$$\begin{bmatrix} 1 & 1 \\ -1 & 3 \end{bmatrix} \begin{bmatrix} a \\ b \end{bmatrix} = 2 \begin{bmatrix} a \\ b \end{bmatrix} \rightarrow \begin{cases} a + b = 2a \\ -a + 3b = 2b \end{cases} \rightarrow b = a$$

Therefore, $x = \begin{bmatrix} 1 \\ 1 \end{bmatrix}$ is the only independent eigenvector.

Example 4.11

Consider the 2×2 matrix

$$A = \begin{bmatrix} 2 & 0 \\ 0 & 2 \end{bmatrix}$$

Show that this matrix has two eigenvalues $\lambda_1 = \lambda_2 = 2$ and two independent eigenvectors $x_1 = \begin{bmatrix} 1 \\ 0 \end{bmatrix}$ and $x_2 = \begin{bmatrix} 0 \\ 1 \end{bmatrix}$.

Solution: The eigenvalues are:

$$P(\lambda) = |\lambda I - A| = \begin{vmatrix} \lambda - 2 & 1 \\ -1 & \lambda - 2 \end{vmatrix} = (\lambda - 2)^2 = 0 \quad \rightarrow \quad \lambda_1 = \lambda_2 = 2$$

The corresponding eigenvectors are:

$$Ax = \lambda x$$

$$\begin{bmatrix} 2 & 0 \\ 0 & 2 \end{bmatrix} \begin{bmatrix} a \\ b \end{bmatrix} = 2 \begin{bmatrix} a \\ b \end{bmatrix} \rightarrow \begin{cases} 2a = 2a \\ 2b = 2b \end{cases} \rightarrow \begin{cases} a = a \\ b = b \end{cases}$$

Since a and b are two arbitrary numbers, $x_1 = \begin{bmatrix} 1 \\ 0 \end{bmatrix}$ and $x_2 = \begin{bmatrix} 0 \\ 1 \end{bmatrix}$ are two independent eigenvectors.

4.3.2.2 Product and Sum of Eigenvalues

The product and sum of the eigenvalues of any matrix are equal to the determinant and trace of that matrix, respectively. We first show that the product of eigenvalues is equal to the determinant of the matrix. Let A be an $n \times n$ matrix with characteristic polynomial $P(\lambda)$. Then:

$$P(\lambda) = |\lambda I - A| = (\lambda - \lambda_1)(\lambda - \lambda_2) \cdots (\lambda - \lambda_n) \qquad (4.35)$$

Setting $\lambda = 0$ yields:

$$|-A| = (-\lambda_1)(-\lambda_2) \cdots (-\lambda_n) = (-1)^n \lambda_1 \lambda_2 \cdots \lambda_n \qquad (4.36)$$

Since $|-A| = (-1)^n |A|$, we have:

$$(-1)^n |A| = (-1)^n \lambda_1 \lambda_2 \cdots \lambda_n \qquad (4.37)$$

Hence:

$$|A| = \lambda_1 \lambda_2 \cdots \lambda_n \qquad (4.38)$$

Eigenvalues and Eigenvectors

Now we show that the sum of eigenvalues is equal to the trace of the matrix. Again, the characteristic polynomial of matrix A is:

$$P(\lambda) = |\lambda I - A| = \begin{bmatrix} \lambda - a_{11} & -a_{12} & \cdots & -a_{1n} \\ -a_{21} & \lambda - a_{22} & \cdots & -a_{2n} \\ \vdots & \vdots & \ddots & \vdots \\ -a_{n1} & -a_{n2} & \cdots & \lambda - a_{nn} \end{bmatrix} \quad (4.39)$$

By expanding the above determinant along the first column, we have:

$$P(\lambda) = (\lambda - a_{11})|M_{11}| - \sum_{i=2}^{n} (-1)^{i+1} a_{i1} |M_{i1}| \quad (4.40)$$

Expanding the determinant of M_{11} results in:

$$|M_{11}| = (\lambda - a_{22})|M'_{11}| - \sum_{i=3}^{n} (-1)^{i+1} a_{i1} |M'_{i1}| \quad (4.41)$$

Continuing these expansions, we will have:

$$P(\lambda) = (\lambda - a_{11})(\lambda - a_{22}) \cdots (\lambda - a_{nn}) + \tilde{P}(\lambda) \quad (4.42)$$

where $\tilde{P}(\lambda)$ is a polynomial of degree $n-2$. Therefore, the leading coefficient of $P(\lambda)$ is one and the second leading coefficient is $a_{11}+a_{22}+\cdots+a_{nm}$. Also, from Equation 4.42, the second leading coefficient of $P(\lambda)$ is $\lambda_1+\lambda_2+\cdots+\lambda_n$. Therefore:

$$\lambda_1 + \lambda_2 + \cdots + \lambda_n = a_{11} + a_{22} + \cdots + a_{nm} = \text{Trace}(A) \quad (4.43)$$

4.3.3 Finding the Characteristic Polynomial of a Matrix

Let A be an $n \times n$ matrix. The characteristic polynomial of matrix A can be found by using the following recursive algorithm. Let $W_k = \text{Trace}(A^K)$, $k = 1, 2, \ldots, n$, then the coefficients of the characteristic equation are:

$$\alpha_1 = -W_1$$

$$\alpha_2 = -\frac{1}{2}(\alpha_1 W_1 + W_2)$$

$$\alpha_3 = -\frac{1}{3}(\alpha_2 W_1 + \alpha_1 W_2 + W_3) \quad (4.44)$$

$$\cdots$$

$$\alpha_n = -\frac{1}{n}(\alpha_{n-1} W_1 + \alpha_{n-2} W_2 + \cdots + \alpha_1 W_{n-1} + W_n)$$

and

$$P(\lambda) = \lambda^n + \alpha_1\lambda^{n-1} + \alpha_2\lambda^{n-2} + \cdots + \alpha_{n-1}\lambda + \alpha_n \qquad (4.45)$$

The proof is left as an exercise (see Problem 4.5).

Example 4.12

Find the characteristic polynomial of the following 3×3 matrix:

$$A = \begin{bmatrix} -1 & -4 & 4 \\ -1 & -5 & 4 \\ 1 & 0 & -2 \end{bmatrix}$$

Solution: We first compute A^2 and A^3.

$$A^2 = A \times A = \begin{bmatrix} -1 & -4 & 4 \\ -1 & -5 & 4 \\ 1 & 0 & -2 \end{bmatrix}\begin{bmatrix} -1 & -4 & 4 \\ -1 & -5 & 4 \\ 1 & 0 & -2 \end{bmatrix} = \begin{bmatrix} 9 & 24 & -28 \\ 10 & 29 & -32 \\ -3 & -4 & 8 \end{bmatrix}$$

$$A^3 = A^2 \times A = \begin{bmatrix} 9 & 24 & -28 \\ 10 & 29 & -32 \\ -3 & -4 & 8 \end{bmatrix}\begin{bmatrix} -1 & -4 & 4 \\ -1 & -5 & 4 \\ 1 & 0 & -2 \end{bmatrix} = \begin{bmatrix} -61 & -156 & 188 \\ -71 & -185 & 220 \\ 15 & 32 & -44 \end{bmatrix}$$

We then find W_1, W_2, and W_3 as follows:

$$W_1 = \text{Trace}(A) = -8$$

$$W_2 = \text{Trace}(A^2) = 46$$

$$W_3 = \text{Trace}(A^3) = -290$$

Using Equation 4.43, we can compute the corresponding characteristic polynomial coefficients as follows:

$$\alpha_1 = -W_1 = 8$$

$$\alpha_2 = -\frac{1}{2}(a_1 W_1 + W_2) = -\frac{1}{2}(-8\times 8 + 46) = 9$$

$$\alpha_3 = -\frac{1}{3}(\alpha_2 W_1 + \alpha_1 W_2 + W_3) = -\frac{1}{3}(-9\times 8 + 8\times 46 - 290) = -2$$

Hence, the characteristic polynomial of matrix A is:

$$P(\lambda) = \lambda^3 + \alpha_1\lambda^2 + \alpha_2\lambda + \alpha_3 = \lambda^3 + 8\lambda^2 + 9\lambda - 2$$

Eigenvalues and Eigenvectors

4.3.4 Modal Matrix

Let $x_1, x_2, \ldots, x_n$ represent the n independent eigenvectors of the $n \times n$ matrix A. The $n \times n$ matrix M formed by the side-by-side stacking of the eigenvectors of A, as shown in Equation 4.46, is called the modal matrix of A.

$$M = \begin{bmatrix} x_1 & x_2 & \cdots & x_n \end{bmatrix} \qquad (4.46)$$

Since the eigenvectors are independent, the n columns of M are linearly independent and the modal matrix is full rank and nonsingular.

Example 4.13

Find the modal matrix of:

$$A = \begin{bmatrix} 8.25 & -9.75 & -7.5 \\ 2.25 & 2.25 & -1.5 \\ 2.25 & -12.75 & -4.5 \end{bmatrix}$$

Solution: Matrix A has three independent eigenvectors (see Example 4.9). They are:

$$x_1 = \begin{bmatrix} 1 \\ 0 \\ 1.5 \end{bmatrix}, \text{ and } x_2 = \begin{bmatrix} 1 \\ -1 \\ 2 \end{bmatrix} \text{ and } x_3 = \begin{bmatrix} 1 \\ 1 \\ -1 \end{bmatrix}$$

Therefore, the modal matrix is:

$$M = \begin{bmatrix} x_1 & x_2 & \cdots & x_n \end{bmatrix} = \begin{bmatrix} 1 & 1 & 1 \\ 0 & -1 & 1 \\ 1.5 & 2 & -1 \end{bmatrix}$$

Note that the determinant of M is equal to two indicating that the matrix is nonsingular.

4.4 Matrix Diagonalization

4.4.1 Distinct Eigenvalues

Let A be an $n \times n$ matrix with n distinct eigenvalues $\lambda_1, \lambda_2, \ldots, \lambda_n$. Let the corresponding independent eigenvectors be $x_1, x_2, \ldots, x_n$. Therefore:

$$Ax_1 = \lambda_1 x_1, \, Ax_2 = \lambda_2 x_2, \ldots, Ax_n = \lambda_n x_n \qquad (4.47)$$

These equations can be concatenated together in matrix form as follows:

$$\begin{bmatrix} Ax_1 & Ax_2 & \cdots & Ax_n \end{bmatrix} = \begin{bmatrix} \lambda_1 x_1 & \lambda_2 x_2 & \cdots & \lambda_n x_n \end{bmatrix} \qquad (4.48)$$

Equation 4.48 can be written as:

$$A[x_1 \ x_2 \ \cdots \ x_n] = [x_1 \ x_2 \ \cdots \ x_n] \begin{bmatrix} \lambda_1 & 0 & \cdots & 0 \\ 0 & \lambda_2 & \cdots & 0 \\ \vdots & \vdots & \ddots & \vdots \\ 0 & 0 & \cdots & \lambda_n \end{bmatrix} \quad (4.49)$$

Let Λ represent the diagonal matrix of eigenvalues shown on the right side in Equation 4.49 above. The above matrix equation can now be written in terms of Λ and modal matrix M as:

$$AM = M\Lambda \quad (4.50)$$

This equation is valid regardless of whether the eigenvectors are linearly independent or not. However, if the eigenvectors are linearly independent, then M is of full rank and its inverse is well defined. In this case, we can postmultiply the above equation by M^{-1} to obtain:

$$A = M\Lambda M^{-1} \quad (4.51)$$

or

$$\Lambda = M^{-1}AM \quad (4.52)$$

Example 4.14

Diagonalize the 3×3 matrix A given by:

$$A = \begin{bmatrix} 8.25 & -9.75 & -7.5 \\ 2.25 & 2.25 & -1.5 \\ 2.25 & -12.75 & -4.5 \end{bmatrix}$$

Solution: The eigenvalues, eigenvectors and modal matrix of A were computed earlier (see Examples 4.9 and 4.13) yielding:

$$\lambda_1 = -3, \lambda_2 = 3, \text{ and } \lambda_3 = 6 \quad M = \begin{bmatrix} 1 & 1 & 1 \\ 0 & -1 & 1 \\ 1.5 & 2 & -1 \end{bmatrix}$$

Thus:

$$M^{-1}AM = \begin{bmatrix} 1 & 1 & 1 \\ 0 & -1 & 1 \\ 1.5 & 2 & -1 \end{bmatrix}^{-1} \begin{bmatrix} 8.25 & -9.75 & -7.5 \\ 2.25 & 2.25 & -1.5 \\ 2.25 & -12.75 & -4.5 \end{bmatrix} \begin{bmatrix} 1 & 1 & 1 \\ 0 & -1 & 1 \\ 1.5 & 2 & -1 \end{bmatrix}$$

$$= \begin{bmatrix} -0.5 & 1.5 & 1 \\ 0.75 & -1.25 & -0.5 \\ 0.75 & -0.25 & -0.5 \end{bmatrix} \begin{bmatrix} 8.25 & -9.75 & -7.5 \\ 2.25 & 2.25 & -1.5 \\ 2.25 & -12.75 & -4.5 \end{bmatrix} \begin{bmatrix} 1 & 1 & 1 \\ 0 & -1 & 1 \\ 1.5 & 2 & -1 \end{bmatrix}$$

$$= \begin{bmatrix} -3 & 0 & 0 \\ 0 & 3 & 0 \\ 0 & 0 & 6 \end{bmatrix} = \begin{bmatrix} \lambda_1 & 0 & 0 \\ 0 & \lambda_2 & 0 \\ 0 & 0 & \lambda_3 \end{bmatrix} = \Lambda$$

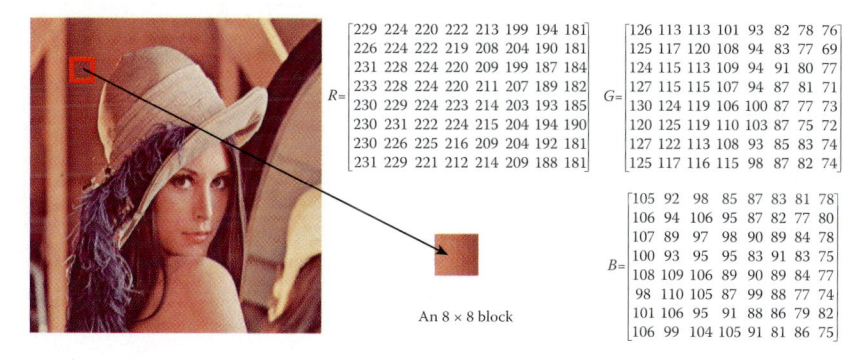

COLOR FIGURE 1.2
RGB LENA image.

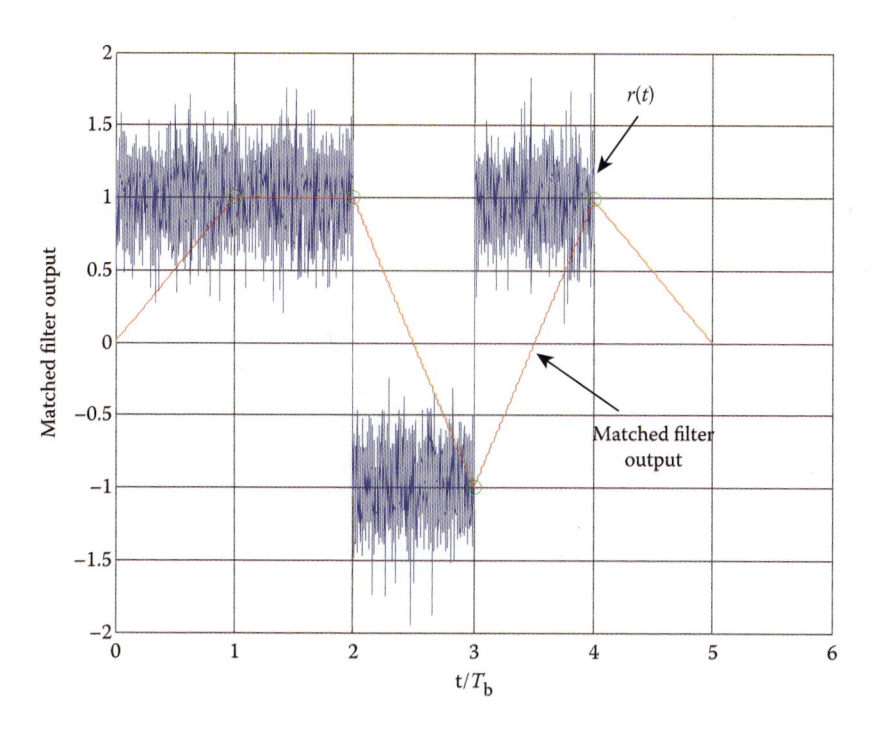

$$R = \begin{bmatrix} 229 & 224 & 220 & 222 & 213 & 199 & 194 & 181 \\ 226 & 224 & 222 & 219 & 208 & 204 & 190 & 181 \\ 231 & 228 & 224 & 220 & 209 & 199 & 187 & 184 \\ 233 & 228 & 224 & 220 & 211 & 207 & 189 & 182 \\ 230 & 229 & 224 & 223 & 214 & 203 & 193 & 185 \\ 230 & 231 & 222 & 224 & 215 & 204 & 194 & 190 \\ 230 & 226 & 225 & 216 & 209 & 204 & 192 & 181 \\ 231 & 229 & 221 & 212 & 214 & 209 & 188 & 181 \end{bmatrix}$$

$$G = \begin{bmatrix} 126 & 113 & 113 & 101 & 93 & 82 & 78 & 76 \\ 125 & 117 & 120 & 108 & 94 & 83 & 77 & 69 \\ 124 & 115 & 113 & 109 & 94 & 91 & 80 & 77 \\ 127 & 115 & 115 & 107 & 94 & 87 & 81 & 71 \\ 130 & 124 & 119 & 106 & 100 & 87 & 77 & 73 \\ 120 & 125 & 119 & 110 & 103 & 87 & 75 & 72 \\ 127 & 122 & 113 & 108 & 93 & 85 & 83 & 74 \\ 125 & 117 & 116 & 115 & 98 & 87 & 82 & 74 \end{bmatrix}$$

$$B = \begin{bmatrix} 105 & 92 & 98 & 85 & 87 & 83 & 81 & 78 \\ 106 & 94 & 106 & 95 & 87 & 82 & 77 & 80 \\ 107 & 89 & 97 & 98 & 90 & 89 & 84 & 78 \\ 100 & 93 & 95 & 95 & 83 & 91 & 83 & 75 \\ 108 & 109 & 106 & 89 & 90 & 89 & 84 & 77 \\ 98 & 110 & 105 & 87 & 99 & 88 & 77 & 74 \\ 101 & 106 & 95 & 91 & 88 & 86 & 79 & 82 \\ 106 & 99 & 104 & 105 & 91 & 81 & 86 & 75 \end{bmatrix}$$

An 8 × 8 block

COLOR FIGURE 3.7
The received signal and the output of the matched filter.

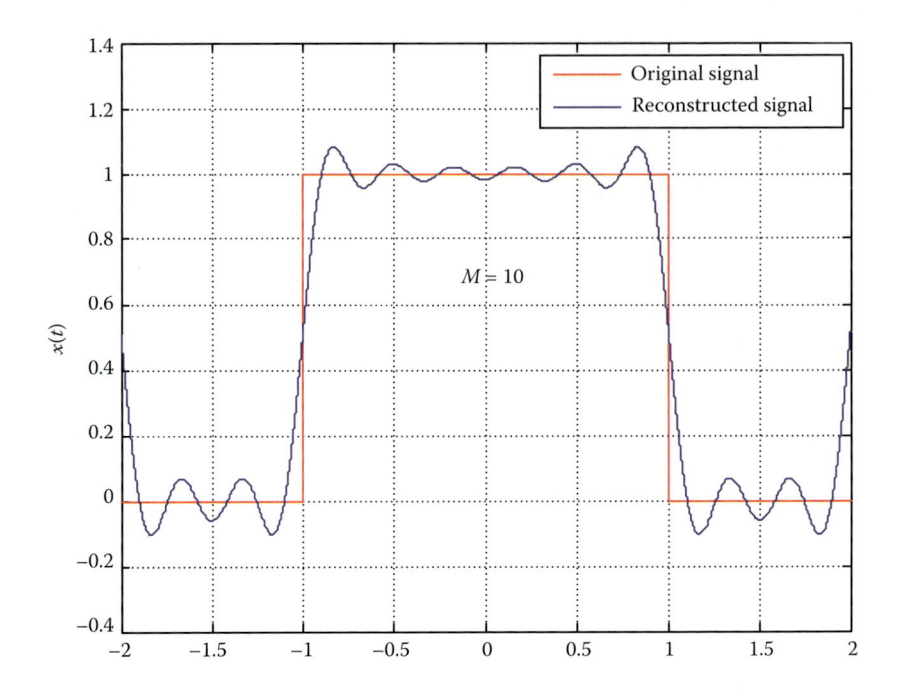

COLOR FIGURE 3.10
Fourier series expansion of $x(t)$ with $M=10$.

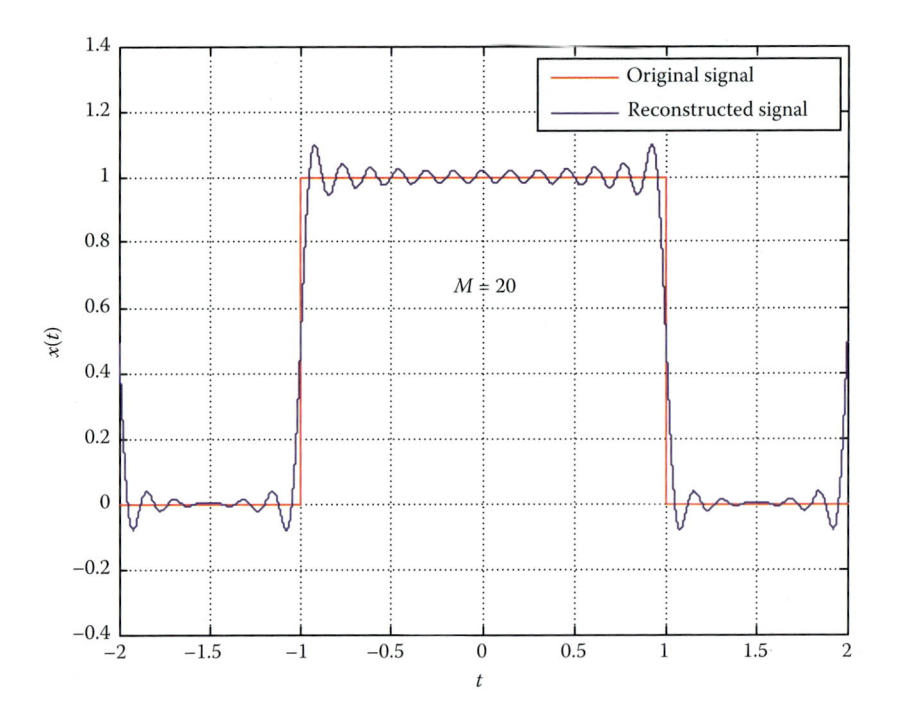

COLOR FIGURE 3.11
Fourier series expansion of $x(t)$ with $M=20$.

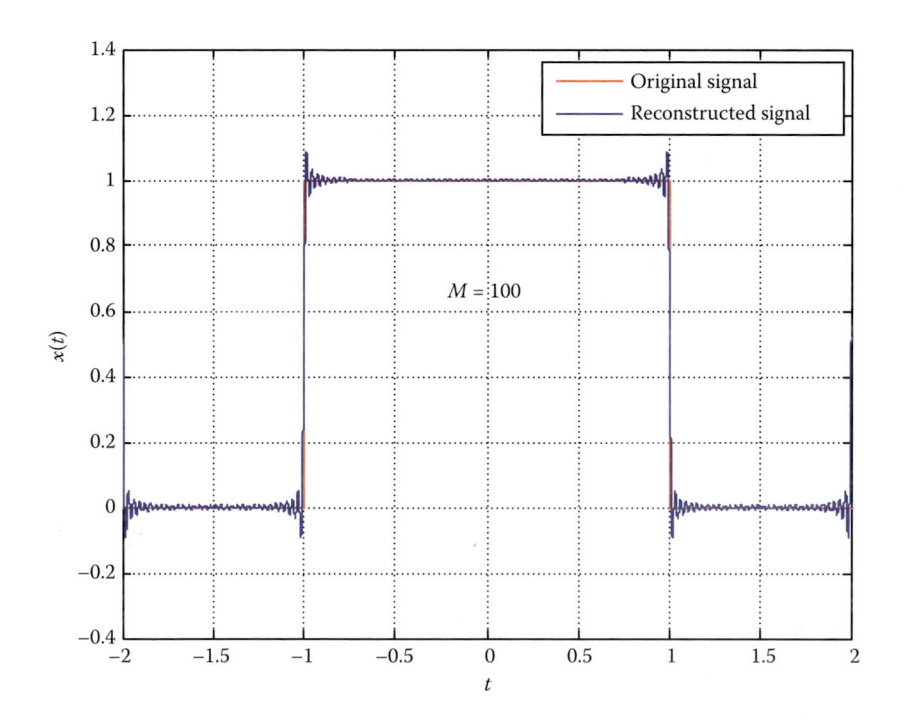

COLOR FIGURE 3.12
Fourier series expansion of $x(t)$ with $M=100$.

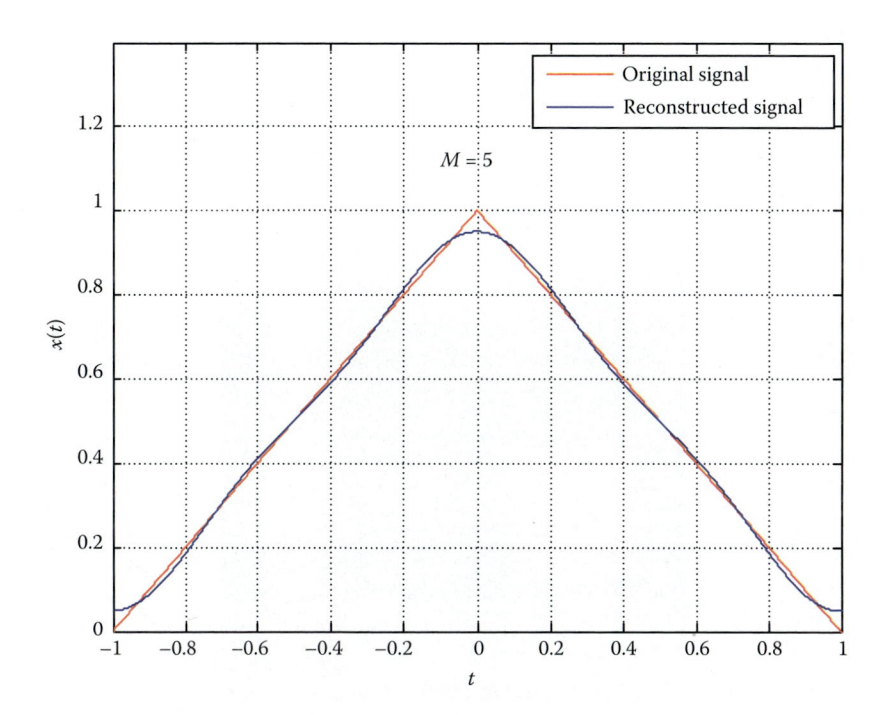

COLOR FIGURE 3.14
Fourier series expansion of $x(t)$ with $M=5$.

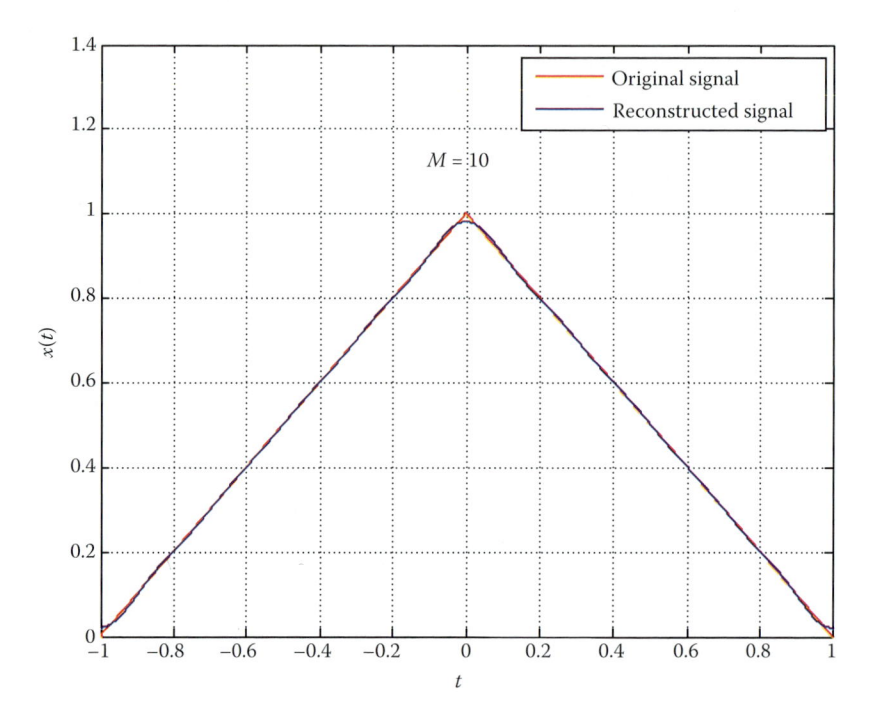

COLOR FIGURE 3.15
Fourier series expansion of $x(t)$ with $M=10$.

COLOR FIGURE 4.17
Computer generated image.

COLOR FIGURE 4.19
LENA image.

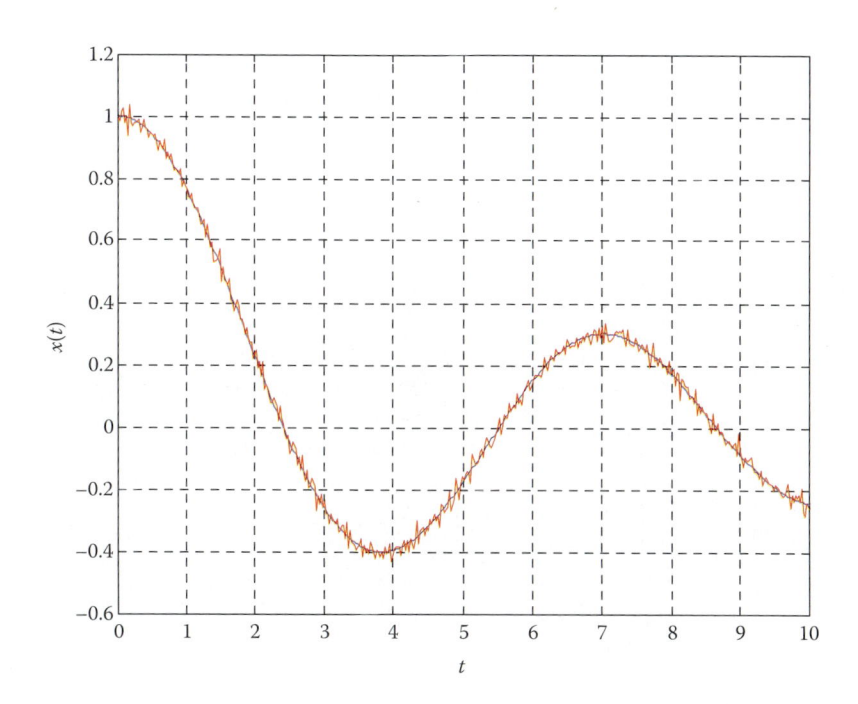

COLOR FIGURE 6.4
Deconvolved signal in presence of noise.

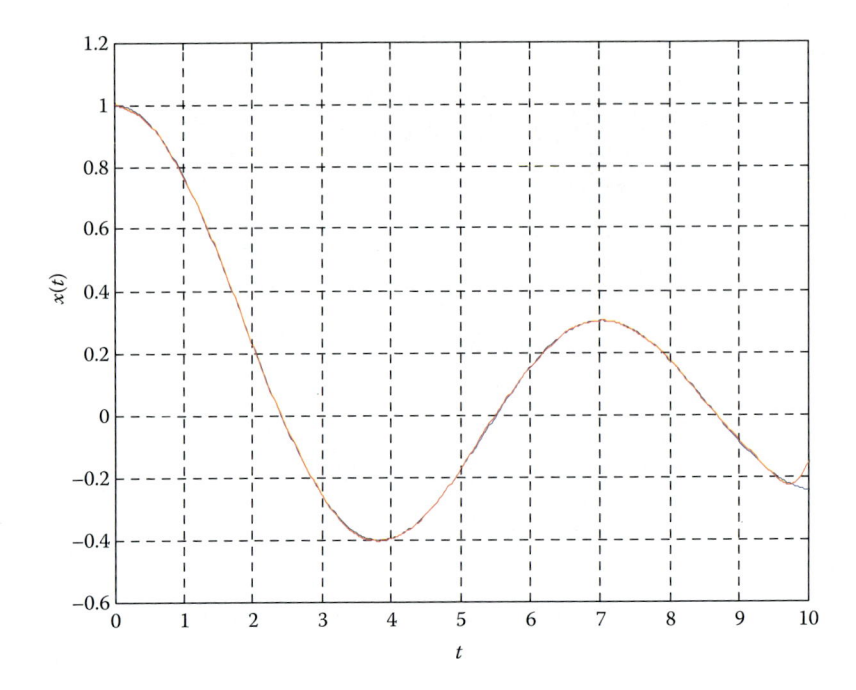

COLOR FIGURE 6.5
Deconvolution with a smooth operator.

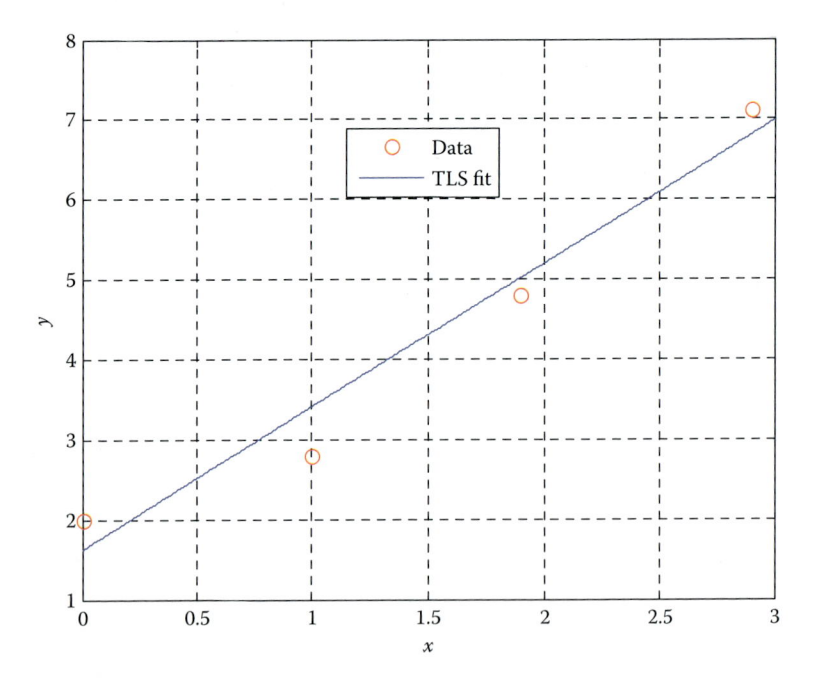

COLOR FIGURE 6.8
TLS linear fit. The original data is shown by the red circles and the least square linear fit is illustrated by the blue line.

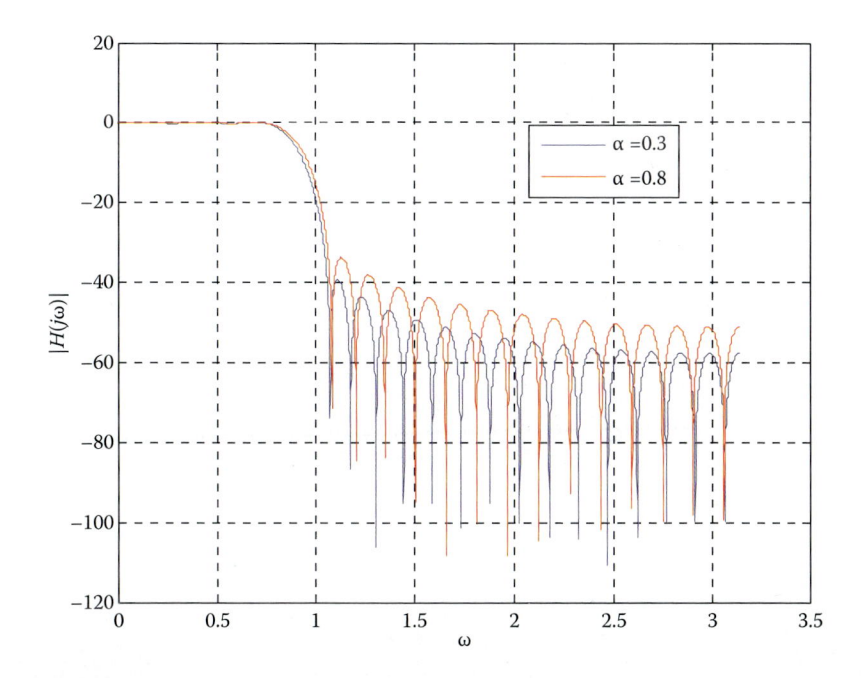

COLOR FIGURE 6.11
Frequency response of the designed low pass filter.

Eigenvalues and Eigenvectors

Example 4.15

Diagonalize matrix A given by:

$$A = \begin{bmatrix} 5 & 1 & 0 \\ 2 & 4 & 0 \\ -1 & 1 & 6 \end{bmatrix}$$

Solution: The eigenvalues of A are: $\lambda_1 = 3$ and $\lambda_2 = \lambda_3 = 6$. Note that A has one distinct and one repeat eigenvalue. However, it possesses three corresponding independent eigenvectors as follows:

$$x_1 = \begin{bmatrix} 1 \\ -2 \\ 1 \end{bmatrix}, \; x_2 = \begin{bmatrix} 0 \\ 0 \\ 1 \end{bmatrix} \text{ and } x_3 = \begin{bmatrix} 23 \\ 23 \\ 18 \end{bmatrix}$$

Since, the repeated eigenvalues have independent eigenvectors, matrix A is diagonalizable. Therefore:

$$M^{-1}AM = \begin{bmatrix} 1 & 0 & 23 \\ -2 & 0 & 23 \\ 1 & 1 & 18 \end{bmatrix}^{-1} \begin{bmatrix} 5 & 1 & 0 \\ 2 & 4 & 0 \\ -1 & 1 & 6 \end{bmatrix} \begin{bmatrix} 1 & 0 & 23 \\ -2 & 0 & 23 \\ 1 & 1 & 18 \end{bmatrix} = \begin{bmatrix} 3 & 0 & 0 \\ 0 & 6 & 0 \\ 0 & 0 & 6 \end{bmatrix}$$

Note that if matrix A has distinct eigenvalues, then it is always diagonalizable. However, if it possesses repeated eigenvalues, it is not always possible to diagonalize it, as shown by the following example.

Example 4.16

Consider the 3×3 matrix A.

$$A = \begin{bmatrix} 2 & 1 & -4 \\ 0 & 2 & 1 \\ 0 & 0 & 2 \end{bmatrix}$$

Matrix A has one eigenvalue repeated three times (multiplicity of three), that is $\lambda_1 = \lambda_2 = \lambda_3 = 2$, and one independent eigenvector $x_1 = \begin{bmatrix} 1 \\ 0 \\ 0 \end{bmatrix}$. Hence, it cannot be diagonalized.

In summary, if the $n \times n$ matrix A has n independent eigenvectors, then it is always diagonalizable regardless whether it has distinct or repeat eigenvalues. On the other hand, if A has less than n independent eigenvectors, then it is not diagonalizable.

4.4.2 Jordan Canonical Form

If the $n \times n$ matrix A has repeated eigenvalues, then it is not always possible to diagonalize it. It can be diagonalized only if it possesses n independent eigenvectors. If some of the eigenvectors become linearly dependent, then matrix A cannot be diagonalized. In this case, it can be transformed into a Jordan canonical form, which is related to matrix A through a similarity transformation. Here, we state a theorem known as the Jordan form.

Theorem 4.5: Let A be an $n \times n$ matrix with $k \leq n$ linearly independent eigenvectors. Then, there is a nonsingular transformation T such that:

$$A = TJT^{-1} \tag{4.53}$$

where matrix J is block diagonal and is given by:

$$J = \begin{bmatrix} J_1 & & & \\ & J_2 & & \\ & & \ddots & \\ & & & J_k \end{bmatrix} \tag{4.54}$$

Each block is known as a Jordan block. Each Jordan block is a square matrix of size $m_i \times m_i$ of the form:

$$J_i = \begin{bmatrix} \lambda_i & 1 & 0 & \cdots & 0 \\ 0 & \lambda_i & 1 & \cdots & 0 \\ \vdots & \vdots & \ddots & \ddots & \vdots \\ 0 & 0 & \cdots & \lambda_i & 1 \\ 0 & 0 & \cdots & 0 & \lambda_i \end{bmatrix} \tag{4.55}$$

where m_i represents the multiplicity of the ith eigenvalue λ_i.

Proof: Without loss of generality, we assume that the $n \times n$ matrix A has one eigenvalue of multiplicity m with one independent eigenvector x_1 and $n-m$ distinct eigenvalues. That is, the eigenvalues of A are:

$$\underbrace{\lambda_1, \lambda_1, \cdots, \lambda_1}_{m}, \lambda_{m+1}, \lambda_{m+2}, \cdots, \lambda_n \tag{4.56}$$

Let us define the vectors $x_2, x_3, \ldots, x_m$ as:

$$Ax_1 = \lambda_1 x_1 \tag{4.57}$$

$$Ax_2 = \lambda_1 x_2 + x_1 \tag{4.58}$$

$$Ax_3 = \lambda_1 x_3 + x_2 \tag{4.59}$$

$$\vdots$$

$$Ax_m = \lambda_1 x_m + x_{m-1} \tag{4.60}$$

Eigenvalues and Eigenvectors

The eigenvectors $x_{m+1}, x_{m+2}, \ldots, x_n$ corresponding to distinct eigenvalues $\lambda_{m+1}, \lambda_{m+2}, \ldots, \lambda_n$ are determined from:

$$Ax_{m+1} = \lambda_{m+1} x_{m+1} \tag{4.61}$$

$$Ax_{m+2} = \lambda_{m+2} x_{m+2} \tag{4.62}$$

$$\vdots$$

$$Ax_n = \lambda_n x_n \tag{4.63}$$

Let us define the transformation T as:

$$T = \begin{bmatrix} x_1 \vdots & x_2 \vdots & x_3 \vdots & \cdots & \vdots x_n \end{bmatrix} \tag{4.64}$$

The n vectors $x_1, x_2, \ldots, x_n$ are linearly independent and therefore matrix T is nonsingular. It is clear that:

$$A\begin{bmatrix} x_1 \vdots & x_2 \vdots & \cdots & \vdots x_m \vdots & x_{m+1} \vdots & \cdots & \vdots x_n \end{bmatrix}$$

$$= \begin{bmatrix} x_1 \vdots & x_2 \vdots & \cdots & \vdots x_m \vdots & x_{m+1} \vdots & \cdots & \vdots x_n \end{bmatrix} \begin{bmatrix} \lambda_1 & 1 & 0 & \cdots & 0 & 0 & \cdots & 0 \\ 0 & \lambda_1 & 1 & \cdots & 0 & 0 & \cdots & 0 \\ \vdots & \vdots & \ddots & \ddots & \vdots & \vdots & \vdots & \vdots \\ 0 & 0 & \cdots & \lambda_1 & 1 & 0 & \cdots & 0 \\ 0 & 0 & \cdots & 0 & \lambda_1 & 0 & \cdots & 0 \\ 0 & 0 & \cdots & 0 & 0 & \lambda_{m+1} & \cdots & 0 \\ \vdots & \vdots & \cdots & \vdots & \vdots & \vdots & \ddots & \vdots \\ 0 & 0 & \cdots & 0 & 0 & 0 & \cdots & \lambda_n \end{bmatrix} \tag{4.65}$$

Therefore:

$$AT = T \begin{bmatrix} \lambda_1 & 1 & 0 & \cdots & 0 & 0 & \cdots & 0 \\ 0 & \lambda_1 & 1 & \cdots & 0 & 0 & \cdots & 0 \\ \vdots & \vdots & \ddots & \ddots & \vdots & \vdots & \vdots & \vdots \\ 0 & 0 & \cdots & \lambda_1 & 1 & 0 & \cdots & 0 \\ 0 & 0 & \cdots & 0 & \lambda_1 & 0 & \cdots & 0 \\ 0 & 0 & \cdots & 0 & 0 & \lambda_{m+1} & \cdots & 0 \\ \vdots & \vdots & \cdots & \vdots & \vdots & \vdots & \ddots & \vdots \\ 0 & 0 & \cdots & 0 & 0 & 0 & \cdots & \lambda_n \end{bmatrix} \tag{4.66}$$

Premultiplying both sides of the above equation by T^{-1} results in:

$$T^{-1}AT = \begin{bmatrix} \lambda_1 & 1 & 0 & \cdots & 0 & 0 & \cdots & 0 \\ 0 & \lambda_1 & 1 & \cdots & 0 & 0 & \cdots & 0 \\ \vdots & \vdots & \ddots & \ddots & \vdots & \vdots & \vdots & \vdots \\ 0 & 0 & \cdots & \lambda_1 & 1 & 0 & \cdots & 0 \\ 0 & 0 & \cdots & 0 & \lambda_1 & 0 & \cdots & 0 \\ 0 & 0 & \cdots & 0 & 0 & \lambda_{m+1} & \cdots & 0 \\ \vdots & \vdots & \cdots & \vdots & \vdots & \vdots & \ddots & \vdots \\ 0 & 0 & \cdots & 0 & 0 & 0 & \cdots & \lambda_n \end{bmatrix} \qquad (4.67)$$

The MATLAB® command to compute the Jordan canonical form is [V,J]=jordan(A).

Example 4.17

Transform the following matrix into its Jordan form:

$$A = \begin{bmatrix} -5 & 6.75 & 5.5 & 12.75 \\ 1 & -2.25 & -0.5 & -0.25 \\ -2 & 2.75 & 0.5 & 3.75 \\ -1 & 1.75 & 1.5 & 1.75 \end{bmatrix}$$

Solution: The characteristic equation of matrix A is:

$$P(\lambda) = |\lambda I - A| = (\lambda + 1)^3(\lambda + 2)$$

There are four eigenvalues: $\lambda_1 = -1$ (repeat eigenvalue) with a multiplicity of three (i.e. $m_1 = 3$) and $\lambda_4 = -2$ (distinct eigenvalue) with a multiplicity of one. The eigenvector corresponding to λ_4 is denoted by x_4 and is computed as:

$$Ax_4 = \lambda_4 x_4$$

$$\begin{bmatrix} -5 & 6.75 & 5.5 & 12.75 \\ 1 & -2.25 & -0.5 & -0.25 \\ -2 & 2.75 & 0.5 & 3.75 \\ -1 & 1.75 & 1.5 & 1.75 \end{bmatrix} \begin{bmatrix} a \\ b \\ c \\ d \end{bmatrix} = -2 \begin{bmatrix} a \\ b \\ c \\ d \end{bmatrix} \rightarrow \begin{bmatrix} a \\ b \\ c \\ d \end{bmatrix} = \begin{bmatrix} 1 \\ -2 \\ 3 \\ 0 \end{bmatrix} \rightarrow x_4 = \begin{bmatrix} 1 \\ -2 \\ 3 \\ 0 \end{bmatrix}$$

The eigenvectors associated with the repeat eigenvalue $\lambda_1 = -1$ are:

$$Ax_1 = \lambda_1 x_1$$

$$\begin{bmatrix} -5 & 6.75 & 5.5 & 12.75 \\ 1 & -2.25 & -0.5 & -0.25 \\ -2 & 2.75 & 0.5 & 3.75 \\ -1 & 1.75 & 1.5 & 1.75 \end{bmatrix} \begin{bmatrix} a \\ b \\ c \\ d \end{bmatrix} = - \begin{bmatrix} a \\ b \\ c \\ d \end{bmatrix} \rightarrow \begin{cases} -5a + 6.75b + 5.5c + 12.75d = -a \\ a - 2.25b - 0.5c - 0.25d = -b \\ -2a + 2.75b + 0.5c + 3.75d = -c \\ -a + 1.75b + 1.5c + 1.75d = -d \end{cases}$$

Eigenvalues and Eigenvectors

Hence:

$$\begin{bmatrix} -4 & 6.75 & 5.5 & 12.75 \\ 1 & -1.25 & -0.5 & -0.25 \\ -2 & 2.75 & 1.5 & 3.75 \\ -1 & 1.75 & 1.5 & 2.75 \end{bmatrix} \begin{bmatrix} a \\ b \\ c \\ d \end{bmatrix} = 0$$

Since rank $(A - \lambda_1 I) = 3$, the above set of equations has one independent solution. The independent solution is:

$$x_1 = \begin{bmatrix} 2 \\ 2 \\ -1 \\ 0 \end{bmatrix}$$

The second vector is obtained by solving the following equation:

$$Ax_2 = \lambda_1 x_2 + x_1$$

or

$$\begin{bmatrix} -5 & 6.75 & 5.5 & 12.75 \\ 1 & -2.25 & -0.5 & -0.25 \\ -2 & 2.75 & 0.5 & 3.75 \\ -1 & 1.75 & 1.5 & 1.75 \end{bmatrix} \begin{bmatrix} a \\ b \\ c \\ d \end{bmatrix} = - \begin{bmatrix} a \\ b \\ c \\ d \end{bmatrix} + \begin{bmatrix} 2 \\ 2 \\ -1 \\ 0 \end{bmatrix}$$

which can be written as:

$$\begin{bmatrix} -4 & 6.75 & 5.5 & 12.75 \\ 1 & -1.25 & -0.5 & -0.25 \\ -2 & 2.75 & 1.5 & 3.75 \\ -1 & 1.75 & 1.5 & 2.75 \end{bmatrix} \begin{bmatrix} a \\ b \\ c \\ d \end{bmatrix} = \begin{bmatrix} 2 \\ 2 \\ -1 \\ 0 \end{bmatrix}$$

There is one independent solution x_2 given by:

$$x_2 = \begin{bmatrix} 3 \\ 1 \\ -1 \\ 1 \end{bmatrix}$$

The third vector is obtained by solving the following equation:

$$Ax_3 = \lambda_1 x_3 + x_2$$

or

$$\begin{bmatrix} -4 & 6.75 & 5.5 & 12.75 \\ 1 & -1.25 & -0.5 & -0.25 \\ -2 & 2.75 & 1.5 & 3.75 \\ -1 & 1.75 & 1.5 & 2.75 \end{bmatrix} \begin{bmatrix} a \\ b \\ c \\ d \end{bmatrix} = \begin{bmatrix} 3 \\ 1 \\ -1 \\ 1 \end{bmatrix}$$

The solution is:

$$x_3 = \begin{bmatrix} 0 \\ -2 \\ 3 \\ 0 \end{bmatrix}$$

Therefore, the transformation matrix T is given by:

$$T = \begin{bmatrix} x_1 & x_2 & x_3 & x_4 \end{bmatrix} = \begin{bmatrix} 2 & 3 & 0 & 1 \\ 2 & 1 & -2 & -2 \\ -1 & -1 & 3 & 3 \\ 0 & 1 & 0 & 0 \end{bmatrix}$$

and

$$T^{-1}AT = \begin{bmatrix} 0 & 0.75 & 0.50 & -0.25 \\ 0 & 0 & 0 & 1 \\ -1 & 1.75 & 1.50 & 2.75 \\ 1 & -1.50 & -1 & -2.50 \end{bmatrix} \begin{bmatrix} -5 & 6.75 & 5.5 & 12.75 \\ 1 & -2.25 & -0.5 & -0.25 \\ -2 & 2.75 & 0.5 & 3.75 \\ -1 & 1.75 & 1.5 & 1.75 \end{bmatrix} \begin{bmatrix} 2 & 3 & 0 & 1 \\ 2 & 1 & -2 & -2 \\ -1 & -1 & 3 & 3 \\ 0 & 1 & 0 & 0 \end{bmatrix}$$

$$= \begin{bmatrix} -1 & 1 & 0 & 0 \\ 0 & -1 & 1 & 0 \\ 0 & 0 & -1 & 0 \\ 0 & 0 & 0 & -2 \end{bmatrix}$$

4.5 Special Matrices

4.5.1 Unitary Matrices

The $n \times n$ matrix A is said to be unitary or orthogonal if:

$$AA^H = A^H A = I \tag{4.68}$$

An important property of unitary matrices is that the inverse of the matrix is equal to its conjugate transpose. In addition, the columns of a unitary matrix are pairwise orthogonal and form a set of orthonormal vectors. A linear transformation defined by a unitary matrix is a norm preserving transform. The following theorem proves the norm preserving property of unitary matrices.

Theorem 4.6: Any unitary transformation is a norm preserving transform.

Proof: Let A be a linear unitary transformation from R^n onto itself, that is $y = Ax$ where $x, y \in R^n$. Then:

$$\|y\|^2 = \|Ax\|^2 = (Ax)^H (Ax) = x^H A^H A x = x^H x = \|x\|^2 \tag{4.69}$$

Eigenvalues and Eigenvectors 181

Therefore:

$$\|y\| = \|x\| \tag{4.70}$$

Hence, A is a norm preserving transform. Most of the frequency domain transforms such as the Continuous Fourier Transform (CFT), the Discrete Time Fourier Transform (DTFT) and the Discrete Fourier Transform (DFT) are norm preserving transforms where the above identity is referred to as the Parseval's identity.

Example 4.18

The N-point DFT of a given discrete signal x is defined as:

$$X = Ax \tag{4.71}$$

where

$$x = \begin{bmatrix} x(0) & x(1) & \cdots & x(N-1) \end{bmatrix}^T, X = \begin{bmatrix} X(0) & X(1) & \cdots & X(N-1) \end{bmatrix}^T \tag{4.72}$$

$$A = \begin{bmatrix} W_N^0 & W_N^0 & W_N^0 & \cdots & W_N^0 \\ W_N^0 & W_N^1 & W_N^2 & \cdots & W_N^{N-1} \\ \vdots & \vdots & \vdots & \cdots & \\ W_N^0 & W_N^{N-2} & W_N^{2(N-2)} & \cdots & W_N^{(N-2)(N-1)} \\ W_N^0 & W_N^{N-1} & W_N^{2(N-1)} & \cdots & W_N^{(N-1)(N-1)} \end{bmatrix} \tag{4.73}$$

and $W_N = e^{j\frac{2\pi}{N}}$. Matrix A is a unitary matrix. To show this, let us define $B = AA^H$. Then:

$$b_{mn} = \sum_{l=1}^{N} (A)_{ml} (A^H)_{ln} = \sum_{l=1}^{N} (A)_{ml} (\bar{A})_{nl} = \sum_{l=1}^{N} W_N^{(m-1)(l-1)} \bar{W}_N^{(n-1)(l-1)}$$

$$= \sum_{l=0}^{N-1} W_N^{(m-1)l} \bar{W}_N^{(n-1)l} = \sum_{l=0}^{N-1} e^{j\frac{2\pi}{N}(m-1)l} e^{-j\frac{2\pi}{N}(n-1)l}$$

$$= \sum_{l=0}^{N-1} e^{j\frac{2\pi}{N}(m-n)l} = \frac{1 - e^{j2\pi(m-n)}}{1 - e^{j\frac{2\pi}{N}(m-n)}} = \begin{cases} N & m = n \\ 0 & m \neq n \end{cases} \tag{4.74}$$

Therefore:

$$AA^H = NI \tag{4.75}$$

where I represents the identity matrix. Hence, A^{-1} is equal to:

$$A^{-1} = \frac{1}{N} A^H \tag{4.76}$$

From Equation 4.71, we have:

$$x = A^{-1}X = \frac{1}{N}A^H X \qquad (4.77)$$

The DFT is a norm (energy) preserving transform, which implies that:

$$\|x\|^2 = \frac{1}{N}\|X\|^2 \qquad (4.78)$$

or

$$\sum_{n=0}^{N-1} |x(n)|^2 = \frac{1}{N}\sum_{k=0}^{N-1} |X(k)|^2 \qquad (4.79)$$

Equation 4.79 is known as the Parseval's identity.

Example 4.19

Find the four-point inverse Discrete Fourier Transform (IDFT) of the complex sequence X defined as follows:

$$X = \begin{bmatrix} 10 \\ -2+j2 \\ -2 \\ -2-j2 \end{bmatrix}$$

and show that it is a norm preserving transform.

Solution: Since $N=4$, then:

$$A = \begin{bmatrix} W_4^0 & W_4^0 & W_4^0 & W_4^0 \\ W_4^0 & W_4^1 & W_4^2 & W_4^3 \\ W_4^0 & W_4^2 & W_4^4 & W_4^6 \\ W_4^0 & W_4^3 & W_4^6 & W_4^9 \end{bmatrix} = \begin{bmatrix} 1 & 1 & 1 & 1 \\ 1 & -j & -1 & j \\ 1 & -1 & 1 & -1 \\ 1 & j & -1 & -j \end{bmatrix}$$

The IDFT of X can be computed by premultiplying vector X by $(1/N)A^H$ as shown below:

$$x = \frac{1}{N}A^H X = \frac{1}{4}\begin{bmatrix} 1 & 1 & 1 & 1 \\ 1 & j & -1 & -j \\ 1 & -1 & 1 & -1 \\ 1 & -j & -1 & j \end{bmatrix}\begin{bmatrix} 10 \\ -2+j2 \\ -2 \\ -2-j2 \end{bmatrix} = \begin{bmatrix} 1 \\ 2 \\ 3 \\ 4 \end{bmatrix}$$

$$\|X\|^2 = 100 + (4+4) + 4 + (4+4) = 120$$

$$\|x\|^2 = 1 + 4 + 9 + 16 = 30 = \frac{1}{4}\|X\|^2$$

Eigenvalues and Eigenvectors

4.5.2 Hermitian Matrices

Hermitian matrices play an important role in communication, signal processing and control. In this section, we provide a definition for Hermitian matrices and describe their important properties. An $n \times n$ complex matrix is said to be Hermitian if $A = A^H$, where H stands for the conjugate transpose. It should be noted that symmetric matrices are a special case of Hermitian matrices. The following theorem states an important property related to eigenvalue/eigenvector structure of Hermitian matrices.

Theorem 4.7: If A is an $n \times n$ Hermitian matrix, then the eigenvalues of A are real and the eigenvectors corresponding to the distinct eigenvalues are orthogonal.

Proof: First we show that the eigenvalues are real. Let λ be an eigenvalue of A with corresponding eigenvector x. Then:

$$Ax = \lambda x \tag{4.80}$$

Premultiply both sides of Equation 4.80 by x^H, the conjugate transpose of x. This yields the following:

$$x^H A x = x^H \lambda x = \lambda x^H x = \lambda \|x\|^2 \tag{4.81}$$

Next, postmultiply the conjugate transpose of Equation 4.80 by x. This yields:

$$x^H A^H x = \bar{\lambda} x^H x = \bar{\lambda} \|x\|^2 \tag{4.82}$$

But since $A^H = A$ and A is symmetric, Equation 4.82 can be transformed into:

$$x^H A x = \bar{\lambda} x^H x = \bar{\lambda} \|x\|^2 \tag{4.83}$$

Comparing Equations 4.81 and 4.83, we have:

$$\bar{\lambda} \|x\|^2 = \lambda \|x\|^2 \tag{4.84}$$

Since x is a nonzero vector, then $\bar{\lambda} = \lambda$ and λ is real.

Now we prove that the eigenvectors corresponding to the distinct eigenvalues are orthogonal. Let λ_1 and λ_2 be two distinct eigenvalues with corresponding eigenvectors x_1 and x_2. Then:

$$Ax_1 = \lambda_1 x_1 \tag{4.85}$$

and

$$Ax_2 = \lambda_2 x_2 \tag{4.86}$$

Premultiplying both sides of Equation 4.86 by x_2^H results in:

$$x_2^H A x_1 = \lambda_1 x_2^H x_1 \tag{4.87}$$

Now, let us take the conjugate transpose of Equation 4.86 and postmultiply both sides by x_1 to obtain:

$$x_2^H A^H x_1 = \bar{\lambda}_2 x_2^H x_1 \tag{4.88}$$

Since A is Hermitian (i.e. $A = A^H$) and the eigenvalues are real ($\bar{\lambda}_2 = \lambda_2$) as shown earlier, Equation 4.88 is reduced to:

$$x_2^H A x_1 = \lambda_2 x_2^H x_1 \tag{4.89}$$

Comparing Equations 4.87 and 4.89, we have:

$$\lambda_1 x_2^H x_1 = \lambda_2 x_2^H x_1 \Leftrightarrow (\lambda_1 - \lambda_2) x_2^H x_1 = 0 \tag{4.90}$$

Since $\lambda_1 \neq \lambda_2$, we must have $x_2^H x_1 = 0$. That is, x_1 is orthogonal to x_2.

Another important property of Hermitian matrices is that the modal matrix M of a Hermitian matrix is unitary. That is:

$$MM^H = M^H M = I \tag{4.91}$$

Therefore, Hermitian matrices with distinct eigenvalues can be decomposed as:

$$A = M \Lambda M^{-1} = M \Lambda M^H \tag{4.92}$$

Example 4.20

Consider the 2×2 Hermitian matrix A given by:

$$A = \begin{bmatrix} 2 & 1-j \\ 1+j & 4 \end{bmatrix}$$

The eigenvalues and corresponding eigenvectors of A are:

$$\lambda_1 = 3 + \sqrt{3}, \ x_1 = \begin{bmatrix} 1 \\ \dfrac{1+\sqrt{3}}{2}(1+j) \end{bmatrix}$$

and

$$\lambda_2 = 3 - \sqrt{3}, \ x_2 = \begin{bmatrix} 1 \\ \dfrac{1-\sqrt{3}}{2}(1+j) \end{bmatrix}$$

Eigenvalues and Eigenvectors

Note that the two eigenvalues are real and the corresponding eigenvectors are orthogonal since:

$$< x_1, x_2 > = x_1^H x_2 = 1 + \frac{1+\sqrt{3}}{2} \times \frac{1-\sqrt{3}}{2}(1-j)(1+j) = 1 - 1 = 0$$

4.5.3 Definite Matrices

Positive and negative definite (semidefinite) matrices are an important class of matrices with applications in different engineering fields. They are particularly useful in optimization type problems. There are four types of definite matrices:

4.5.3.1 Positive Definite Matrices

The $n \times n$ Hermitian matrix A is said to be positive definite if for any nonzero vector $x \in C^n$, the quantity $x^H A x > 0$.

4.5.3.2 Positive Semidefinite Matrices

The $n \times n$ Hermitian matrix A is said to be positive semidefinite if for any nonzero vector $x \in C^n$, the quantity $x^H A x \geq 0$.

4.5.3.3 Negative Definite Matrices

The $n \times n$ Hermitian matrix A is said to be negative definite if for any nonzero vector $x \in C^n$, the quantity $x^H A x < 0$.

4.5.3.4 Negative Semidefinite Matrices

The $n \times n$ Hermitian matrix A is said to be negative semidefinite if for any nonzero vector $x \in C^n$, the quantity $x^H A x \leq 0$.

4.5.3.5 Test for Matrix Positiveness

The following theorem states the necessary and sufficient condition for positive definiteness of a Hermitian matrix.

Theorem 4.8: An $n \times n$ Hermitian matrix is positive definite if and only if all its eigenvalues are positive.

Proof: First assume that A is Hermitian and all of its eigenvalues are positive. Then:

$$x^H A x = x^H M \Lambda M^H x \tag{4.93}$$

Let $y = M^H x$. Then Equation 4.93 can be written as:

$$x^H A x = x^H M \Lambda M^H x = y^H \Lambda y = \sum_{i=1}^{n} \lambda_i |y_i|^2 > 0 \tag{4.94}$$

Next we need to show that if A is Hermitian and positive definite, then all of its eigenvalues are positive. Let x_i be the eigenvector corresponding to the ith eigenvalue λ_i of A. Then:

$$0 < x_i^H A x_i = x_i^H \lambda_i x_i = \lambda_i \|x_i\|^2 \tag{4.95}$$

Therefore, $\lambda_i > 0$. Similar theorems can be stated for positive semidefinite, negative definite and negative semidefinite matrices. These are summarized below:

(a) Positive definite: → All eigenvalues are positive.
(b) Negative definite: → All eigenvalues are negative.
(c) Positive semidefinite: → All eigenvalues are nonnegative (zero or positive).
(d) Negative semidefinite: → All eigenvalues are nonpositive (zero or negative).
(e) Indefinite: → Matrix A has positive and negative eigenvalues.

Example 4.21

State whether the following matrices are positive definite/semidefinite or negative definite/semidefinite.

a. $A = \begin{bmatrix} 3 & 2 \\ 2 & 3 \end{bmatrix}$

b. $B = \begin{bmatrix} 1 & 2 \\ 2 & 4 \end{bmatrix}$

c. $C = \begin{bmatrix} 1 & -2 \\ -2 & -6 \end{bmatrix}$

d. $D = \begin{bmatrix} 1 & 1-j \\ 1+j & 5 \end{bmatrix}$

Solution: The eigenvalues of A are 5 and 1. Both are positive, hence matrix A is a positive definite matrix. The eigenvalues of matrix B are 0 and 5, therefore B is positive semidefinite. The eigenvalues of matrix C are 1.5311 and −6.5311, therefore C is indefinite. Finally, the eigenvalues of matrix D are 5.4495 and 0.5505 which makes it positive definite.

However, since the computation of eigenvalues is generally nontrivial, a test which does not require eigenvalue computation is sometimes preferred. The following theorem provides a computationally easier approach to test matrices for positiveness.

Eigenvalues and Eigenvectors

Theorem 4.9: A necessary and sufficient condition for an $n \times n$ matrix A to be positive definite is that the following n submatrices have a positive determinant.

$$a_{11} > 0$$

$$\begin{vmatrix} a_{11} & a_{12} \\ a_{21} & a_{22} \end{vmatrix} > 0$$

$$\begin{vmatrix} a_{11} & a_{12} & a_{13} \\ a_{21} & a_{22} & a_{23} \\ a_{31} & a_{32} & a_{33} \end{vmatrix} > 0 \quad (4.96)$$

$$\vdots$$

$$|A| > 0$$

Proof: The $n \times n$ matrix A is positive definite if $x^H A x > 0$ for any $x \in R^n$. First, assume that $n=2$ and let $x = [1 \ \ 0]^T$. Thus:

$$x^H A x = a_{11} > 0 \quad (4.97)$$

and since the product of the eigenvalues is equal to the determinant of the matrix, then:

$$\lambda_1 \lambda_2 = |A| = \begin{vmatrix} a_{11} & a_{12} \\ a_{21} & a_{22} \end{vmatrix} > 0 \quad (4.98)$$

Now assume that the matrix is 3×3 and let $x = [x_1 \ \ x_2 \ \ 0]^T$. Then:

$$x^H A x = \begin{bmatrix} x_1 & x_2 \end{bmatrix} \begin{bmatrix} a_{11} & a_{12} \\ a_{21} & a_{22} \end{bmatrix} \begin{bmatrix} x_1 \\ x_2 \end{bmatrix} > 0 \quad (4.99)$$

for any x_1 and x_2. This means that the 2×2 matrix $\begin{bmatrix} a_{11} & a_{12} \\ a_{21} & a_{22} \end{bmatrix}$ must be positive definite. Similarly, since the eigenvalues of A are positive, therefore the determinant of A must be positive. Hence:

$$a_{11} > 0 \quad \begin{vmatrix} a_{11} & a_{12} \\ a_{21} & a_{22} \end{vmatrix} > 0 \quad \begin{vmatrix} a_{11} & a_{12} & a_{13} \\ a_{21} & a_{22} & a_{23} \\ a_{31} & a_{32} & a_{33} \end{vmatrix} > 0 \quad (4.100)$$

Using induction, assume that the conditions are met for the $(n-1) \times (n-1)$ matrix A, and let B be an $n \times n$ matrix defined by:

$$B = \begin{bmatrix} A & \vdots & b \\ \cdots & \cdots & \cdots \\ c & \vdots & b_{nn} \end{bmatrix} \quad (4.101)$$

where b is an $(n-1)\times 1$ column vector, c is a $1 \times (n-1)$ row vector and b_{nm} is scalar. Since matrix B is positive definite, then $y^T By$ must be positive for any $y \in R^n$. Now let $y = [x\ \ 0]^T$, where $x \in R^{n-1}$. Then, since $y^T By = x^T Ax$, A must be positive definite and all the $n-1$ submatrices of A must have positive determinants. Also, since the eigenvalues of matrix B are all positive, the product of these eigenvalues (which is equal to the determinant of B) has to be positive. This completes the proof of the theorem.

Example 4.22

Is the following matrix positive definite?

$$A = \begin{bmatrix} 3 & 1 & -2 \\ 1 & 6 & -4 \\ -2 & -4 & 6 \end{bmatrix}$$

Solution: Using the above theorem, we have:

$$a_{11} = 3 > 0,$$

$$\begin{vmatrix} a_{11} & a_{12} \\ a_{21} & a_{22} \end{vmatrix} = \begin{vmatrix} 3 & 1 \\ 1 & 6 \end{vmatrix} = 17 > 0,$$

and

$$\begin{vmatrix} a_{11} & a_{12} & a_{13} \\ a_{21} & a_{22} & a_{23} \\ a_{31} & a_{32} & a_{33} \end{vmatrix} = \begin{vmatrix} 3 & 1 & -2 \\ 1 & 6 & -4 \\ -2 & -4 & 6 \end{vmatrix} = 46 > 0$$

Therefore, it is positive definite.

4.6 Singular Value Decomposition (SVD)

One of the most important tools in signal processing and numerical linear algebra is the singular value decomposition (SVD) of matrices. SVD was developed for square matrices by Beltrami and Jordan in the 18th century. The theory for general matrices was established by Eckart and Young. We first state the SVD theorem and then explore its use in several engineering applications.

4.6.1 Definition of SVD

Singular value decomposition of a real or complex $m \times n$ matrix is defined through the following theorem.

Theorem 4.10: Let A be an $m \times n$ real or complex matrix with rank r. Then, there exist unitary matrices U ($m \times m$) and V ($n \times n$) such that:

$$A = U\Sigma V^H \qquad (4.102)$$

Eigenvalues and Eigenvectors 189

where Σ is an $m \times n$ matrix with entries:

$$\Sigma_{ij} = \begin{cases} \sigma_i & \text{if } i = j \\ 0 & \text{if } i \neq j \end{cases} \quad (4.103)$$

The quantities $\sigma_1 \geq \sigma_2 \geq \cdots \geq \sigma_r > \sigma_{r+1} = \sigma_{r+2} = \cdots = \sigma_n = 0$ are called the singular values of A.

Proof: Let $S = A^H A$. Matrix S is an $n \times n$ Hermitian and positive semidefinite with rank r. Therefore, it has nonnegative eigenvalues. Let the eigenvalues and eigenvectors of S be $\sigma_1^2 \geq \sigma_2^2 \geq \cdots \geq \sigma_r^2 > \sigma_{r+1}^2 = \sigma_{r+2}^2 = \cdots = \sigma_n^2 = 0$ and $v_1, v_2, \ldots, v_n$, respectively. The eigenvectors form an orthonormal set. Let $V_1 = [v_1, v_2, \ldots, v_r]$, $V_2 = [v_{r+1}, v_{r+2}, \ldots, v_n]$ and $\Lambda = \text{diag}(\sigma_1, \sigma_2, \ldots, \sigma_r)$. Then:

$$A^H A V_1 = V_1 \Lambda^2 \quad (4.104)$$

Premultiplying both sides of Equation 4.104 by V_1^H followed by post and premultiplications by Λ^{-1} result in:

$$\Lambda^{-1} V_1^H A^H A V_1 \Lambda^{-1} = I \quad (4.105)$$

Let $U_1 = A V_1 \Lambda^{-1}$. Then, using Equation 4.105, we have $U_1^H U_1 = I$. Notice that U_1 is a unitary matrix of size $m \times r$. Let U_2 be another unitary matrix of size $m \times (m-r)$ orthogonal to U_1 and define $U = [U_1 \quad U_2]$ and $V = [V_1 \quad V_2]$. Then:

$$U^H A V = [U_1 \quad U_2]^H A [V_1 \quad V_2] = \begin{bmatrix} U_1^H A V_1 & U_1^H A V_2 \\ U_2^H A V_1 & U_2^H A V_2 \end{bmatrix} = \begin{bmatrix} \Lambda & 0 \\ 0 & 0 \end{bmatrix} = \Sigma \quad (4.106)$$

Premultiply and postmultiply both sides of Equation 4.104 by U and V^H. This yields:

$$A = U \Sigma V^H \quad (4.107)$$

The nonnegative numbers $\sigma_1 \geq \sigma_2 \geq \cdots \geq \sigma_r \geq \sigma_{r+1} = \sigma_{r+2} = \cdots = \sigma_n = 0$ are called singular values of A and they are the square roots of the eigenvalues of $A^H A$ or $A A^H$.

Example 4.23

Find the singular value decomposition of matrix A given by:

$$A = \begin{bmatrix} 1 & 2 & 3 \\ -1 & 3 & -2 \end{bmatrix}$$

Solution:

$$S = A^T A = \begin{bmatrix} 2 & -1 & 5 \\ -1 & 13 & 0 \\ 5 & 0 & 13 \end{bmatrix}$$

The eigenvalues of S are $\sigma_1^2 = 15$, $\sigma_2^2 = 13$ and $\sigma_3^2 = 0$. The corresponding eigenvectors are:

$$v_1 = \begin{bmatrix} -0.3651 \\ 0.1826 \\ -0.9129 \end{bmatrix}, v_2 = \begin{bmatrix} 0 \\ -0.9806 \\ 0.1961 \end{bmatrix} \text{ and } v_3 = \begin{bmatrix} -0.9309 \\ -0.0716 \\ 0.3581 \end{bmatrix}$$

Therefore:

$$V_1 = \begin{bmatrix} -0.3651 & 0 \\ 0.1826 & -0.9806 \\ -0.9129 & 0.1961 \end{bmatrix} \text{ and } V_2 = \begin{bmatrix} -0.9309 \\ -0.0716 \\ 0.3581 \end{bmatrix}$$

and

$$U_1 = AV_1\Lambda^{-1} = \begin{bmatrix} 1 & 2 & 3 \\ -1 & 3 & -2 \end{bmatrix} \begin{bmatrix} -0.3651 & 0 \\ 0.1826 & -0.9806 \\ -0.9129 & 0.1961 \end{bmatrix} \begin{bmatrix} \sqrt{15} & 0 \\ 0 & \sqrt{13} \end{bmatrix}^{-1}$$

$$= \begin{bmatrix} -0.707 & -0.707 \\ 0.707 & -0.707 \end{bmatrix}$$

Since $m=2$ and $r=2$, we have $U=U_1$. Therefore:

$$A = U\Sigma V^H = \begin{bmatrix} -0.707 & -0.707 \\ 0.707 & -0.707 \end{bmatrix} \begin{bmatrix} \sqrt{15} & 0 & 0 \\ 0 & \sqrt{13} & 0 \end{bmatrix} \begin{bmatrix} -0.3651 & 0 & -0.9309 \\ 0.1826 & -0.9806 & -0.0716 \\ -0.9129 & 0.1961 & 0.3581 \end{bmatrix}^T$$

Example 4.24

Consider the 256×256 LENA image. If we decompose the image into its SVD components, each SVD component is an image of size 256×256. The ith component of the image is $\sigma_i^2 u_i v_i$ and the image itself can be expressed as:

$$A = U\Sigma V^T = \sum_{i=1}^{256} \sigma_i u_i v_i^T$$

The plot of the 256 singular values of the image normalized with respect to the largest singular value is shown in Figure 4.6. As can be seen, most of the singular

Eigenvalues and Eigenvectors 191

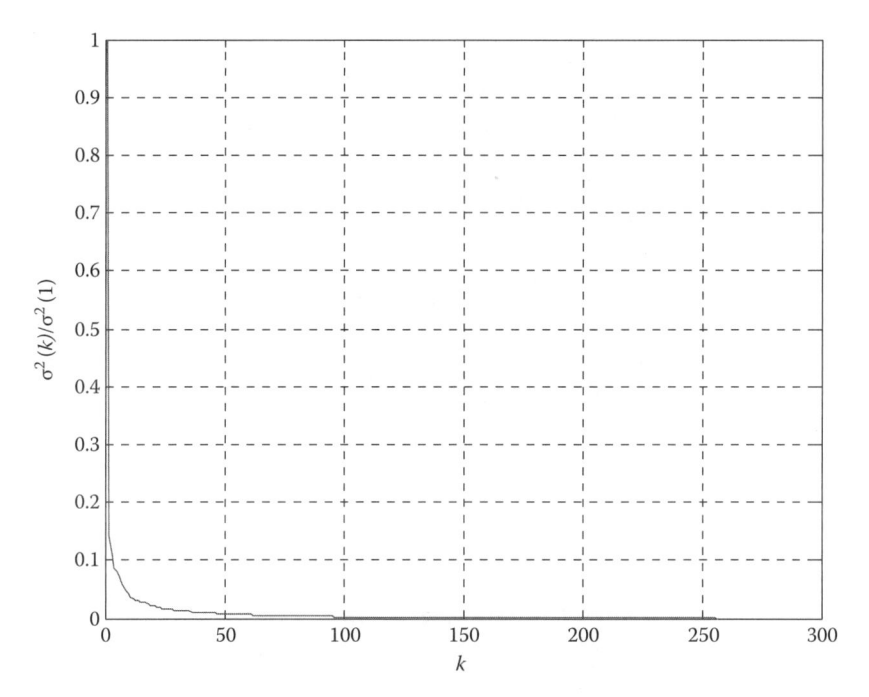

FIGURE 4.6
Singular values normalized with respect to the largest singular value.

FIGURE 4.7
LENA image.

values are small. The original image and the images reconstructed using the first 10 and 50 singular values are shown in Figures 4.7 through 4.9, respectively. The MATLAB® code to compute and approximate an image by its eigenimages is shown in Table 4.1.

FIGURE 4.8
Image reconstructed using the first 10 eigenimages.

FIGURE 4.9
Image reconstructed using the first 50 eigenimages.

4.6.2 Matrix Norm

Matrix norm, like other vector space norms, must satisfy the properties of norms as stated in Section 4 of Chapter 3. Let A be an $m \times n$ matrix mapping vector space R^n to vector space R^m. By definition, the p matrix norm is defined as:

$$\|A\|_p = \sup_{\substack{x \in R^n \\ \|x\|_p = 1}} \|Ax\|_p \qquad (4.108)$$

Eigenvalues and Eigenvectors

TABLE 4.1

MATLAB® Code to Find Eigenimages

```
% f = input image
% fh1 = image reconstructed by using the first M SVD
% MSE = resulting normalized mean square error
f = imread('lena.tif')
[u,s,v] = svd(f);
fh = u*s*v';
sv = diag(s);
fh1 = zeros(size(f));
M = 50;
for i = 1:M
fh1 = fh1 + u(:,i)*s(i,i)*v(:,i)';
end
MSE = 100*sum(sum((f-fh1).^2))/sum(sum(f.^2));
figure(1)
imshow(f/255);
figure(2)
imshow(fh1/255);
```

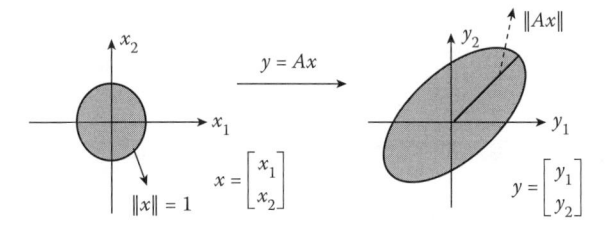

FIGURE 4.10
Concept of matrix norm.

Matrix norm is a measure of the boundness of a matrix. The concept of a matrix norm for a 2×2 matrix is illustrated in Figure 4.10.

The matrix norm depends on the vector norm. For example, if the vector norm utilized is the l_1 norm, then the matrix norm is based on vector l_1 norm. We consider three cases corresponding to $p=1$, $p=2$, and $p=\infty$.

Case I: $p=1$

In this case, the matrix norm becomes:

$$\|A\|_1 = \max_{\|x\|_1=1} \|Ax\|_1 = \max_j \sum_{i=1}^{m} |a_{ij}| \qquad (4.109)$$

Therefore, $\|A\|_1$ is equal to the longest column sum, which means that to find the $p=1$ norm, it is necessary to compute the sum of absolute values of each column and pick the maximum.

Case II: $p=\infty$

In this case, the matrix norm becomes:

$$\|A\|_\infty = \max_{\|x\|_\infty=1} \|Ax\|_\infty = \max_i \sum_{j=1}^m |a_{ij}| \qquad (4.110)$$

Therefore, $\|A\|_\infty$ is equal to the longest row sum, which means that to find the $p=\infty$ norm, you must compute the sum of absolute values of each row and pick the maximum.

Case III: $p=2$

In this case, the matrix norm becomes:

$$\|A\|_2 = \max_{\|x\|_2=1} \|Ax\|_2 \qquad (4.111)$$

To find the $p=2$ matrix norm, we must solve the following optimization problem:

$$\begin{array}{l}\displaystyle\max_x \|Ax\|_2 = x^H A^H A x \\ \text{subject to: } x^H x = 1\end{array} \qquad (4.112)$$

Using the Lagrange multiplier technique, this is equivalent to the following optimization problem:

$$\max_x J = x^H A^H A x - \lambda(x^H x - 1) \qquad (4.113)$$

Setting the gradient of J with respect to x equal to zero, we obtain the equation:

$$\frac{\partial J}{\partial x} = 2A^H A x - 2\lambda x = 0 \rightarrow A^H A x = \lambda x \qquad (4.114)$$

Therefore, the solution for vector x must be an eigenvector of the square matrix $A^H A$ corresponding to eigenvalue λ and the resulting norm is:

$$\|Ax\|_2 = x^H A^H A x = \lambda x^H x = \lambda \qquad (4.115)$$

Since we are maximizing the norm, λ must be chosen to be the maximum eigenvalue of the positive definite matrix $A^H A$. Therefore:

$$\|A\|_2 = \lambda_{\max}(A^H A) \qquad (4.116)$$

The matrix p norm has the characteristic that, for any two matrices A and B, the following properties hold:

$$\|Ax\|_p \leq \|A\|_p \|x\|_p \qquad (4.117)$$

Eigenvalues and Eigenvectors

and

$$\|AB\|_p \leq \|A\|_p \|B\|_p \tag{4.118}$$

4.6.3 Frobenius Norm

The Frobenius norm is another matrix norm that is not a p norm. It is defined as:

$$\|A\|_F = \left(\sum_{i=1}^{m}\sum_{j=1}^{n}|a_{ij}|^2\right)^{\frac{1}{2}} \tag{4.119}$$

The Frobenius norm is also called the Euclidean norm. As a simple example, the Frobenius norm of an $n \times n$ identity matrix is $\|I\|_F = \sqrt{n}$. It can also be expressed as:

$$\|A\|_F = \sqrt{\text{Trace}(A^H A)} \tag{4.120}$$

The MATLAB® command to compute the norm of matrix A is **norm(A,p)**. The command to compute the Frobenius norm is **norm(A,'fro')**.

Example 4.25

Find the $p=1$, $p=2$, $p=\infty$, and Frobenius norm of the following matrix:

$$A = \begin{bmatrix} 2 & 1 \\ -3 & 5 \end{bmatrix}$$

Solution: The $p=1$ norm is:

$$\|A\|_1 = \max_j \sum_{i=1}^{m}|a_{ij}| = \max_j \left[\sum_{i=1}^{m}|a_{i1}| \quad \sum_{i=1}^{m}|a_{i2}|\right] = \max_j [5 \quad 6] = 6$$

The $p=\infty$ norm is:

$$\|A\|_\infty = \max_i \sum_{j=1}^{m}|a_{ij}| = \max_i \left[\sum_{j=1}^{m}|a_{1j}| \quad \sum_{j=1}^{m}|a_{2j}|\right] = \max_i [3 \quad 8] = 8$$

The $p=2$ norm is:

$$\|A\|_2 = \sqrt{\max \lambda(A^H A)} = \sqrt{\max \lambda \left(\begin{bmatrix} 2 & -3 \\ 1 & 5 \end{bmatrix}\begin{bmatrix} 2 & 1 \\ -3 & 5 \end{bmatrix}\right)}$$

$$= \sqrt{\max \lambda \left(\begin{bmatrix} 13 & -13 \\ -13 & 26 \end{bmatrix}\right)} = \sqrt{34.0344} = 5.8339$$

The Frobenius norm is:

$$\|A\|_F = \left(\sum_{i=1}^{2}\sum_{j=1}^{2}|a_{ij}|^2\right)^{\frac{1}{2}} = \sqrt{4+1+9+25} = \sqrt{39} = 6.245$$

4.6.4 Matrix Condition Number

Consider the linear equation $Ax=b$. If A is nonsingular, then the solution is $x_0 = A^{-1}b$. If we perturb A by εB, where εB is small (ε is a small scalar and B is a matrix), then the solution to $(A+\varepsilon B)x=b$ is:

$$x_\varepsilon = (A+\varepsilon B)^{-1}b = [A(I+\varepsilon A^{-1}B)]^{-1}b = (I+\varepsilon A^{-1}B)^{-1}A^{-1}b \qquad (4.121)$$

Using the expansion $(I + \varepsilon A^{-1}B)^{-1} = I - \varepsilon A^{-1}B + O(\varepsilon^2 \|B\|^2)$, we have:

$$x_\varepsilon = A^{-1}b - \varepsilon A^{-1}B(A^{-1}b) + O(\varepsilon^2 \|B\|^2)A^{-1}b = x_0 - \varepsilon A^{-1}Bx_0 + O(\varepsilon^2 \|B\|^2)x_0 \qquad (4.122)$$

The normalized error between the exact solution x_0 and the perturbed solution x_ε is

$$\frac{\|x_0 - x_\varepsilon\|}{\|x_0\|} = \frac{\|[\varepsilon A^{-1}B - O(\varepsilon^2 \|B\|^2)]x_0\|}{\|x_0\|} \le \varepsilon \|A^{-1}\|\|B\| + O(\varepsilon^2 \|B\|^2) \qquad (4.123)$$

Defining the condition number $c(A) = \|A\|\|A^{-1}\|$ and $\rho = \varepsilon \|B\|/\|A\|$, the relative error is bounded by:

$$\frac{\|x_0 - x_\varepsilon\|}{\|x_0\|} \le c(A)\rho + O(\varepsilon^2 \|B\|^2) \qquad (4.124)$$

Equation 4.124 indicates that the relative error due to a small perturbation of matrix A is small if the condition number of matrix A is small. A matrix with a large condition number is said to be ill conditioned and Equation 4.124 suggests that, for an ill conditioned matrix, the relative error in the solution of the linear equation $Ax=b$ using matrix inversion may be large, even for a small error in matrix A. Using L_2 norm, we have:

$$\|A\| = \sqrt{\lambda_{\max}(A^T A)} \qquad (4.125)$$

and

$$\|A^{-1}\| = \sqrt{\lambda_{\max}[(A^T A)^{-1}]} = \frac{1}{\sqrt{\lambda_{\min}(A^T A)}} \qquad (4.126)$$

Eigenvalues and Eigenvectors

Hence, the condition number of matrix A is:

$$c(A) = \|A\| \|A^{-1}\| = \sqrt{\frac{\lambda_{max}(A^T A)}{\lambda_{min}(A^T A)}} = \frac{\sigma_{max}}{\sigma_{min}} \qquad (4.127)$$

where σ_{max} and σ_{min} are the maximum and minimum singular values of matrix A. If the condition number of matrix A is large, it will be difficult to find its inverse. The MATLAB® command to find the condition number of matrix A is **c=cond(A)**.

Example 4.26

a. Find the condition number of the 3×3 matrix A.

$$A = \begin{bmatrix} 23 & -18 & -52 \\ 13 & -10 & -29 \\ 28 & -21 & -64 \end{bmatrix}$$

b. Solve the linear equation $Ax=b$, where $b=[169 \quad 94 \quad 26]^T$, and denote the solution by x_0.

c. Solve the perturbed equation $(A+\varepsilon B)x=b$, where:

$$\varepsilon B = \begin{bmatrix} 0.0003 & -0.0195 & 0.0182 \\ 0.0209 & -0.0310 & -0.0197 \\ -0.0071 & -0.0307 & 0.0042 \end{bmatrix}$$

Solution:

a. Use the MATLAB command **s=svd(A)** to find the singular values of matrix A. They are $\sigma_1 = 99.937$, $\sigma_2 = 0.7483$, and $\sigma_3 = 0.1471$. The matrix condition number is:

$$c(A) = \frac{\sigma_{max}}{\sigma_{min}} = \frac{99,937}{0.1471} = 679.796$$

Matrix A is ill conditioned since its condition number is large.

b.
$$x_0 = A^{-1}b = \begin{bmatrix} 23 & -18 & -52 \\ 13 & -10 & -29 \\ 28 & -21 & -64 \end{bmatrix}^{-1} \begin{bmatrix} 169 \\ 94 \\ 26 \end{bmatrix} = \begin{bmatrix} -1 \\ -2 \\ -3 \end{bmatrix}$$

c.
$$x_\varepsilon = (A+\varepsilon B)^{-1}b = \begin{bmatrix} 23.0003 & -18.0195 & -51.9818 \\ 13.0209 & -10.031 & -29.0197 \\ 27.9929 & -21.0307 & -63.9958 \end{bmatrix}^{-1} \begin{bmatrix} 169 \\ 94 \\ 26 \end{bmatrix} = \begin{bmatrix} -1.5686 \\ -2.2247 \\ -3.174 \end{bmatrix}$$

Since matrix A is ill-conditioned, a small perturbation of A results in a large error. This is illustrated by the fact that the first component of x_0 is significantly different from the first component of x_e.

Example 4.27

Consider a set of linear equations of the form $Ax=b$, where A is a 100×100 ill-conditioned matrix with condition number 10^3, b is 100×1 and x is a 100×1 unknown vector. To solve this set of equations in MATLAB, we can use the command $x=A^{-1}b$ or the command $x=A\backslash b$. The first approach is based on matrix inversion, which is not stable if matrix A is ill-conditioned. However, the second approach utilizes the Gauss elimination technique and is more robust, especially when matrix A is ill conditioned. The MATLAB script to create matrix A and simulate the solution is given in Table 4.2.

If we execute the above MATLAB script, we will have:

$$E_1 = 0.0047 \text{ and } E_2 = 1.4904 \times 10^{-14} \qquad (4.128)$$

Therefore, the second approach $x=A\backslash b$ is more robust and produces mean square error (MSE) equal to 1.4904×10^{-14}. The MSE between the true solution and the first and second solutions are $MSE_1=1.092$ and $MSE_2=0.8417$, respectively.

TABLE 4.2

MATLAB Code for Example 4.27

```
% Create data
% Create ill conditioned matrix A
M=rand(100,100);
lamda=diag([1:1:100]);
lamda(1) =10^12;
A=M*lamda*inv(M);
A=A/norm(A);
% True solution x
x=[1:1:100]';
% Create vector b
b=A*x;
% Solve Ax=b using matrix inversion
x1=inv(A)*b;
% Compute MSE between Ax1 and b
E1=norm(A*x1-b);
% Solve Ax=b using the second technique
x2=A\b;
% Compute MSE between Ax2 and b
E2=norm(A*x2-b);
% Compute the MSE between the true solution and x1
MSE1=norm(x-x1);
% Compute the MSE between the true solution and x2
MSE2=norm(x-x2);
```

4.7 Numerical Computation of Eigenvalues and Eigenvectors

Let A be an $n \times n$ matrix. The eigenvalues of A are the n roots of the characteristic polynomial $P(\lambda)$. Since the characteristic polynomial is of degree n, the eigenvalues must, in general, be found numerically when the value of n exceeds 4 (i.e. $n \geq 4$). A possible technique is to apply a numerical method such as Newton's technique to find the roots of the characteristic polynomial $P(\lambda)$. This is not a suitable approach, since a small change in the coefficients of $P(\lambda)$ can lead to a large change in the roots of the characteristic polynomial. As an example, consider the polynomial of degree 12 given by:

$$P(\lambda) = \lambda^{12} - a \tag{4.129}$$

If $a=0$, then this polynomial has 12 roots that are all equal to zero. Now, assuming that $a=10^{-12}$, the 12 roots of $P(\lambda)$ are:

$$\lambda_k = 0.1\, e^{j\frac{\pi}{6}k} \quad k = 0, 1, 2, ..., 11 \tag{4.130}$$

Therefore, the roots are very sensitive to variations in the coefficients of the polynomial. Generally, it is very difficult to compute the eigenvalues and eigenvectors of matrices with linearly dependent or almost linearly dependent eigenvectors. Hence the power method described below assumes that the matrix has n independent eigenvectors.

4.7.1 Power Method

Assume that the $n \times n$ matrix A has n linearly independent eigenvectors $x_1, x_2, ..., x_n$. These eigenvectors form a basis for R^n. Now let u_0 be an arbitrary vector in R^n. Then:

$$u_0 = \sum_{i=1}^{n} \alpha_i x_i \tag{4.131}$$

Define:

$$u_k = A^k u_0 = \sum_{i=1}^{n} \alpha_i A^k x_i = \sum_{i=1}^{n} \alpha_i A^{k-1} \lambda_i x_i = \cdots = \sum_{i=1}^{n} \alpha_i \lambda_i^k x_i \tag{4.132}$$

Assume that $|\lambda_1| > |\lambda_2| \geq \cdots \geq |\lambda_n|$. Then:

$$\frac{u_k}{\lambda_1^k} = \alpha_1 x_1 + \sum_{i=2}^{n} \alpha_i \left(\frac{\lambda_i}{\lambda_1}\right)^k x_i \tag{4.133}$$

Now, take the limit as $k \to \infty$ for both sides of the above equation. This yields:

$$\lim_{k \to \infty} \frac{u_k}{\lambda_1^k} = \alpha_1 x_1 \qquad (4.134)$$

Hence, u_k/λ_1^k converges to the eigenvector corresponding to the eigenvalue with the largest magnitude λ_1. If $\lambda_1 \neq 1$, the limit of the quantity λ_1^k as $k \to \infty$ approaches zero or infinity dependent on the value of λ_1. Thus, the components of the vector u_k, given that fact that each of them is individually divided by λ_1^k, will either grow without bound or decrease to zero as $k \to \infty$. To avoid this situation as k is incremented for each iteration, we divide all the components of u_k by a scale factor g such that one component of u_k is equal to one. The scale factor $g = u_k(I)$, where I represents the index corresponding to the maximum absolute value of the components of u_k. A summary of the power method algorithm is shown below:

Step 1: Pick any starting vector u_0 and a stopping criterion such as the maximum number of iterations (K_{max}).

Step 2: For $k = 1, 2, \ldots, K_{max}$, compute:

$$v = A u_k$$

$$[\max, \ I] = \max(v)$$

$$u_k = \frac{v}{v(I)}$$

Step 3: $x_1 = u_K$ and $\lambda_1 = \dfrac{x_1^H A x_1}{x_1^H x_1}$

Once the dominant eigenvalue and the corresponding eigenvector are found, they are removed from matrix A. This yields an $(n-1) \times (n-1)$ matrix A_1 with eigenvalues and eigenvectors equal to the remainder of the eigenvalues and eigenvectors of matrix A. Let:

$$H = I - 2\frac{ww^H}{w^H w}, \text{ where } w = x_1 + \|x_1\| e_1 \text{ and } e_1 = \begin{bmatrix} 1 & 0 & \cdots & 0 \end{bmatrix}^T \qquad (4.135)$$

Let us form the matrix B given by:

$$B = HAH = \begin{bmatrix} \lambda_1 & a^H \\ 0 & A_1 \end{bmatrix} \qquad (4.136)$$

Then, the matrix A_1 can be obtained by deleting the first row and first column of matrix B.

Eigenvalues and Eigenvectors 201

We now apply the power method to A_1 and compute x_2 and λ_2. This process is repeated until all eigenvalues and eigenvectors of matrix A have been computed. The steps of the algorithm are clearly illustrated in following examples.

Example 4.28

Use the power method to compute the eigenvalues and eigenvectors of the following matrix:

$$A = \begin{bmatrix} 5 & -6 \\ 4 & -6 \end{bmatrix}$$

Solution: Let $u_0 = \begin{bmatrix} 1 \\ 1 \end{bmatrix}$, then:

$$v = Au_0 = \begin{bmatrix} 5 & -6 \\ 4 & -6 \end{bmatrix}\begin{bmatrix} 1 \\ 1 \end{bmatrix} = \begin{bmatrix} -1 \\ -2 \end{bmatrix}, \quad u_1 = \frac{v}{-2} = \begin{bmatrix} 0.5 \\ 1 \end{bmatrix},$$

Note that the components of u_1 were divided by -2 so that the largest component is now equal to one.

$$\lambda_1(1) = \frac{u_1^H A u_1}{u_1^H u_1} = -4.6$$

$$v = Au_1 = \begin{bmatrix} 5 & -6 \\ 4 & -6 \end{bmatrix}\begin{bmatrix} 0.5 \\ 1 \end{bmatrix} = \begin{bmatrix} -3.5 \\ -4 \end{bmatrix}, \quad u_2 = \frac{v}{-4} = \begin{bmatrix} 0.875 \\ 1 \end{bmatrix}, \quad \lambda_1(2) = \frac{u_2^H A u_2}{u_2^H u_2} = -2.2212$$

$$v = Au_2 = \begin{bmatrix} 5 & -6 \\ 4 & -6 \end{bmatrix}\begin{bmatrix} 0.875 \\ 1 \end{bmatrix} = \begin{bmatrix} -1.625 \\ -2.5 \end{bmatrix},$$

$$u_3 = \frac{v}{-2.5} = \begin{bmatrix} 0.65 \\ 1 \end{bmatrix}, \quad \lambda_1(3) = \frac{u_3^H A u_3}{u_3^H u_3} = -3.6467$$

$$v = Au_3 = \begin{bmatrix} 5 & -6 \\ 4 & -6 \end{bmatrix}\begin{bmatrix} 0.65 \\ 1 \end{bmatrix} = \begin{bmatrix} -2.75 \\ -3.4 \end{bmatrix},$$

$$u_4 = \frac{v}{-3.4} = \begin{bmatrix} 0.808 \\ 1 \end{bmatrix}, \quad \lambda_1(4) = \frac{u_4^H A u_4}{u_4^H u_4} = -2.6277$$

This process is continued until convergence.

A plot of $\lambda_1(k)$ versus iteration index k is shown in Figure 4.11.

As can be seen from Figure 4.11, $\lambda_1 = \lim_{k \to \infty} \lambda_1(k) = -3$. The eigenvector corresponding to λ_1 is $x_1 = [0.75 \ \ 1]$. To find the second eigenvalue and eigenvector, we form matrix A_1 as follows:

$$w = x_1 + \|x_1\|e_1 = \begin{bmatrix} 0.75 \\ 1 \end{bmatrix} + 1.25\begin{bmatrix} 1 \\ 0 \end{bmatrix} = \begin{bmatrix} 2 \\ 1 \end{bmatrix}$$

$$H = I - 2\frac{ww^H}{w^H w} = \begin{bmatrix} 1 & 0 \\ 0 & 1 \end{bmatrix} - \frac{2}{2^2 + 1^2}\begin{bmatrix} 2 \\ 1 \end{bmatrix}\begin{bmatrix} 2 & 1 \end{bmatrix} = \begin{bmatrix} -0.6 & -0.8 \\ -0.8 & 0.6 \end{bmatrix}$$

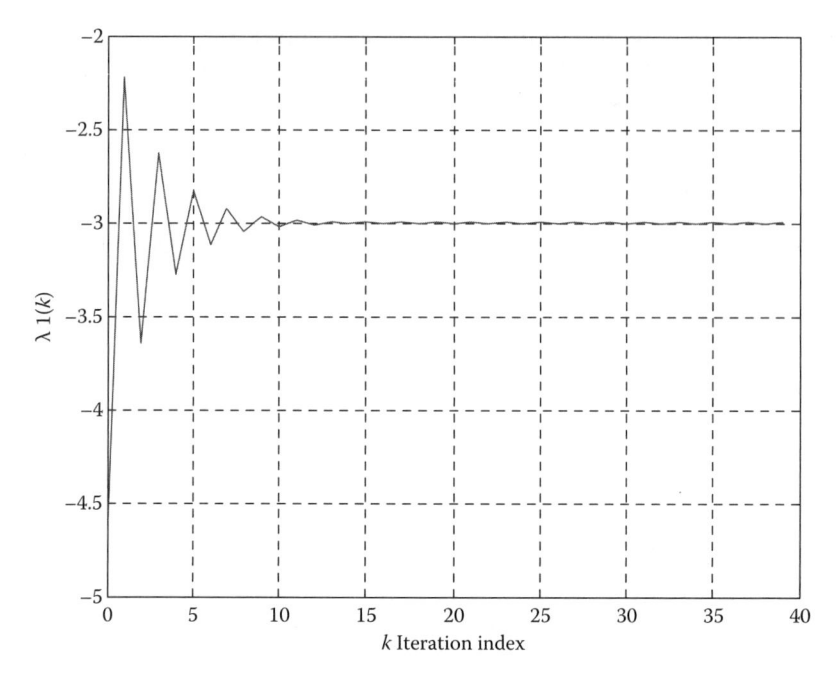

FIGURE 4.11
Convergence of the dominant eigenvalue.

$$HAH = \begin{bmatrix} -3 & 10 \\ 0 & 2 \end{bmatrix} = \begin{bmatrix} \lambda_1 & a^H \\ 0 & A_1 \end{bmatrix}$$

Therefore:

$$A_1 = 2$$

And proceed to apply the above algorithm yielding:

$$\lambda_2 = 2$$

MATLAB® code to implement the power method

The MATLAB function file shown in Table 4.3 can be used to numerically compute the eigenvalues using the power method algorithm.

Example 4.29

Using the power method, compute the dominant eigenvalue and eigenvector of the following matrix:

$$A = \begin{bmatrix} -2.4 & -13.8 & 11 \\ -1.8 & -4.10 & 4.5 \\ -4.6 & -14.7 & 13.5 \end{bmatrix}$$

Eigenvalues and Eigenvectors

TABLE 4.3

MATLAB Code to Compute the Eigenvalues Using the Power Method

```
function [lam] = eigenval(A,K)
[n,m] = size(A);
for i=1:n
  u = rand(n-i+1,1);
  for k=1:K
    v = A*u;
    [vmax,I] = max(abs(v));
    u = v/v(I);
    a(k) = (u'*A*u)/(u'*u);
  end
  lam(i) = a(K);
  x1 = u;
  e1 = zeros(n-i+1,1);
  e1(1) = 1;
  w = u+norm(u)*e1;

  H = eye(n-i+1) - (2*w*w')/((w'*w));
  B = H*A*H;
  A = B(2:end,2:end);
end
```

Solution: The first iteration with initial condition $u_0 = \begin{bmatrix} 1 \\ 1 \\ 0 \end{bmatrix}$ is:

$$k = 1$$

$$v = Au_0 = \begin{bmatrix} -2.4 & -13.8 & 11 \\ -1.8 & -4.10 & 4.5 \\ -4.6 & -14.7 & 13.5 \end{bmatrix} \begin{bmatrix} 1 \\ 1 \\ 0 \end{bmatrix} = \begin{bmatrix} -16.2 \\ -5.9 \\ -19.3 \end{bmatrix}, \quad u_1 = \frac{v}{-19.3} = \begin{bmatrix} 0.8394 \\ 0.3057 \\ 1 \end{bmatrix},$$

$$\lambda_1(1) = \frac{u_1^H A u_1}{u_1^H u_1} = 5.382$$

$$k = 2$$

$$v = Au_1 = \begin{bmatrix} -2.4 & -13.8 & 11 \\ -1.8 & -4.10 & 4.5 \\ -4.6 & -14.7 & 13.5 \end{bmatrix} \begin{bmatrix} 0.8394 \\ 0.3057 \\ 1 \end{bmatrix} = \begin{bmatrix} 4.76682 \\ 1.7358 \\ 5.1451 \end{bmatrix}, \quad u_2 = \frac{v}{5.1451} = \begin{bmatrix} 0.9265 \\ 0.3374 \\ 1 \end{bmatrix},$$

$$\lambda_1(2) = \frac{u_1^H A u_2}{u_2^H u_2} = 4.3534$$

$$k = 3$$

$$v = Au_2 = \begin{bmatrix} -2.4 & -13.8 & 11 \\ -1.8 & -4.10 & 4.5 \\ -4.6 & -14.7 & 13.5 \end{bmatrix} \begin{bmatrix} 0.9265 \\ 0.3374 \\ 1 \end{bmatrix} = \begin{bmatrix} 4.1202 \\ 1.4491 \\ 4.279 \end{bmatrix}, \quad u_3 = \frac{v}{4.279} = \begin{bmatrix} 0.9631 \\ 0.3387 \\ 1 \end{bmatrix},$$

$$\lambda_1(3) = \frac{u_3^H A u_3}{u_3^H u_3} = 4.1255$$

$$k = 4$$

$$v = Au_3 = \begin{bmatrix} -2.4 & -13.8 & 11 \\ -1.8 & -4.10 & 4.5 \\ -4.6 & -14.7 & 13.5 \end{bmatrix} \begin{bmatrix} 0.9631 \\ 0.3387 \\ 1 \end{bmatrix} = \begin{bmatrix} 4.0151 \\ 1.378 \\ 4.0916 \end{bmatrix}, \quad u_4 = \frac{v}{4.0916} = \begin{bmatrix} 0.9813 \\ 0.3368 \\ 1 \end{bmatrix},$$

$$\lambda_1(4) = \frac{u_4^H A u_4}{u_4^H u_4} = 4.0519$$

A plot of $\lambda_1(k)$ versus iteration index k is shown in Figure 4.12.

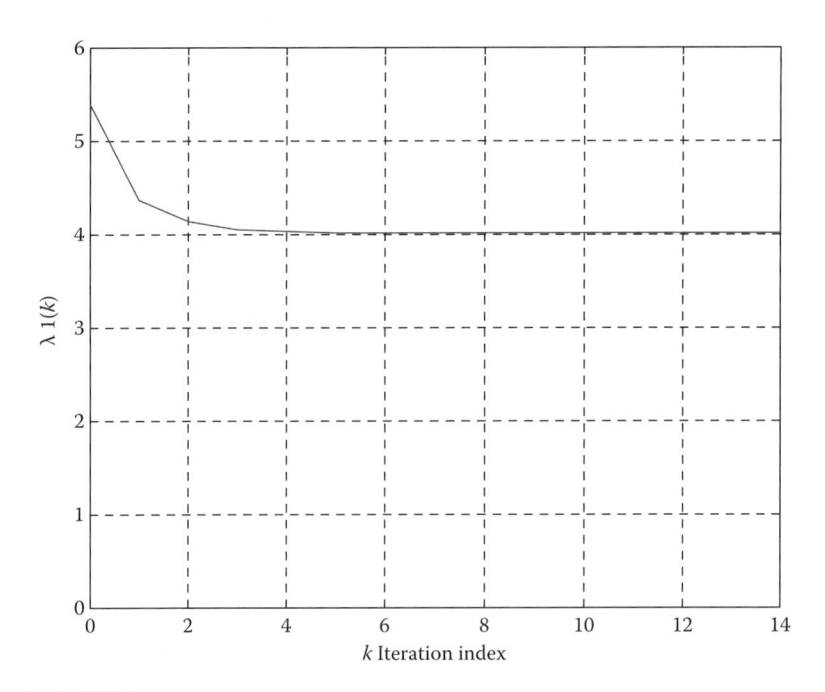

FIGURE 4.12
Plot of $\lambda_1(k)$ versus iteration index k.

Eigenvalues and Eigenvectors

As can be seen, the first eigenvalue is $\lambda_1=4$. The eigenvector corresponding to λ_1 is $x_1 = [1 \quad 0.3333 \quad 1]^T$. To find the second eigenvalue and eigenvector, we form matrix A_1 as follows:

$$w = x_1 + \|x_1\|e_1 = \begin{bmatrix} 1 \\ \frac{1}{3} \\ 1 \end{bmatrix} + 1.453 \begin{bmatrix} 1 \\ 0 \\ 0 \end{bmatrix} = \begin{bmatrix} 2.453 \\ \frac{1}{3} \\ 1 \end{bmatrix}$$

$$H = I - 2\frac{ww^H}{w^H w} = \begin{bmatrix} 1 & 0 & 0 \\ 0 & 1 & 0 \\ 0 & 0 & 1 \end{bmatrix} - \frac{2}{2.453^2 + \left(\frac{1}{3}\right)^2 + 1^2} \begin{bmatrix} 2.453 \\ \frac{1}{3} \\ 1 \end{bmatrix} \begin{bmatrix} 2.453 & \frac{1}{3} & 1 \end{bmatrix}$$

$$= \begin{bmatrix} -0.6882 & -0.2294 & -0.6883 \\ -0.2294 & -0.9688 & -0.0935 \\ -0.6883 & -0.0935 & 0.7194 \end{bmatrix}$$

$$B = HAH = \begin{bmatrix} 4 & 20.3884 & -18.3961 \\ 0 & 0.6723 & 0.8846 \\ 0 & -0.4918 & 2.3276 \end{bmatrix} = \begin{bmatrix} \lambda_1 & a^H \\ 0 & A_1 \end{bmatrix}$$

Therefore:

$$A_1 = \begin{bmatrix} 0.6723 & 0.8846 \\ -0.4918 & 2.3276 \end{bmatrix}$$

Applying the power methods to A_1, we can compute the other two eigenvalues as:

$$\lambda_2 = 2 \text{ and } \lambda_3 = 1.$$

4.8 Properties of Eigenvalues and Eigenvectors of Different Classes of Matrices

The properties of the eigenvalues and eigenvectors of special matrices are summarized in Table 4.4. Some of these have already been covered in the previous sections; the others are left as an exercise (see Problem 4.9).

TABLE 4.4
Structure of Eigenvalues and Eigenvectors of Certain Class of Matrices

Matrix Class	Definition	Eigenvalues	Eigenvectors
Hermitian	$A = A^H$	Real	Orthogonal
Orthogonal	$A^{-1} = A^H$	All $\|\lambda_i\| = 1$	Orthogonal
Skew-symmetric	$A^T = -A$	Imaginary	Orthogonal
Positive definite	$x^H A x > 0$	Positive	Orthogonal
Markov	$a_{ij} > 0$ $\sum_{i=1}^{n} a_{ij} = 1$	$\lambda_{max} = 1$	$x > 0$ Components of eigenvectors are positive
Similar	$B = T^{-1} A T$	$\lambda(B) = \lambda(A)$	$x_B = T^{-1} x_A$
Projection	$A = A^2 = A^T$	$\lambda = 1, 0$	Column space; null space
Reflection	$I - 2uu^T$	$\lambda = -1; 1, 1, \ldots, 1$	$u; u^\perp$
Inverse	A^{-1}	$1/\lambda(A)$	Eigenvectors of A
Shift	$A + \alpha I$	$\lambda(A) + \alpha$	Eigenvectors of A
Scale	αA	$\alpha \lambda$	Eigenvectors of A
Stable power	$A^n \underset{n \to \infty}{\to} 0$	$\|\lambda_i\| < 1$	No structure
Stable exponential	$e^{At} \underset{t \to \infty}{\to} 0$	Real $\lambda_i < 0$	No structure
Circulant	i^{th} row is circular shift of the $(i-1)^{th}$ row	$\lambda_k = \exp\left(j \frac{2\pi}{n} k\right)$ $k = 1, 2, \ldots, n$	$x_k = \begin{bmatrix} 1 & \lambda_k & \cdots & \lambda_k^{n-1} \end{bmatrix}$ $k = 1, 2, \ldots, n$
Tridiagonal	$-1, 2, -1$ on Diagonals	$\lambda_k = 2 - 2\cos\left(\frac{\pi k}{n+1}\right)$	$x_k = \begin{bmatrix} \sin\frac{k\pi}{n+1} & \cdots & \sin\frac{nk\pi}{n+1} \end{bmatrix}$

4.9 Applications

Eigenvalues and eigenvectors have applications in many engineering disciplines. Control system theory, signal/image processing, vibration, advanced circuit theory, and quantum mechanics represent a small sample of the application areas that have benefited from the computation of eigenvalues and eigenvectors to solve real world scenarios. In this section, we cover a few applications in the areas of signal/image processing, control and vibration.

4.9.1 Image Edge Detection

Edge detection often represents the primary step in image analysis and understanding applications. In this section, we first discuss edge detection

Eigenvalues and Eigenvectors

for gray scale images and then extend the results to RGB color images. A common approach for edge detection is through the use of the derivative operator. Consider the synthetic images with perfect edges as shown in Figure 4.13. The horizontal profiles of these two images are denoted by $f(x)$ and $g(x)$. The first and second derivatives of these two profiles are shown in Figure 4.14. The first derivative of $f(x)$ contains an impulse at the edge point and the second derivative shows a double impulse at the same edge point. In Figure 4.15, we have an image where the transition from black to white has a smooth edge. The first and second derivatives of the horizontal profile $f(x)$ of this image are shown in Figure 4.16. As can be seen, the first derivative reaches its maximum at the edge point while the second derivative is zero. From these simple examples, it is obvious that the first and second derivatives can be used to locate edge points in images. In the following paragraphs, we

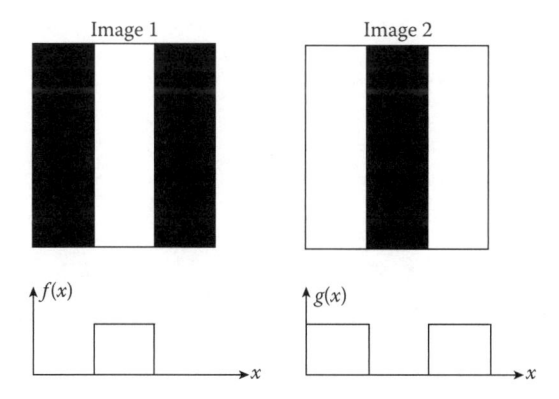

FIGURE 4.13
Two images with sharp edges.

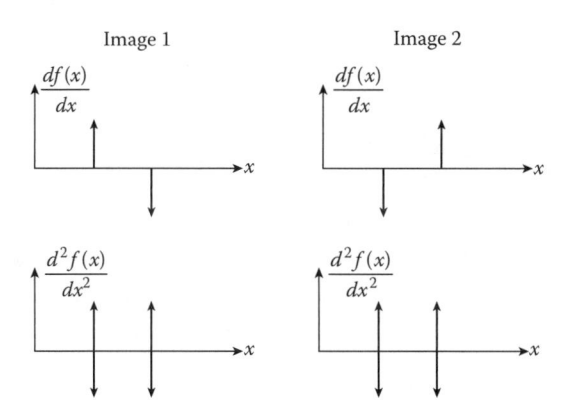

FIGURE 4.14
First and second derivatives of $f(x)$ and $g(x)$.

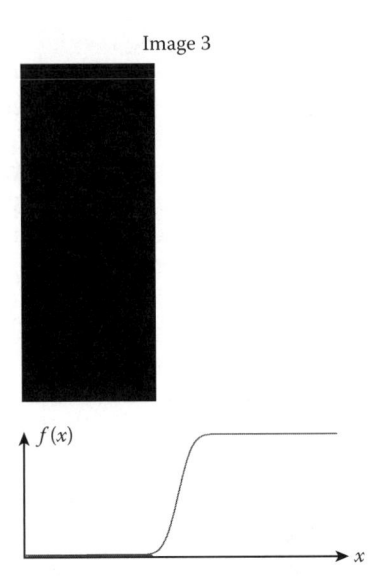

FIGURE 4.15
Image with a smooth edge.

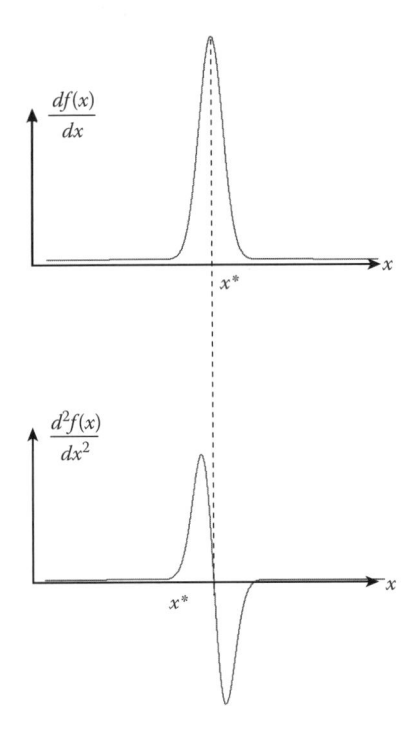

FIGURE 4.16
First and second derivatives of $f(x)$.

Eigenvalues and Eigenvectors

outline the use of gradient based edge detection technique for gray scale as well as RGB type color images.

4.9.1.1 Gradient Based Edge Detection of Gray Scale Images

Using this approach, the gradient (first derivative) of the image is computed and its magnitude is compared with a user selected threshold. If the magnitude of the gradient is greater than a certain threshold, it is considered an edge point otherwise it is classified as a nonedge point.

The gradient of a function $f(x, y)$ is a vector defined by:

$$\nabla f(x, y) = \begin{bmatrix} \dfrac{\partial f(x, y)}{\partial x} \\ \dfrac{\partial f(x, y)}{\partial y} \end{bmatrix} \tag{4.137}$$

and its magnitude is:

$$|\nabla f(x, y)| = \sqrt{\left(\dfrac{\partial f(x, y)}{\partial x}\right)^2 + \left(\dfrac{\partial f(x, y)}{\partial y}\right)^2} \tag{4.138}$$

In practice, horizontal and vertical derivatives are estimated numerically using finite difference approximation. They are implemented using 3×3 Finite Impulse Response (FIR) filters as follows:

$$\dfrac{\partial f(x, y)}{\partial x} \approx f(x, y) * h_1(x, y)$$

$$\dfrac{\partial f(x, y)}{\partial y} \approx f(x, y) * h_2(x, y) \tag{4.139}$$

where $h_1(x, y)$ and $h_2(x, y)$ are defined as:

$$h_1(x, y) = \begin{bmatrix} 1 & 2 & 1 \\ 0 & 0 & 0 \\ -1 & -2 & -1 \end{bmatrix} \text{ and } h_2(x, y) = \begin{bmatrix} 1 & 0 & -1 \\ 2 & 0 & -2 \\ 1 & 0 & -1 \end{bmatrix} \tag{4.140}$$

A pixel at location (x, y) is an edge point if:

$$|\nabla f(x, y)| \geq T \tag{4.141}$$

where T is a predefined threshold. The MATLAB® implementation of this approach is shown in Table 4.5.

TABLE 4.5

MATLAB Code to Create an Edge Map for a Gray Scale Image

```
% image = Input image
% T = Threshold
function [em] = edgemap(image,T)
[n,m] = size(image);
h = [-1 0 1;-2 0 2; -1 0 1];
Gx = filter2(h,image);
Gy = filter2(h',image);
G = sqrt(Gx.^2+Gy.^2);
EM = ones(n,m);
[I] = find(G>=T);
em(I) = 0
```

4.9.1.2 Gradient Based Edge Detection of RGB Images

The above approach can be extended to color images. Let $R(x, y)$, $G(x, y)$, and $B(x, y)$ denote the three channels of an RGB color image. The 3×2 gradient matrix $D(x, y)$ at spatial location x, y is defined as:

$$D(x, y) = \begin{bmatrix} \dfrac{\partial R(x, y)}{\partial x} & \dfrac{\partial R(x, y)}{\partial y} \\ \dfrac{\partial G(x, y)}{\partial x} & \dfrac{\partial G(x, y)}{\partial y} \\ \dfrac{\partial B(x, y)}{\partial x} & \dfrac{\partial B(x, y)}{\partial y} \end{bmatrix} \qquad (4.142)$$

The image formed by computing the norm of the gradient matrix at spatial location (x, y) is the edge map of the image. The norm of the gradient matrix $D(x, y)$ is the largest singular value of $D(x, y)$ or the square root of the largest eigenvalue of the 2×2 positive definite matrix $K(x, y)$ defined by:

$$K(x, y) = D^T(x, y)D(x, y) = \begin{bmatrix} k_{11}(x, y) & k_{12}(x, y) \\ k_{21}(x, y) & k_{22}(x, y) \end{bmatrix} \qquad (4.143)$$

where:

$$k_{11}(x, y) = \left(\frac{\partial R(x, y)}{\partial x}\right)^2 + \left(\frac{\partial G(x, y)}{\partial x}\right)^2 + \left(\frac{\partial B(x, y)}{\partial x}\right)^2, \qquad (4.144)$$

Eigenvalues and Eigenvectors

$$k_{12}(x, y) = k_{21}(x, y) = \frac{\partial R(x, y)}{\partial x}\frac{\partial R(x, y)}{\partial y} + \frac{\partial G(x, y)}{\partial x}\frac{\partial G(x, y)}{\partial y}$$

$$+ \frac{\partial B(x, y)}{\partial x}\frac{\partial B(x, y)}{\partial y}, \quad (4.145)$$

and

$$k_{22}(x, y) = \left(\frac{\partial R(x, y)}{\partial y}\right)^2 + \left(\frac{\partial G(x, y)}{\partial y}\right)^2 + \left(\frac{\partial B(x, y)}{\partial y}\right)^2 \quad (4.146)$$

The largest eigenvalue of $K(x, y)$ is given by:

$$\lambda_{max} = 0.5\left(k_{11}(x, y) + k_{22}(x, y) + \sqrt{(k_{11}(x, y) - k_{22}(x, y))^2 + 4k_{12}^2(x, y)}\right) \quad (4.147)$$

Therefore:

$$G(x, y) = \left[0.5\left(k_{11}(x, y) + k_{22}(x, y) + \sqrt{(k_{11}(x, y) - k_{22}(x, y))^2 + 4k_{12}^2(x, y)}\right)\right]^{\frac{1}{2}}$$

$$(4.148)$$

The MATLAB® code implementation of this algorithm is shown in Table 4.6.

TABLE 4.6

MATLAB Code to Create an Edge Map for an RGB Image

```
% image = input RGB image (nxmx3)
% edge = output edge image
%
function [edge] = edgecolr(image)
image = double(image);
h = [-1 0 1; -2 0 2; -1 0 1];
GRx = filter2(h, image(:,:,1));
GRy = filter2(h', image(:,:,1));
GGx = filter2(h, image(:,:,2));
GGy = filter2(h', image(:,:,2));
GBx = filter2(h, image(:,:,3));
GBy = filter2(h', image(:,:,3));
A = GRx.^2 + GGx.^2 + GBx.^2;
B = GRx.*GRy + GGx.*GGy + GBx.*GBy;
C = GRy.^2 + GGy.^2 + GBy.^2;
edge = 0.5*(A+C) + 0.5*sqrt((A-C).^2 + 4*B.^2);
edge = 255*sqrt(edge/max(max(edge)));
```

Example 4.30

Consider the computer generated image with sharp edges shown in Figure 4.17. The gradient image using the eigenvalue technique is shown in Figure 4.18. Since, the edges in this image are "perfect", i.e. noise free, the eigen based detector provides an excellent representation of the edges found in the image.

FIGURE 4.17
(See color insert following page 174.) Computer generated image.

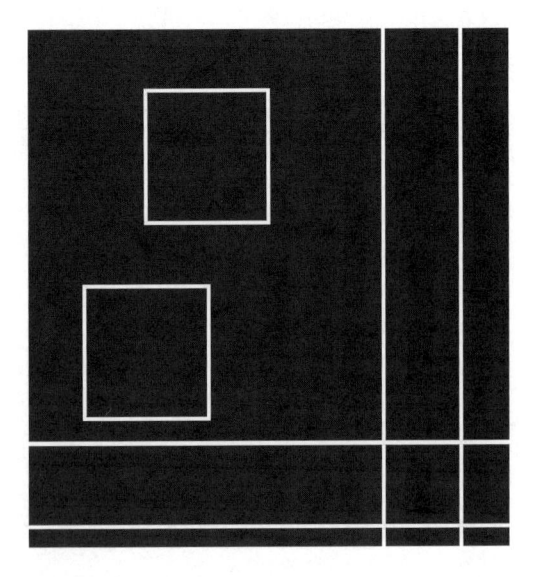

FIGURE 4.18
Gradient of computer generated image.

Example 4.31

Consider the RGB LENA image shown in Figure 4.19. The gradient image using the eigenvalue based technique is shown in Figure 4.20. In this case, the gradient

FIGURE 4.19
(See color insert following page 174.) LENA image.

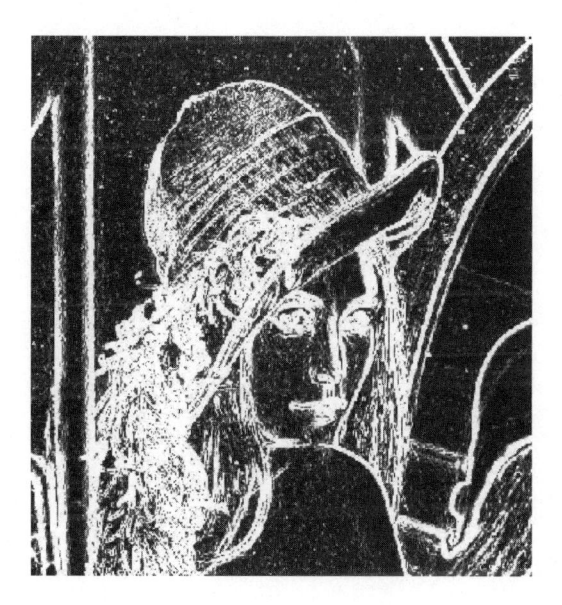

FIGURE 4.20
Gradient of LENA image.

4.9.2 Vibration Analysis

Eigenvalue/eigenvector techniques have also been utilized extensively to analyze vibration systems. In this section, we examine their use in the analysis of a frictionless vibration system. Consider the mass-spring system shown in Figure 4.21. The system has two degrees of freedom: $x_1(t)$ and $x_2(t)$. The equations of motion of this system are given by:

$$m_1 \frac{d^2 x_1(t)}{dt^2} + (k_1 + k_2)x_1(t) - k_2 x_2(t) = 0 \qquad (4.149)$$

$$m_2 \frac{d^2 x_2(t)}{dt^2} - k_2 x_1(t) + (k_2 + k_3)x_2(t) = 0 \qquad (4.150)$$

where m_1 and m_2 are the masses, k_1, k_2 and k_3 are constants that represent the spring stiffness. Let us define:

$$X(t) = \begin{bmatrix} x_1(t) \\ x_2(t) \end{bmatrix}, \; M = \begin{bmatrix} m_1 & 0 \\ 0 & m_2 \end{bmatrix}, \text{ and } K = \begin{bmatrix} k_1 + k_2 & -k_2 \\ -k_2 & k_2 + k_3 \end{bmatrix}$$

Thus, Equations 4.149 and 4.150 can be written in matrix form as:

$$M \frac{d^2 X(t)}{dt^2} + KX(t) = 0 \qquad (4.151)$$

Let us assume that a solution to Equation 4.151 has the form:

$$X(t) = X e^{j\omega t} \qquad (4.152)$$

where X is a 2×1 column vector. Substituting Equation 4.152 into Equation 4.151 results in:

$$-\omega^2 M X e^{j\omega t} + K X e^{j\omega t} = 0 \qquad (4.153)$$

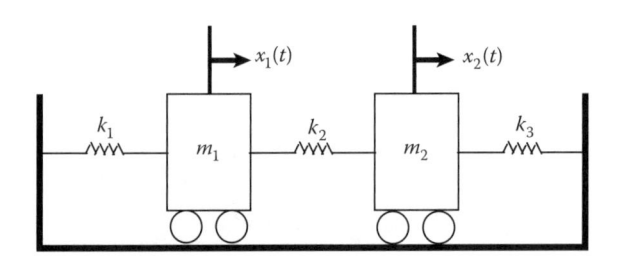

FIGURE 4.21
A second order vibration system.

Eigenvalues and Eigenvectors 215

or

$$(-\omega^2 M + K)X = 0 \tag{4.154}$$

Multiplying both sides of Equation 4.154 by M^{-1}, we have:

$$M^{-1}KX = \omega^2 X \tag{4.155}$$

Equation 4.155 is an eigenvalue/eigenvector problem. That is, X and ω^2 are the eigenvector and eigenvalue of the 2×2 matrix $A = M^{-1}K$, respectively. As an example, assume that $m_1=1$, $m_2=2$, $k_1=k_2=0.3$, $k_3=0.5$, $x_1(0)=0$, $x_2(0)=1$, and $dx_1(t)/dt\,|_{t=0} = dx_2(t)/dt\,|_{t=0} = 0$. Then:

$$B = M^{-1}K = \begin{bmatrix} m_1 & 0 \\ 0 & m_2 \end{bmatrix}^{-1} \begin{bmatrix} k_1+k_2 & -k_2 \\ -k_2 & k_2+k_3 \end{bmatrix} \tag{4.156}$$

$$= \begin{bmatrix} 1 & 0 \\ 0 & 2 \end{bmatrix}^{-1} \begin{bmatrix} 0.6 & -0.3 \\ -0.3 & 0.8 \end{bmatrix} = \begin{bmatrix} 0.6 & -0.3 \\ -0.15 & 0.4 \end{bmatrix}$$

The eigenvalues and eigenvectors of matrix B are:

$$\lambda_1 = \omega_1^2 = 0.7345,\ X_1 = \begin{bmatrix} 1 \\ -0.448 \end{bmatrix} \tag{4.157}$$

and

$$\lambda_2 = \omega_2^2 = 0.2655,\ X_2 = \begin{bmatrix} 1 \\ 1.115 \end{bmatrix} \tag{4.158}$$

There are four eigenfunctions (modes of the system). The corresponding frequencies and amplitudes of these eigenfunctions are:

$$\omega_1 = \sqrt{0.734} = 0.857,\ X_1 = \begin{bmatrix} 1 \\ -0.448 \end{bmatrix},\ \omega_1 = -\sqrt{0.734} = -0.857,$$

$$X_1 = \begin{bmatrix} 1 \\ -0.448 \end{bmatrix} \tag{4.159}$$

$$\omega_2 = \sqrt{0.265} = 0.515,\ X_2 = \begin{bmatrix} 1 \\ 1.115 \end{bmatrix},\ \omega_2 = -\sqrt{0.265} = -0.515,$$

$$X_2 = \begin{bmatrix} 1 \\ 1.115 \end{bmatrix} \tag{4.160}$$

The solution $X(t)$ is a linear combination of the four modes (eigenfunctions) of the systems. That is:

$$X(t) = c_1 X_1 e^{j\omega_1 t} + c_2 X_1 e^{-j\omega_1 t} + c_3 X_2 e^{j\omega_2 t} + c_4 X_2 e^{-j\omega_2 t} \qquad (4.161)$$

Substituting the numerical values of ω_1, ω_2, X_1, and X_2 into Equation 4.161 results in:

$$X(t) = c_1 \begin{bmatrix} 1 \\ -0.448 \end{bmatrix} e^{j0.857t} + c_2 \begin{bmatrix} 1 \\ -0.448 \end{bmatrix} e^{-j0.857t}$$

$$+ c_3 \begin{bmatrix} 1 \\ 1.115 \end{bmatrix} e^{j0.515t} + c_4 \begin{bmatrix} 1 \\ 1.115 \end{bmatrix} e^{-j0.515t} \qquad (4.162)$$

or

$$X(t) = \begin{bmatrix} c_1 e^{j0.857t} + c_2 e^{-j0.857t} + c_3 e^{j0.515t} + c_4 e^{-j0.515t} \\ -0.448 c_1 e^{j0.857t} - 0.448 c_2 e^{-j0.857t} + 1.115 c_3 e^{j0.515t} + 1.115 c_4 e^{-j0.515t} \end{bmatrix} \qquad (4.163)$$

Therefore:

$$x_1(t) = c_1 e^{j0.857t} + c_2 e^{-j0.857t} + c_3 e^{j0.515t} + c_4 e^{-j0.515t} \qquad (4.164)$$

and

$$x_2(t) = -0.448 c_1 e^{j0.857t} - 0.448 c_2 e^{-j0.857t} + 1.115 c_3 e^{j0.515t} + 1.115 c_4 e^{-j0.515t} \qquad (4.165)$$

Applying the initial conditions, we have:

$$x_1(0) = c_1 + c_2 + c_3 + c_4 = 0$$

$$x_2(0) = -0.448 c_1 - 0.448 c_2 + 1.115 c_3 + 1.115 c_4 = 1$$

$$\left.\frac{dx_1(t)}{dt}\right|_{t=0} = j0.857 c_1 - j0.857 c_2 + j0.515 c_3 - j0.515 c_4 = 0 \qquad (4.166)$$

$$\left.\frac{dx_2(t)}{dt}\right|_{t=0} = -j0.3839 c_1 + j0.3839 c_2 + j0.5742 c_3 - j0.5742 c_4 = 0$$

Therefore:

$$\begin{bmatrix} 1 & 1 & 1 & 1 \\ -0.448 & -0.448 & 1.115 & 1.115 \\ 0.857 & -0.857 & 0.515 & -0.515 \\ -0.3839 & 0.3839 & 0.5742 & -0.5742 \end{bmatrix} \begin{bmatrix} c_1 \\ c_2 \\ c_3 \\ c_4 \end{bmatrix} = \begin{bmatrix} 0 \\ 1 \\ 0 \\ 0 \end{bmatrix} \qquad (4.167)$$

Eigenvalues and Eigenvectors 217

Solving the above set of linear equations results in:

$$\begin{bmatrix} c_1 \\ c_2 \\ c_3 \\ c_4 \end{bmatrix} = \begin{bmatrix} -0.3199 \\ -0.3199 \\ 0.3199 \\ 0.3199 \end{bmatrix} \qquad (4.168)$$

Then:

$$x_1(t) = -0.3199e^{j0.857t} - 0.3199e^{-j0.857t} + 0.3199e^{j0.515t} + 0.3199e^{-j0.515t}$$
$$= -0.6398\cos(0.857t) + 0.6398\cos(0.515t) \qquad (4.169)$$

and

$$x_2(t) = 0.1433e^{j0.857t} + 0.1433e^{-j0.857t} + 0.3567e^{j0.515t} + 0.3567e^{-j0.515t}$$
$$= 0.2866\cos(0.857t) + 0.7134\cos(0.515t) \qquad (4.170)$$

4.9.3 Signal Subspace Decomposition

Another application of eigen type representation is described through the use of SVD to extract the signal information by decomposing the available (signal plus noise) waveform into its corresponding signal and noise subspaces. Since the noise subspace is orthogonal to the signal subspace, the signal parameters are easily extracted from this information. This is demonstrated in the following subsections.

4.9.3.1 Frequency Estimation

As the first example of subspace decomposition, we consider the problem of estimating frequencies of a narrowband signal $s(n)$ buried in white noise $v(n)$. Suppose that signal $s(n)$ consists of the sum of P complex exponentials (signal) buried in zero mean white noise. The observed signal $x(n)$ is:

$$x(n) = s(n) + v(n) = \sum_{k=1}^{P} A_k e^{j\omega_k n} + v(n) \qquad (4.171)$$

where A_k, ω_k are the complex amplitude and frequency of the kth signal, respectively, and $v(n)$ is zero mean additive white noise independent of the signal. The autocorrelation function of $x(n)$ is:

$$r_x(l) = E[x(n)\bar{x}(n-l)] = \sum_{k=1}^{P} |A_k|^2 e^{j\omega_k l} + \sigma^2 \delta(l) \qquad (4.172)$$

where E is the expectation operator, σ^2 is the variance of the additive noise and $\delta(.)$ is the discrete delta function. The $(P+1)\times(P+1)$ autocorrelation matrix of signal $x(n)$ is:

$$R_x = \begin{bmatrix} r_x(0) & r_x(-1) & \cdots & r_x(-P) \\ r_x(1) & r_x(0) & \cdots & r_x(-P+1) \\ \vdots & \vdots & & \vdots \\ r_x(P) & r_x(P-1) & \cdots & r_x(0) \end{bmatrix} \quad (4.173)$$

The autocorrelation matrix R_x is:

$$R_x = \sum_{k=1}^{P} |A_k|^2 S_k S_k^H + \sigma^2 I \quad (4.174)$$

where $S_k = [1 \ e^{j\omega_k} \ e^{j2\omega_k} \ \cdots \ e^{jP\omega_k}]^T$. Equation 4.174 indicates that the rank of $(P+1)\times(P+1)$ matrix R_x is equal to P if $\sigma=0$ and $P+1$ if $\sigma \neq 0$. The eigenvalue-eigenvector (or SVD) decomposition of positive definite matrix R_x is:

$$R_x = \sum_{i=1}^{P+1} \lambda_i u_i u_i^T \quad (4.175)$$

where $\lambda_1 > \lambda_2 > \cdots > \lambda_p > \lambda_{p+1} = \sigma^2$. Since eigenvectors are pair wise orthogonal, then u_{p+1} is orthogonal to the signal subspace. The signal subspace is the span of vectors $u_1, u_2, \ldots, u_p$. Since the vector $S(e^{j\omega}) = [1 \ e^{j\omega} \ e^{j2\omega} \ \cdots \ e^{jP\omega}]^T$ is in the signal subspace, it is therefore orthogonal to u_{p+1}. Hence:

$$f(e^{j\omega}) = \langle S(e^{j\omega}), u_{P+1} \rangle = S^H(e^{j\omega}) u_{P+1} = 0 \quad (4.176)$$

Equation 4.176 is a polynomial of degree P in $z = e^{j\omega}$. The roots of this polynomial are complex numbers located on the boundary of the unit circle in the complex z-plane. The angles of these complex roots are the frequencies of the narrowband signal $s(n)$. The variance of the additive noise is the smallest eigenvalues of R_x. In practice, the autocorrelation matrix R_x is not available and has to be estimated from the observed signal $x(n)$. The estimated value of $R_x(l, m)$ is given by:

$$\hat{R}_x(l, m) = \frac{1}{N} \sum_{n=0}^{N-1} x(n) \bar{x}(n-l+m) \quad (4.177)$$

Example 4.32

Let $x(n)$ be written as follows:

$$x(n) = Ae^{j\omega_1 n} + v(n)$$

…where A represents the amplitude and $v(n)$ the noise. Assume that the 2×2 data autocorrelation matrix R_x is estimated to be:

$$R_x = \begin{bmatrix} 4.396 & 2.014 - j3.456 \\ 2.014 + j3.456 & 4.396 \end{bmatrix}$$

Estimate the frequency of the signal.

Solution: Using $[u, D] = \text{SVD}(R_x)$, we obtain:

$$u = \begin{bmatrix} -0.7071 & 0.7071 \\ -0.3560 - j0.6109 & -0.3560 - j0.6109 \end{bmatrix} \text{ and } D = \begin{bmatrix} 8.3959 & 0 \\ 0 & 0.3961 \end{bmatrix}$$

Then:

$$f(e^{j\omega}) = \langle S(e^{j\omega}), u_{P+1} \rangle = \begin{bmatrix} 1 & e^{-j\omega} \end{bmatrix} \begin{bmatrix} 0.7071 \\ -0.356 - j0.6109 \end{bmatrix} = 0$$

or

$$f(z) = 0.7071 + (-0.356 - j0.6109)z^{-1} = 0 \quad \rightarrow \quad z = \frac{0.356 + j0.6109}{0.7071} = e^{j0.332\pi}$$

Therefore, the frequency of the signal is estimated to be $\hat{\omega}_1 = 0.332\pi$. The variance of the noise is estimated to be $\sigma^2 = \lambda_{\min} = 0.3961$.

4.9.3.2 Direction of Arrival Estimation

The second example of subspace decomposition is the angle of arrival estimation problem encountered in radar and wireless communication systems. The multiple signal classification (MUSIC) algorithm can be used to estimate the angles of arrival of different signals. MUSIC is based on signal subspace decomposition. Consider a linear array of M sensors as shown in Figure 4.22.

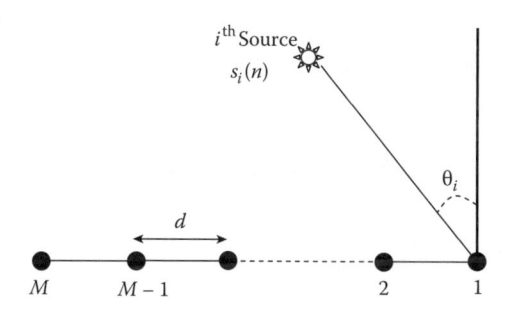

FIGURE 4.22
Linear array of sensors.

Assume that I far field sources are impending on this array. The angle of arrival of the ith signal is θ_i. The received signal at the mth sensor at time n is given by:

$$x_m(n) = \sum_{i=1}^{I} S_i(n)\exp\left(-j\omega_0(m-1)\frac{\sin\theta_i}{C}d\right) + v_m(n) \qquad (4.178)$$

$$n = 0, 1, 2, ..., N-1, \quad m = 1, 2, ..., M$$

where d is the distance between sensors, ω_0 is the frequency of the signal, $S_i(n)$ is the signal emitted from the ith source, θ_i is the angle of arrival of the ith signal, C is the speed of light, and $v_m(n)$ is the measurement noise at the mth sensor at time n. The MUSIC algorithm is based on the fact that the signal subspace is orthogonal to the noise subspace. The algorithm is implemented using the following steps:

Step 1: Use the N data snapshots to estimate the $M \times M$ data covariance matrix R as:

$$\hat{R} = \frac{1}{N}\sum_{n=0}^{N-1} X(n)X^H(n) \qquad (4.179)$$

where $X(n) = [x_1(n) \ x_2(n) \ ... \ x_M(n)]^T$ is a data snapshot vector at time n.

Step 2: Perform SVD on covariance matrix $\hat{R}$. Then:

$$\hat{R} = \sum_{i=1}^{M} \lambda_i u_i u_i^H \qquad (4.180)$$

where

$$\lambda_i = \begin{cases} \lambda_{is} + \sigma^2 & 1 \le i \le I \\ \sigma^2 & i > I \end{cases} \qquad (4.181)$$

Therefore:

$$\hat{R} = \sum_{i=1}^{I}(\lambda_{is} + \sigma^2)u_i u_i^H + \sum_{i=I+1}^{M}\sigma^2 u_i u_i^H \qquad (4.182)$$

The first term in the expansion for $\hat{R}$ is the signal subspace and the second term is the noise subspace.

Step 3: Since the signal subspace is orthogonal to the noise subspace, the $M \times 1$ vector

Eigenvalues and Eigenvectors

$$A(\theta) = \begin{bmatrix} 1 & \exp\left(j\omega_0 \frac{d}{C}\sin\theta\right) & \exp\left(j2\omega_0 \frac{d}{C}\sin\theta\right) & \cdots & \exp\left(j(M-1)\omega_0 \frac{d}{C}\sin\theta\right) \end{bmatrix}^T$$

(4.183)

is orthogonal to any vector in the noise subspace, where θ is one of the angles of arrival. Therefore, it should be orthogonal to $u_{I+1}, u_{I=1}, \ldots, u_M$. Hence, $P(\theta)$ is defined as:

$$P(\theta) = \frac{1}{\sum_{i=I+1}^{M} |A^H(\theta) u_i|^2}$$

(4.184)

and has peaks at $\theta_1, \theta_2, \ldots, \theta_J$. By locating the peaks of $P(\theta)$, the angles of arrival are estimated from the locations of these peaks.

Example 4.33

Consider a linear array of M=8 elements. Assume that there are three uncorrelated Gaussian sources of equal power arriving at the angles of −60, 30 and 60 degrees. Let the SNR be 10 dB. Assume that $d=\lambda/2$, where λ is the wavelength. N=64 data snapshots are used to estimate the data covariance matrix. The plot of $P(\theta)$ as a function of θ is shown in Figure 4.23. The three peaks indicate that the directions of arrival are approximately −60, 30 and 60 degrees. The MATLAB code implementing the MUSIC algorithm is shown in Table 4.7.

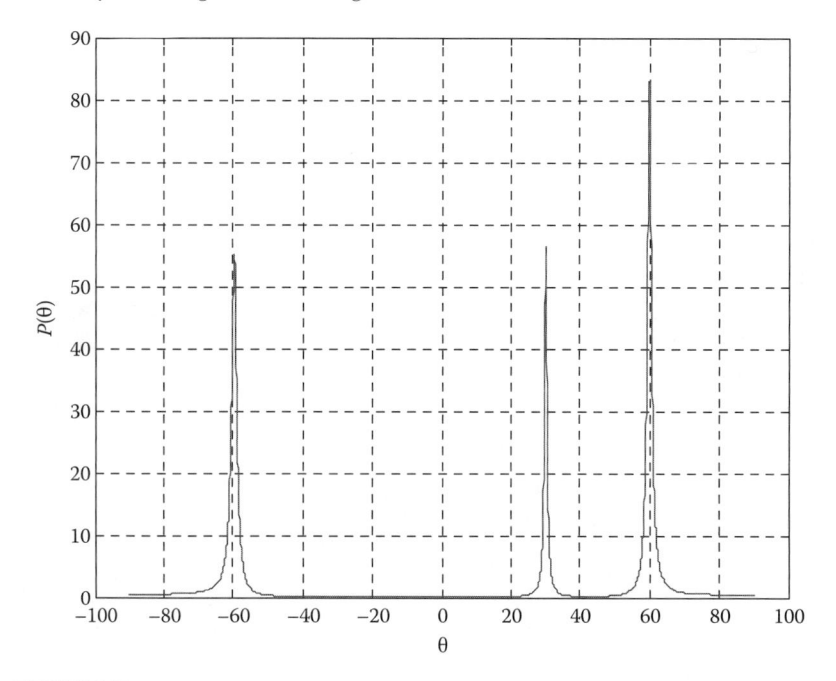

FIGURE 4.23
$P(\theta)$ as a function of θ.

TABLE 4.7
MATLAB Code Implementation of MUSIC Algorithm

```
% function [teta] =music(X,dn,I)
% X=Data matrix (NxM)
% dn=Distance between sensors normalized with
  respect to λ
% I=Number of sources
% teta=Output vector= [θ1 θ2 … θI]
%
function [teta] =music(X,dn,I)
L=4000;
[N,M] = size(X);
R=X'*X/N;
[U,S,V] =svd(R);
theta = linspace(-90,90,L);
m = 0:1:M-1;
C=exp(j*2*pi*dn*kron(m',sin(theta*pi/180)));
D=abs(C'*U(:,I+1:M));
P=1./sum(D.^2,2);
plot(theta,P);
xlabel('\theta')
ylabel('P(\theta)')
grid on
[X,Y] =GINPUT(I);
teta=X;
```

Problems

PROBLEM 4.1
Find the eigenvalues and eigenvectors of matrix A:

$$A = \begin{bmatrix} 1 & 2 \\ 3 & -1 \end{bmatrix}$$

PROBLEM 4.2
Find the eigenvalues and eigenvectors of the following matrices:

$$A = \begin{bmatrix} 2 & -1 \\ 0 & 3 \end{bmatrix}, \quad B = \begin{bmatrix} \cos(\theta) & -\sin(\theta) \\ \sin(\theta) & \cos(\theta) \end{bmatrix}, \quad C = \begin{bmatrix} 0 & 0 & 2 \\ 0 & 2 & 0 \\ -2 & 0 & 0 \end{bmatrix}, \text{ and}$$

$$D = \begin{bmatrix} 2 & 1-j & 3 \\ 1+j & 4 & j \\ 3 & -j & 2 \end{bmatrix}$$

Eigenvalues and Eigenvectors

PROBLEM 4.3
Let A be a 3×3 matrix with eigenvalues $-1, 2, 3$. Find the following (if possible):

a. $\det(A^T)$
b. $\text{trace}(A^{-1})$
c. $\det(A - 2I)$

PROBLEM 4.4
Let A be a 4×4 matrix with eigenvalues $-1, -2, -3$, and -4. Find the following quantities (if possible):

a. $\det(A^T)$
b. $\text{trace}(A^{-1})$
c. $\det(A + 4I)$

PROBLEM 4.5
Let A be an $n \times n$ matrix. The characteristic polynomial of matrix A is given by: $P(\lambda) = \lambda^n + \alpha_1 \lambda^{n-1} + \alpha_2 \lambda^{n-2} + \cdots + \alpha_{n-1} \lambda + \alpha_n$

Show that the coefficients of the characteristic polynomial can be computed by the following recursive equations:

$$\alpha_1 = -W_1$$

$$\alpha_2 = -\frac{1}{2}(a_1 W_1 + W_2)$$

$$\alpha_3 = -\frac{1}{3}(a_2 W_1 + a_1 W_2 + W_3)$$

$$\cdots$$

$$\alpha_n = -\frac{1}{n}(a_{n-1} W_1 + a_{n-2} W_2 + \cdots + a_1 W_{n-1} + W_n)$$

where $W_k = \text{Trace}(A^k)$, $k = 1, 2, \ldots, n$.

PROBLEM 4.6
Compute the eigenvalues and eigenvectors of the matrix:

$$A = \begin{bmatrix} 0 & 1 & 1 \\ 1 & 0 & 1 \\ 1 & 1 & 0 \end{bmatrix}$$

Is A symmetric? Justify your answer. Find, if possible, the nonsingular matrix T such that $T^{-1}AT$ is a diagonal matrix.

PROBLEM 4.7

Compute the eigenvalues and eigenvectors of the matrix:

$$A = \begin{bmatrix} 1 & 1 & 1 \\ 0 & 2 & 1 \\ 0 & 0 & 3 \end{bmatrix}$$

Find, if possible, the matrix T such that $T^{-1}AT$ is a diagonal matrix.

PROBLEM 4.8

a. If A and B are square matrices, show that AB and BA have the same eigenvalues.
b. What about if A is $m \times n$ and B is $n \times m$?

PROBLEM 4.9

Prove the eigenvalue and eigenvector properties shown in Table 4.4.

PROBLEM 4.10

Show that if the $n \times n$ matrices A and B have a common set of n linearly independent eigenvectors, then $AB = BA$. Is the converse true?

PROBLEM 4.11

Show that the eigenvalues and eigenvectors of an $N \times N$ circulant matrix with l^{th}, m^{th} element $\exp(j(2\pi/N)(l-1)(m-1))$ are:

$$\lambda_k = \exp\left(j\frac{2\pi}{N}k\right) \text{ and } x_k = \begin{bmatrix} 1 & \lambda_k & \cdots & \lambda_k^{n-1} \end{bmatrix}$$

PROBLEM 4.12

Show that the eigenvalues of $A + \alpha I$ are related to the eigenvalues of A by:

$$\lambda(A + \alpha I) = \lambda(A) + \alpha$$

PROBLEM 4.13

Show that the algebraic matrix equation:

$$AX + XB = C$$

where A, B, and C are matrices of appropriate dimensions, has a unique solution if the sum of the ith eigenvalue of A and j^{th} eigenvalue of B is different from zero for all values of i and j.

PROBLEM 4.14

Find the SVD of the following matrices:

$$A = \begin{bmatrix} 1 & 2 \\ 2 & 4 \end{bmatrix}, \quad B = \begin{bmatrix} 1 & -1 & 2 \\ 3 & 2 & 0 \\ 1 & -3 & 4 \end{bmatrix} \text{ and } C = \begin{bmatrix} 1 & -1 \\ 2 & -3 \\ 3 & 4 \end{bmatrix}$$

Eigenvalues and Eigenvectors

PROBLEM 4.15
Use the power method algorithm to compute the eigenvalue with the largest magnitude of the following matrix:

$$A = \begin{bmatrix} 2 & 3 \\ 3 & -4 \end{bmatrix}$$

PROBLEM 4.16
Let $A = B + jC$, where B and C are real matrices defined as:

$$B = \begin{bmatrix} 2 & 3 \\ 3 & -4 \end{bmatrix}, \quad C = \begin{bmatrix} 1 & -1 \\ 1 & 4 \end{bmatrix}$$

a. Find the SVD of A, B, and C.
b. Find the SVD of A in terms of the SVD of matrix T given by

$$T = \begin{bmatrix} B & -C \\ C & B \end{bmatrix}$$

PROBLEM 4.17
Consider the symmetric square matrix:

$$A = \begin{bmatrix} b & 1 \\ 1 & a \end{bmatrix} \begin{bmatrix} 4 & 0 \\ 0 & -2 \end{bmatrix} \begin{bmatrix} b & 1 \\ 1 & a \end{bmatrix}^{-1}$$

a. Find the eigenvalues of $A + 3I$.
b. Find b if $a = -1$.
c. For what value of α is the matrix $A + \alpha I$ singular?

PROBLEM 4.18
Determine whether the following matrices are (1) positive definite, (2) positive semidefinite, (3) negative definite, or (4) negative semidefinite.

$$A = \begin{bmatrix} 2 & 1 & 1 \\ 1 & 3 & 0 \\ 1 & 0 & 1 \end{bmatrix} \quad B = \begin{bmatrix} -1 & -1 & -1 \\ -1 & -4 & -1 \\ -1 & -1 & -2 \end{bmatrix} \quad C = \begin{bmatrix} 2 & 4 \\ 1 & 2 \end{bmatrix}$$

PROBLEM 4.19
Let T be a nonsingular $n \times n$ matrix. Show that A and TAT^{-1} have the same eigenvalues. What can you say about their eigenvectors?

PROBLEM 4.20
Let $A = \begin{bmatrix} 2 & 3 \\ 1 & -2 \end{bmatrix}$ and $B = \begin{bmatrix} 0 & 1 \\ 2 & 3 \end{bmatrix}$. Show that the eigenvalues of $C = A \otimes B$ are $\{\lambda_i(A) \lambda_j(B)\}$, $i, j = 1, 2$. Can you generalize the results if A and B are $n \times n$ matrices?

PROBLEM 4.21
Let A be an $n \times n$ matrix such that $A^k = 0$ for some $k > 1$. Show that the characteristic polynomial of A is given by:

$$P(\lambda) = \lambda^n$$

PROBLEM 4.22
Let A be an $n \times n$ matrix and $f(x)$ be a monic polynomial (a polynomial is monic if the leading coefficient is one). The monic polynomial $f(x)$ of smallest degree such that $f(A) = 0$ is called the minimal polynomial of A. Find the minimal polynomial of the following matrices:

$$A = \begin{bmatrix} -2 & 1 \\ 0 & -2 \end{bmatrix}, \quad B = \begin{bmatrix} -1 & 1 & 0 \\ 0 & -1 & 1 \\ 0 & 0 & -1 \end{bmatrix}$$

PROBLEM 4.23
Show that the following function is always positive for any vector $x = [x_1 \quad x_2 \quad x_3]^T$

$$f(x) = 5x_1^2 + 9x_2^2 + 6x_3^2 + 12x_1x_2 + 8x_1x_3 + 14x_2x_3$$

PROBLEM 4.24
Show that if A is a symmetric $n \times n$ positive definite matrix, then

$$\lambda_{\min} \|x\|^2 \leq x^H A x \leq \lambda_{\max} \|x\|^2$$

PROBLEM 4.25
Let A be an $n \times n$ skew symmetric ($A^T = -A$) with n odd. Show that:

a. A is singular.
b. $B = (I+A)(I-A)^{-1}$ is an orthogonal matrix.

PROBLEM 4.26 (MATLAB)
a. Let A be a symmetric positive definite matrix. It can be shown that A can be factored as $A = BB^T$ for some symmetric matrix B. Matrix B is called the square root of A. Develop an algorithm to compute the square root of a positive definite matrix A.
b. Test your algorithm using the following matrix:

$$A = \begin{bmatrix} 6 & 5 & 4 \\ 5 & 6 & 6 \\ 4 & 6 & 8 \end{bmatrix}$$

c. Develop a MATLAB code to implement the developed algorithm.
d. Use the code to generate 400 samples of a zero mean Gaussian random vector of size 2×1 with covariance matrix

Eigenvalues and Eigenvectors

$$R = \begin{bmatrix} 1 & 0.5 \\ 0.5 & 2 \end{bmatrix}$$

e. Use the generated data, estimate the covariance matrix and compare it with the true covariance matrix.

PROBLEM 4.27 (MATLAB)

Select an RGB image of your choice (e.g. natural scene). Perform the following:

a. Convert the image from RGB to gray and display it.
b. Perform gradient based edge detection on the gray image.
c. Perform Laplacian based edge detection on the same gray image.
d. Discuss your observations between the two techniques.
e. Can you combine the gradient and Laplacian based techniques together, building on their collective strength to establish a better edge map?
f. Perform color based edge detection on the RGB image and discuss your results as compared to the gray scale version.

PROBLEM 4.28 (MATLAB)

Consider the following two matrices:

$$A = \begin{bmatrix} 1 & 3 \\ 4 & 5 \end{bmatrix} \text{ and } B = \begin{bmatrix} 1 & 2 \\ 0 & 3 \\ 2 & 4 \end{bmatrix}$$

a. Find the eigenvalues and eigenvectors of A, BB^T and B^TB.
b. Compute the SVD of matrix B.
c. Compute the modal matrix of A.

PROBLEM 4.29 (MATLAB)

Consider the following two matrices:

$$A = \begin{bmatrix} 1 & 2 & 3 & -1 & 2 \\ 2 & 3 & 0 & 3 & 4 \\ -4 & 3 & 6 & 2 & 8 \\ 2 & 3 & 5 & 7 & 8 \\ 1 & 2 & 0 & -1 & -1 \end{bmatrix} \text{ and } B = \begin{bmatrix} 1 & 2 \\ 3 & 4 \end{bmatrix}$$

a. Find eigenvalues and eigenvectors of A.
b. Find eigenvalues and eigenvectors of B.
c. Find eigenvalues and eigenvectors of $A \otimes B$.
d. Can you establish a relationship between the eigenvalues of $A \otimes B$ and the eigenvalues of A and B?
e. Compute the SVD of matrix A.
f. Compute the SVD of matrix B.
g. Compute the SVD of matrix $A \otimes B$.
h. Is there a relationship between the SVDs of A, B, and $A \otimes B$?

PROBLEM 4.30 (MATLAB)

A 2-d FIR filter $h(n, m)$ is said to be separable if it can be written as product of two 1-d filters, that is:

$$h(n,m) = h_1(n)h_2(m)$$

a. Find the necessary and sufficient condition for FIR filter $h(n, m)$ to be separable.
b. If $h(n, m)$ is not separable, it can be approximated by a separable filter. Write a MATLAB script to approximate a nonseparable filter by a separable filter.
c. Test your code using the following two filters:

$$h_I(n, m) = \frac{1}{7}\begin{bmatrix} 1 & -1 & 2 \\ -1 & 5 & -1 \\ 2 & -1 & 1 \end{bmatrix} \text{ and } h_{II}(n, m) = \begin{bmatrix} 1 & 0 & -1 \\ 2 & 0 & -2 \\ 1 & 0 & -1 \end{bmatrix}$$

PROBLEM 4.31 (MATLAB)

Let $r(n)$, $n = 0, \pm 1, \pm 2, \ldots, \pm m$ be the autocorrelation function of a zero mean real stationary process $x(n)$. The power spectral density of this process is defined as:

$$S(\omega) = \sum_{n=-m}^{m} r(n)e^{-jn\omega}$$

a. If $m = 7$ and $r(n) = 1 - |n|/m$, show that $S(\omega) \geq 0$ and the $(m+1) \times (m+1)$ Toeplitz autocorrelation matrix R defined as $R(i, j) = r(i-j)$ is positive definite.
b. Show that in general if $S(\omega) \geq 0$ then the $(m+1) \times (m+1)$ Toeplitz autocorrelation matrix R defined as $R(i, j) = r(i-j)$ is positive semidefinite.

5

Matrix Polynomials and Functions of Square Matrices

5.1 Introduction

Matrix polynomials and functions of matrices have extensive applications in different engineering problems such as modern control system design, vibration analysis, queueing theory and communication systems. In this chapter, we focus on matrix polynomials, special functions of square matrices and discuss their analytical and numerical computation.

5.2 Matrix Polynomials

Consider the polynomial $f(x)$ of degree m given by:

$$f(x) = a_0 x^m + a_1 x^{m-1} + a_2 x^{m-2} + \cdots + a_{m-1} x + a_m \tag{5.1}$$

If the scalar variable x is replaced by the $n \times n$ matrix A, then the corresponding matrix polynomial $f(A)$ is defined by:

$$f(A) = a_0 A^m + a_1 A^{m-1} + a_2 A^{m-2} + \cdots + a_{m-1} A + a_m I \tag{5.2}$$

where $A^m = \underbrace{A \times A \times \cdots \times A}_{m \text{ times}}$ and I is the $n \times n$ identity matrix. The polynomial $f(x)$ can be written in factor form as:

$$f(x) = a_0 (x - \alpha_1)(x - \alpha_2) \cdots (x - \alpha_m) \tag{5.3}$$

where $\alpha_1, \alpha_2, \ldots, \alpha_m$ are the roots of the polynomial $f(x)$. Similarly, the matrix polynomial $f(A)$ can be factored as:

$$f(A) = a_0 (A - \alpha_1 I)(A - \alpha_2 I) \cdots (A - \alpha_m I) \tag{5.4}$$

Note that m, the degree of the polynomial $f(x)$, can be greater than, equal to or less than n, the size of matrix A.

The MATLAB® command to factor polynomial $f(x)$ is **p=roots(a)**, where $a=[a_0\ a_1\ \dots\ a_m]$ is the vector of polynomial coefficients and $p=[\alpha_0\ \alpha_1\ \dots\ \alpha_m]$ is the vector of the roots of the polynomial $f(x)$.

Example 5.1

If $f(x)=2x^3+x^2+2x+4$ and $A=\begin{bmatrix}1 & -2\\ 3 & 2\end{bmatrix}$
Find $f(A)$.

Solution:

$f(A) = 2A^3 + A^2 + 2A + 4I$

$$= 2\begin{bmatrix}1 & -2\\ 3 & 2\end{bmatrix}\begin{bmatrix}1 & -2\\ 3 & 2\end{bmatrix}\begin{bmatrix}1 & -2\\ 3 & 2\end{bmatrix} + \begin{bmatrix}1 & -2\\ 3 & 2\end{bmatrix}\begin{bmatrix}1 & -2\\ 3 & 2\end{bmatrix} + 2\begin{bmatrix}1 & -2\\ 3 & 2\end{bmatrix} + 4\begin{bmatrix}1 & 0\\ 0 & 1\end{bmatrix}$$

$$= \begin{bmatrix}-46 & -4\\ 6 & -44\end{bmatrix} + \begin{bmatrix}-5 & -6\\ 9 & -2\end{bmatrix} + \begin{bmatrix}2 & -4\\ 6 & 4\end{bmatrix} + \begin{bmatrix}4 & 0\\ 0 & 4\end{bmatrix} = \begin{bmatrix}-45 & -14\\ 21 & -38\end{bmatrix}$$

Example 5.2

If $f(x)=x^2+3x+2$, $g(x)=x^2+2x+2$, and $A=\begin{bmatrix}1 & -2\\ 3 & 2\end{bmatrix}$
a. Find $f(A)$ by first factoring $f(A)$.
b. Find $g(A)$ by first factoring $g(A)$.

Solution:

a. $f(x)=x^2+3x+2=(x+1)(x+2)$

Therefore:

$$f(A) = (A+I)(A+2I) = \left(\begin{bmatrix}1 & -2\\ 3 & 2\end{bmatrix}+\begin{bmatrix}1 & 0\\ 0 & 1\end{bmatrix}\right)\left(\begin{bmatrix}1 & -2\\ 3 & 2\end{bmatrix}+2\begin{bmatrix}1 & 0\\ 0 & 1\end{bmatrix}\right) = \begin{bmatrix}0 & -12\\ 18 & 6\end{bmatrix}$$

b. $g(x)=x^2+2x+2=(x+1+j)(x+1-j)$

Therefore:

$$g(A) = [A+(1+j)I][A+(1-j)I] = \begin{bmatrix}2+j & -2\\ 3 & 3+j\end{bmatrix}\begin{bmatrix}2-j & -2\\ 3 & 3-j\end{bmatrix} = \begin{bmatrix}4 & -4\\ 6 & 6\end{bmatrix}$$

5.2.1 Infinite Series of Matrices

An infinite series of matrix A is defined as:

$$S(A) = a_0 I + a_1 A + a_2 A^2 + \dots = \sum_{k=0}^{\infty} a_k A^k \qquad (5.5)$$

Matrix Polynomials and Functions of Square Matrices

where a_k is a scalar f_n, $k=0, 1, \ldots$ and A is a real or complex $n \times n$ square matrix.

5.2.2 Convergence of an Infinite Matrix Series

It can be shown that the infinite matrix series $S(A)$ given by Equation 5.5 converges if and only if the scalar infinite series $S(\lambda_i)$ converges for all values of i, where λ_i is the ith eigenvalue of matrix A. For example, the geometric matrix series:

$$S(A) = I + aA + a^2 A^2 + \cdots = \sum_{k=0}^{\infty} a^k A^k \tag{5.6}$$

is a convergent series and converges to:

$$S(A) = (I - aA)^{-1} \tag{5.7}$$

if and only if the scalar geometric series:

$$S(\lambda) = 1 + a\lambda + a^2\lambda^2 + \cdots = \sum_{k=0}^{\infty} (a\lambda)^k \tag{5.8}$$

converges when λ is replaced by any eigenvalue of matrix A. Since $S(\lambda)$ converges if $|a\lambda| < 1$, then under the condition that the eigenvalues of A are distinct we must have:

$$|a\lambda_i| < 1 \quad \text{or} \quad -\frac{1}{|a|} < \lambda_i < \frac{1}{|a|} \quad \text{for} \quad i = 1, 2, \ldots, n \tag{5.9}$$

Another example is the exponential matrix polynomial defined by:

$$e^A = I + A + \frac{A^2}{2!} + \frac{A^3}{3!} \cdots = \sum_{k=0}^{\infty} \frac{A^k}{k!} \tag{5.10}$$

Since $1 + \lambda + \lambda^2/2! + \lambda^3/3! \cdots = \sum_{k=0}^{\infty} \lambda^k / k!$ converges for all values of λ, the matrix exponential defined by Equation 5.10 converges for any square matrix A regardless of its eigenvalues.

Example 5.3

Let $A = \begin{bmatrix} 0.25 & 0.5 \\ -0.5 & 1 \end{bmatrix}$ and $f(A) = \sum_{k=1}^{\infty} kA^k$. Is $f(A)$ a convergent series? If so, what does it converge to?

Solution: The scalar series $f(\lambda) = \sum_{k=1}^{\infty} k\lambda^k$ converges to $\lambda/(\lambda-1)^2$ if $|\lambda|<1$. The eigenvalues of matrix A are $\lambda_1 = 5+j\sqrt{7}/8$ and $\lambda_2 = 5-j\sqrt{7}/8$. Since $|\lambda_1|=|\lambda_2|=\sqrt{2}/2 < 1$, the function $f(A)$ converges to $A(A-I)^{-2}$.

$$f(A) = A(A-I)^{-2} = \begin{bmatrix} 0.25 & 0.5 \\ -0.5 & 1 \end{bmatrix} \left\{ \begin{bmatrix} -0.75 & 0.5 \\ -0.5 & 0 \end{bmatrix}^{-1} \right\}^2 = \begin{bmatrix} 0.25 & 0.5 \\ -0.5 & 1 \end{bmatrix} \begin{bmatrix} 0 & -2 \\ 2 & 3 \end{bmatrix}^2$$

$$= \begin{bmatrix} 0.25 & 0.5 \\ -0.5 & 1 \end{bmatrix} \begin{bmatrix} -4 & 6 \\ -6 & 5 \end{bmatrix} = \begin{bmatrix} -4 & 4 \\ -4 & 2 \end{bmatrix}$$

5.3 Cayley–Hamilton Theorem

Theorem 5.1: Any $n \times n$ square matrix satisfies its own characteristic polynomial, that is:

$$P(A) = 0 \qquad (5.11)$$

where $P(\lambda) = |\lambda I - A|$ is the characteristic polynomial of matrix A.

Proof: Let $P(\lambda)$ be the characteristic polynomial of matrix A. Then $P(\lambda)$ is a polynomial of degree n in λ and is given by:

$$P(\lambda) = \lambda^n + \alpha_1 \lambda^{n-1} + \alpha_2 \lambda^{n-2} + \cdots + \alpha_{n-1}\lambda + \alpha_n = \sum_{i=0}^{n} \alpha_i \lambda^{n-i} \qquad (5.12)$$

Given Equation 5.12, $P(A)$ can be written as follows:

$$P(A) = A^n + \alpha_1 A^{n-1} + \alpha_2 A^{n-2} + \cdots + \alpha_{n-1} A + \alpha_n I = \sum_{i=0}^{n} \alpha_i A^{n-i} \qquad (5.13)$$

where $\alpha_0 = 1$. Since $\text{adj}(\lambda I - A)$ is a polynomial in λ of degree $n-1$, let:

$$\text{adj}(\lambda I - A) = I\lambda^{n-1} + B_1 \lambda^{n-2} + B_2 \lambda^{n-3} + \cdots + B_{n-2}\lambda + B_{n-1} = \sum_{i=0}^{n-1} B_i \lambda^{n-1-i} \qquad (5.14)$$

where $B_0 = I$. Then:

$$(\lambda I - A)\text{adj}(\lambda I - A) = (\text{adj}(\lambda I - A))(\lambda I - A) = |\lambda I - A| I = P(\lambda)I \qquad (5.15)$$

As a result:

$$\sum_{i=0}^{n} \alpha_i I \lambda^{n-i} = (\lambda I - A) \sum_{i=0}^{n-1} B_i \lambda^{n-1-i} = \sum_{i=0}^{n-1} B_i \lambda^{n-i} - \sum_{i=0}^{n-1} AB_i \lambda^{n-1-i} \qquad (5.16)$$

Matrix Polynomials and Functions of Square Matrices

This yields:

$$\sum_{i=0}^{n}\alpha_i I\lambda^{n-i} = \sum_{i=0}^{n-1}B_i\lambda^{n-i} - \sum_{i=0}^{n-1}AB_i\lambda^{n-1-i} \tag{5.17}$$

Equation 5.17 can be rewritten as:

$$\sum_{i=0}^{n}\alpha_i I\lambda^{n-i} = \sum_{i=1}^{n-1}(B_i - AB_{i-1})\lambda^{n-i} + B_0\lambda^n - AB_{n-1} \tag{5.18}$$

Comparing the right hand side of Equation 5.18 with the left hand side results in:

$$B_i - AB_{i-1} = \alpha_i I \quad i = 1, 2, \ldots, n-1 \tag{5.19}$$

and

$$-AB_{n-1} = \alpha_n I \tag{5.20}$$

Multiplying the ith equation given by Equation 5.19 by A^{n-i} and adding them together, we have:

$$A^{n-1}(B_1 - AB_0) + A^{n-2}(B_2 - AB_1) + A^{n-3}(B_3 - AB_2) + \cdots + A(B_{n-1} - AB_{n-2})$$

$$= \sum_{i=1}^{n-1}\alpha_i A^{n-i} \tag{5.21}$$

Equation 5.21 can be simplified as:

$$-A^n B_0 + AB_{n-1} = \sum_{i=1}^{n-1}\alpha_i A^{n-i} \tag{5.22}$$

Adding Equations 5.20 through 5.22 and substituting $B_0 = I$ results in the equation:

$$-A^n = \sum_{i=1}^{n-1}\alpha_i A^{n-i} + \alpha_n I \tag{5.23}$$

or

$$P(A) = \sum_{i=0}^{n}\alpha_i A^{n-i} = 0 \tag{5.24}$$

The above proof is independent of whether the eigenvalues of matrix A are distinct or not. However, if the eigenvalues are distinct, an alternative easier proof is:

$$P(A) = \sum_{i=0}^{n} \alpha_i A^{n-i} = \sum_{i=0}^{n} \alpha_i M \Lambda^{n-i} M^{-1} = M \sum_{i=0}^{n} \alpha_i \Lambda^{n-i} M^{-1}$$

$$= M \begin{bmatrix} P(\lambda_1) & 0 & \cdots & 0 \\ 0 & P(\lambda_2) & \cdots & 0 \\ \vdots & & \ddots & \vdots \\ 0 & 0 & \cdots & P(\lambda_n) \end{bmatrix} M^{-1} \qquad (5.25)$$

Since $P(\lambda_1) = P(\lambda_2) = \cdots = P(\lambda_n) = 0$, then:

$$P(A) = 0 \qquad (5.26)$$

Example 5.4

Show that the following 2×2 matrix A satisfies its characteristic polynomial.

$$A = \begin{bmatrix} 1 & -2 \\ 3 & 2 \end{bmatrix}$$

Solution: The characteristic polynomial of A is:

$$P(\lambda) = |\lambda I - A| = \begin{vmatrix} \lambda - 1 & 2 \\ -3 & \lambda - 2 \end{vmatrix} = \lambda^2 - 3\lambda + 8$$

Therefore:

$$P(A) = A^2 - 3A + 8I$$

$$= \begin{bmatrix} 1 & -2 \\ 3 & 2 \end{bmatrix} \begin{bmatrix} 1 & -2 \\ 3 & 2 \end{bmatrix} - 3 \begin{bmatrix} 1 & -2 \\ 3 & 2 \end{bmatrix} + 8 \begin{bmatrix} 1 & 0 \\ 0 & 1 \end{bmatrix}$$

$$= \begin{bmatrix} -5 & -6 \\ 9 & -2 \end{bmatrix} + \begin{bmatrix} -3 & 6 \\ -9 & -6 \end{bmatrix} + \begin{bmatrix} 8 & 0 \\ 0 & 8 \end{bmatrix} = \begin{bmatrix} 0 & 0 \\ 0 & 0 \end{bmatrix} = 0$$

5.3.1 Matrix Polynomial Reduction

Cayley–Hamilton theorem can be used to reduce any matrix polynomial of A of degree $m \geq n$ to a matrix polynomial of degree $n-1$, where n is the size of matrix A. This is shown by the following theorem.

Matrix Polynomials and Functions of Square Matrices

Theorem 5.2: Matrix polynomial reduction theorem

Consider an $n \times n$ matrix A and a matrix polynomial in A of degree $m \geq n$. That is:

$$f(A) = a_m A^m + a_{m-1} A^{m-1} + a_{m-2} A^{m-2} + \cdots + a_1 A + a_0 I \qquad (5.27)$$

Then, $f(A)$ is reducible to a polynomial of degree $n-1$ given by:

$$R(A) = b_{n-1} A^{n-1} + b_{n-2} A^{n-2} + b_{n-3} A^{n-3} + \cdots + b_1 A + b_0 I \qquad (5.28)$$

That is $f(A) = R(A)$.

Proof: Consider the scalar polynomial $f(\lambda)$ in λ of degree m. Dividing this polynomial by $P(\lambda)$, the characteristic polynomial of matrix A results in:

$$f(\lambda) = Q(\lambda) P(\lambda) + R(\lambda) \qquad (5.29)$$

where $Q(\lambda)$ and $R(\lambda)$ are the quotient and the remainder polynomials, respectively that result by dividing $f(\lambda)$ by $P(\lambda)$. The degree of the remainder polynomial is at most $n-1$ since the degree of the characteristic polynomial is n. Since $P(A) = 0$ by the Cayley–Hamilton theorem, we have:

$$f(A) = Q(A) P(A) + R(A) = Q(A) \times 0 + R(A) = R(A) \qquad (5.30)$$

Example 5.5

Find the matrix polynomial $f(A) = A^8 + 6A^6 + A^5 + 2A^3 + A + 4I$ if:

$$A = \begin{bmatrix} 1 & -2 \\ 3 & 2 \end{bmatrix}$$

Solution: The characteristic polynomial of A is:

$$P(\lambda) = |\lambda I - A| = \begin{vmatrix} \lambda - 1 & 2 \\ -3 & \lambda - 2 \end{vmatrix} = \lambda^2 - 3\lambda + 8$$

Dividing $f(\lambda)$ by $P(\lambda)$, we have:

$$\frac{\lambda^8 + 6\lambda^6 + \lambda^5 + 2\lambda^3 + \lambda + 4}{\lambda^2 - 3\lambda + 8}$$

$$= \lambda^6 + 3\lambda^5 + 7\lambda^4 - 2\lambda^3 - 62\lambda^2 - 168\lambda - 8 + \frac{1321\lambda + 68}{\lambda^2 - 3\lambda + 8}$$

Therefore, the remainder polynomial is $R(\lambda) = 1321\lambda + 68$ and

$$f(A) = R(A) = 1321 A + 68 I = 1321 \begin{bmatrix} 1 & -2 \\ 3 & 2 \end{bmatrix} + 68 \begin{bmatrix} 1 & 0 \\ 0 & 1 \end{bmatrix}$$

$$= \begin{bmatrix} 1321 & -2642 \\ 3963 & 2642 \end{bmatrix} + \begin{bmatrix} 68 & 0 \\ 0 & 68 \end{bmatrix} = \begin{bmatrix} 1389 & -2642 \\ 3963 & 2710 \end{bmatrix}$$

Example 5.6

Find the matrix polynomial $f(A)=A^{20}$ if:

$$A = \begin{bmatrix} 0 & 1 \\ 0 & 1 \end{bmatrix}$$

Solution: The characteristic polynomial of A is:

$$P(\lambda) = |\lambda I - A| = \begin{vmatrix} \lambda & -1 \\ 0 & \lambda-1 \end{vmatrix} = \lambda^2 - \lambda$$

Dividing $f(\lambda)$ by $P(\lambda)$, we have:

$$\frac{\lambda^{20}}{\lambda^2 - \lambda} = \sum_{i=0}^{18} \lambda^i + \frac{\lambda}{\lambda^2 - \lambda}$$

Therefore, the remainder polynomial is $R(\lambda)=\lambda$ and $f(A) = R(A) = A = \begin{bmatrix} 0 & 1 \\ 0 & 1 \end{bmatrix}$

5.4 Functions of Matrices

There are several techniques available to find a function of a matrix. They include: (a) use of Sylvester theorem, (b) Cayley–Hamilton technique, and (c) matrix diagonalization.

5.4.1 Sylvester's Expansion

Sylvester's expansion can be used to compute $f(A)$ if it can be expressed as a matrix polynomial.

Theorem 5.3: Sylvester's theorem.

Sylvester's theorem states that if $f(A)$ is a matrix polynomial in A and if the eigenvalues of A are distinct, and then $f(A)$ can be decomposed as:

$$f(A) = \sum_{i=1}^{n} f(\lambda_i) z(\lambda_i) \tag{5.31}$$

where $\{\lambda_i\}_{i=1}^{n}$ are the eigenvalues of the $n \times n$ matrix A and $z(\lambda_i)$ is an $n \times n$ matrix given by:

$$z(\lambda_i) = \frac{\prod_{\substack{k=1 \\ k \neq i}}^{n} (A - \lambda_k I)}{\prod_{\substack{k=1 \\ k \neq i}}^{n} (\lambda_i - \lambda_k)} \tag{5.32}$$

Matrix Polynomials and Functions of Square Matrices

Proof: By Cayley–Hamilton theorem, $f(A)$ can be reduced to a matrix polynomial of degree $n-1$.

$$f(A) = b_{n-1}A^{n-1} + b_{n-2}A^{n-2} + b_{n-3}A^{n-3} + \cdots + b_1 A + b_0 I \tag{5.33}$$

This matrix polynomial can be factored as:

$$f(A) = \sum_{k=1}^{n} w_k \prod_{\substack{i=1 \\ i \neq k}}^{n} (A - \lambda_i I) \tag{5.34}$$

To compute the weights w_k, we post multiply both sides of the above equation by x_j, the jth eigenvector of matrix A. This yields:

$$f(A)x_j = \sum_{k=1}^{n} w_k \prod_{\substack{i=1 \\ i \neq k}}^{n} (A - \lambda_i I) x_j \tag{5.35}$$

Since x_j is an eigenvector, then $(A - \lambda_j I)x_j = 0$. Therefore, all the terms except the ith term in the summation are zero. Hence:

$$f(A)x_j = w_j \prod_{\substack{i=1 \\ i \neq j}}^{n} (A - \lambda_i I) x_j = w_j \prod_{\substack{i=1 \\ i \neq j}}^{n} (Ax_j - \lambda_i x_j)$$

$$= w_j \prod_{\substack{i=1 \\ i \neq j}}^{n} (\lambda_j x_j - \lambda_i x_j) = w_j \prod_{\substack{i=1 \\ i \neq j}}^{n} (\lambda_j - \lambda_i) x_j \tag{5.36}$$

Since the eigenvalues of A are distinct, then $f(A)x_j = f(\lambda_j)x_j$ and

$$w_j = \frac{f(\lambda_j)}{\prod_{\substack{i=1 \\ i \neq j}}^{n} (\lambda_j - \lambda_i)} \tag{5.37}$$

Therefore:

$$f(A) = \sum_{i=1}^{n} f(\lambda_i) \frac{\prod_{\substack{k=1 \\ k \neq i}}^{n} (A - \lambda_k I)}{\prod_{\substack{k=1 \\ k \neq i}}^{n} (\lambda_i - \lambda_k)} \tag{5.38}$$

Example 5.7

Let A be the 2×2 matrix

$$A = \begin{bmatrix} 0 & 1 \\ 0 & -1 \end{bmatrix}$$

Using Sylvester's theorem, find
a. $f(A) = e^{At}$ if $t \geq 0$
b. $f(A) = A^k$
c. $f(A) = \tan(\frac{\pi}{4}A)\cos(\frac{\pi}{3}A)$

Solution: The eigenvalues of A are $\lambda_1 = 0$ and $\lambda_2 = -1$. Therefore:

$$f(A) = \sum_{i=1}^{2} f(\lambda_i) \frac{\prod_{\substack{k=1 \\ k \neq i}}^{2}(A - \lambda_k I)}{\prod_{\substack{k=1 \\ k \neq i}}^{2}(\lambda_i - \lambda_k)} = f(\lambda_1)\frac{A - \lambda_2 I}{\lambda_1 - \lambda_2} + f(\lambda_2)\frac{A - \lambda_1 I}{\lambda_2 - \lambda_1}$$

Therefore:

$$f(A) = f(0)\frac{A - (-1)I}{0 - (-1)} + f(-1)\frac{A - 0I}{-1 - 0} = f(0)(A + I) - f(-1)A$$

a. $f(A) = e^{At}$, hence:

$$f(A) = (A + I) - e^{-t}A = \begin{bmatrix} 1 & 1 \\ 0 & 0 \end{bmatrix} + \begin{bmatrix} 0 & -e^{-t} \\ 0 & e^{-t} \end{bmatrix} = \begin{bmatrix} 1 & 1 - e^{-t} \\ 0 & e^{-t} \end{bmatrix}$$

b. $f(A) = A^k$. In this case $f(0) = (0)^k = \delta(k)$ and $f(-1) = (-1)^k$. Therefore:

$$f(A) = f(0)(A + I) - f(-1)A = \delta(k)(A + I) - (-1)^k A = \begin{bmatrix} \delta(k) & \delta(k) \\ 0 & 0 \end{bmatrix} - \begin{bmatrix} 0 & (-1)^k \\ 0 & -(-1)^k \end{bmatrix}$$

Hence:

$$f(A) = \begin{bmatrix} \delta(k) & \delta(k) - (-1)^k \\ 0 & (-1)^k \end{bmatrix}$$

c. $f(A) = \tan\left(\frac{\pi}{4}A\right)\cos\left(\frac{\pi}{3}A\right)$.

$$f(A) = f(0)(A + I) - f(-1)A = 0 - \tan\left(-\frac{\pi}{4}\right)\cos\left(-\frac{\pi}{3}\right)A = \frac{1}{2}A = \begin{bmatrix} 0 & \frac{1}{2} \\ 0 & -\frac{1}{2} \end{bmatrix}$$

In the case where the eigenvalues are not distinct, Gantmacher's algorithm can be used to expand analytical function $f(A)$ as

$$f(A) = \sum_{j=1}^{m} \sum_{k=1}^{m_j} \frac{1}{(k-1)!} \left.\frac{\partial^{k-1}f(\lambda)}{\partial \lambda^{k-1}}\right|_{\lambda = \lambda_j} Z_{jk} \qquad (5.39)$$

where m is the number of independent eigenvalues, m_j is the multiplicity of the jth eigenvalue, and Z_{jk} are constituent matrices. The following example shows how to apply Gantmacher's algorithm to compute $f(A)$.

Example 5.8

Given the 3×3 matrix A

$$A = \begin{bmatrix} 0 & 2 & -2 \\ 0 & 1 & 0 \\ 1 & -1 & 3 \end{bmatrix}$$

Compute e^{At} using Gantmacher's algorithm.

Solution: The characteristic polynomial of A is $P(\lambda) = (\lambda - 1)^2 (\lambda - 2)$. Therefore, there are two independent eigenvalues: $\lambda_1 = 1$ with multiplicity $m_1 = 2$ and $\lambda_2 = 2$ with multiplicity $m_2 = 1$. Hence:

$$f(A) = \sum_{j=1}^{m} \sum_{k=1}^{m_j} \frac{1}{(k-1)!} \frac{\partial^{k-1} f(\lambda)}{\partial \lambda^{k-1}}\bigg|_{\lambda = \lambda_j} Z_{jk} = \sum_{j=1}^{2} \sum_{k=1}^{m_j} \frac{1}{(k-1)!} \frac{\partial^{k-1} f(\lambda)}{\partial \lambda^{k-1}}\bigg|_{\lambda = \lambda_j} Z_{jk}$$

$$= \sum_{k=1}^{2} \frac{1}{(k-1)!} \frac{\partial^{k-1} f(\lambda)}{\partial \lambda^{k-1}}\bigg|_{\lambda = \lambda_1} Z_{1k} + \sum_{k=1}^{1} \frac{1}{(k-1)!} \frac{\partial^{k-1} f(\lambda)}{\partial \lambda^{k-1}}\bigg|_{\lambda = \lambda_2} Z_{2k}$$

$$= f(\lambda_1) Z_{11} + f'(\lambda_1) Z_{12} + f(\lambda_2) Z_{21}$$

To find the constituent matrices Z_{11}, Z_{12}, and Z_{21}, we need three equations for the three unknowns. Since the above expansion is valid for any analytical function, first let $f(A) = 1$. In this case:

$$f(A) = f(\lambda_1) Z_{11} + f'(\lambda_1) Z_{12} + f(\lambda_2) Z_{21} \rightarrow I = Z_{11} + Z_{21}$$

Next, assume that $f(A) = A$. Then we have:

$$f(A) = f(\lambda_1) Z_{11} + f'(\lambda_1) Z_{12} + f(\lambda_2) Z_{21} \rightarrow A = Z_{11} + Z_{12} + 2Z_{21}$$

Finally, let $f(A) = A^2$. Thus:

$$f(A) = f(\lambda_1) Z_{11} + f'(\lambda_1) Z_{12} + f(\lambda_2) Z_{21} \rightarrow A^2 = Z_{11} + 2Z_{12} + 4Z_{21}$$

The above three equations can be solved for the three unknowns. The solutions are:

$$Z_{11} = A(2I - A) = \begin{bmatrix} 2 & 0 & 2 \\ 0 & 1 & 0 \\ -1 & 0 & -1 \end{bmatrix}, \quad Z_{12} = (A - I)(2I - A) = \begin{bmatrix} 0 & 2 & 0 \\ 0 & 0 & 0 \\ 0 & -1 & 0 \end{bmatrix}, \text{ and}$$

$$Z_{21} = (A - I)^2 = \begin{bmatrix} -1 & 0 & -2 \\ 0 & 0 & 0 \\ 1 & 0 & 2 \end{bmatrix}$$

Therefore:

$$f(A) = f(\lambda_1)Z_{11} + f'(\lambda_1)Z_{12} + f(\lambda_2)Z_{21} = e^{\lambda_1 t}Z_{11} + te^{\lambda_1 t}Z_{12} + e^{\lambda_2 t}Z_{21}$$

$$= e^t \begin{bmatrix} 2 & 0 & 2 \\ 0 & 1 & 0 \\ -1 & 0 & -1 \end{bmatrix} + te^t \begin{bmatrix} 0 & 2 & 0 \\ 0 & 0 & 0 \\ 0 & -1 & 0 \end{bmatrix} + e^{2t} \begin{bmatrix} -1 & 0 & -2 \\ 0 & 0 & 0 \\ 1 & 0 & 2 \end{bmatrix}$$

Simplifying the above equation results in:

$$e^{At} = \begin{bmatrix} 2e^t - e^{2t} & 2te^t & 2e^t - 2e^{2t} \\ 0 & e^t & 0 \\ -e^t + e^{2t} & -te^t & -e^t + 2e^{2t} \end{bmatrix}$$

Note that $e^{A0} = I$.

5.4.2 Cayley–Hamilton Technique

First we assume that the eigenvalues of matrix A are distinct. Using the results of the previous theorem, we have:

$$f(A) = R(A) = b_{n-1}A^{n-1} + b_{n-2}A^{n-2} + b_{n-3}A^{m-3} + \cdots + b_1 A + b_0 I \quad (5.40)$$

where

$$f(\lambda) = Q(\lambda)P(\lambda) + R(\lambda) \quad (5.41)$$

To get the coefficients of the polynomial $R(\lambda)$, we set $\lambda = \lambda_i$, $i = 1, 2, \ldots, n$ in the above equation:

$$f(\lambda_i) = Q(\lambda_i)P(\lambda_i) + R(\lambda_i) = R(\lambda_i) \quad (5.42)$$

This yields a set of n simultaneous linear equations that can be solved for $b_0, b_1, \ldots, b_{n-1}$.

$$\begin{bmatrix} 1 & \lambda_1 & \lambda_1^2 & \cdots & \lambda_1^{n-1} \\ 1 & \lambda_2 & \lambda_2^2 & \cdots & \lambda_2^{n-1} \\ 1 & \lambda_3 & \lambda_3^2 & \cdots & \lambda_3^{n-1} \\ \vdots & \vdots & \vdots & \vdots & \vdots \\ 1 & \lambda_n & \lambda_n^2 & \cdots & \lambda_n^{n-1} \end{bmatrix} \begin{bmatrix} b_0 \\ b_1 \\ b_2 \\ \vdots \\ b_{n-1} \end{bmatrix} = \begin{bmatrix} f(\lambda_1) \\ f(\lambda_2) \\ f(\lambda_3) \\ \vdots \\ f(\lambda_n) \end{bmatrix} \quad (5.43)$$

The process is illustrated in Example 5.9.

Example 5.9

Find $f(A) = e^A$ using Cayley–Hamilton theorem if:

$$A = \begin{bmatrix} 1 & -2 \\ 0 & -3 \end{bmatrix}$$

Matrix Polynomials and Functions of Square Matrices

Solution: The characteristic polynomial of A is:

$$|\lambda I - A| = \begin{vmatrix} \lambda - 1 & -2 \\ 0 & \lambda + 3 \end{vmatrix} = (\lambda - 1)(\lambda + 3) = 0$$

Hence, the eigenvalues of A are: $\lambda_1 = 1$ and $\lambda_2 = -3$
Therefore:

$$\begin{bmatrix} 1 & \lambda_1 \\ 1 & \lambda_2 \end{bmatrix} \begin{bmatrix} b_0 \\ b_1 \end{bmatrix} = \begin{bmatrix} f(\lambda_1) \\ f(\lambda_2) \end{bmatrix} \rightarrow \begin{bmatrix} 1 & 1 \\ 1 & -3 \end{bmatrix} \begin{bmatrix} b_0 \\ b_1 \end{bmatrix} = \begin{bmatrix} e \\ e^{-3} \end{bmatrix} \rightarrow \begin{bmatrix} b_0 \\ b_1 \end{bmatrix} = \begin{bmatrix} 0.75e + 0.25e^{-3} \\ 0.25e - 0.25e^{-3} \end{bmatrix}$$

Then:

$$e^A = b_0 I + b_1 A = \begin{bmatrix} b_0 & 0 \\ 0 & b_0 \end{bmatrix} + \begin{bmatrix} b_1 & -2b_1 \\ 0 & -3b_1 \end{bmatrix} = \begin{bmatrix} b_0 + b_1 & -2b_1 \\ 0 & b_0 - 3b_1 \end{bmatrix} = \begin{bmatrix} e & -0.5e + 0.5e^{-3} \\ 0 & e^{-3} \end{bmatrix}$$

Now let us assume that we have repeated eigenvalues. Without loss of generality, we assume that the $n \times n$ matrix A has one eigenvalue of multiplicity m and $n-m$ distinct eigenvalues. That is, the eigenvalues of A are:

$$\underbrace{\lambda_1, \lambda_1, ..., \lambda_1}_{m}, \lambda_{m+1}, \lambda_{m+2}, ..., \lambda_n \tag{5.44}$$

Since λ_1 is an eigenvalue with a multiplicity of m, then:

$$\frac{d^k Q(\lambda) P(\lambda)}{d\lambda^k}\bigg|_{\lambda_1} = 0, \quad \text{for} \quad k = 1, 2, ..., m \tag{5.45}$$

Therefore, we have the following n equations for the n unknowns $b_0, b_1, ..., b_{n-1}$

$$f(\lambda_1) = R(\lambda_1)$$
$$f'(\lambda_1) = R'(\lambda_1)$$
$$f''(\lambda_1) = R''(\lambda_1)$$
$$\vdots \quad = \quad \vdots$$
$$f^{m-1}(\lambda_1) = R^{m-1}(\lambda_1) \tag{5.46}$$
$$f(\lambda_{m+1}) = R(\lambda_{m+1})$$
$$f(\lambda_{m+2}) = R(\lambda_{m+2})$$
$$\vdots \quad = \quad \vdots$$
$$f(\lambda_n) = R(\lambda_n)$$

Example 5.10

Find $f(A) = e^{At}$ if:

$$A = \begin{bmatrix} 1 & -2 & -1 \\ 2 & -3 & -1 \\ 2 & -2 & -2 \end{bmatrix}$$

Solution: The eigenvalues of A are:

$$|\lambda I - A| = \begin{bmatrix} \lambda-1 & 2 & 1 \\ -2 & \lambda+3 & 1 \\ -2 & 2 & \lambda+2 \end{bmatrix} = (\lambda+1)^2(\lambda+2) = 0$$

$\lambda_1 = \lambda_2 = -1$ and $\lambda_3 = -2$

In this case, we have a simple eigenvalue and an eigenvalue with a multiplicity of two. Therefore, we have the following equations:

$$f(\lambda_1) = R(\lambda_1) = b_0 + b_1\lambda_1 + b_2\lambda_1^2 \qquad b_0 - b_1 + b_2 = e^{-t}$$

$$f'(\lambda_1) = R'(\lambda_1) = b_1 + 2b_2\lambda_1 \quad \rightarrow \quad b_1 - 2b_2 = te^{-t}$$

$$f(\lambda_2) = R(\lambda_2) = b_0 + b_1\lambda_2 + b_2\lambda_2^2 \qquad b_0 - 2b_1 + 4b_2 = e^{-2t}$$

These can be written in matrix form as:

$$\begin{bmatrix} 1 & -1 & 1 \\ 0 & 1 & -2 \\ 1 & -2 & 4 \end{bmatrix} \begin{bmatrix} b_0 \\ b_1 \\ b_2 \end{bmatrix} = \begin{bmatrix} e^{-t} \\ te^{-t} \\ e^{-2t} \end{bmatrix}$$

The solution is:

$$\begin{bmatrix} b_0 \\ b_1 \\ b_2 \end{bmatrix} = \begin{bmatrix} 1 & -1 & 1 \\ 0 & 1 & -2 \\ 1 & -2 & 4 \end{bmatrix}^{-1} \begin{bmatrix} e^{-t} \\ te^{-t} \\ e^{-2t} \end{bmatrix} = \begin{bmatrix} 0 & 2 & 1 \\ -2 & 3 & 2 \\ -1 & 1 & 1 \end{bmatrix} \begin{bmatrix} e^{-t} \\ te^{-t} \\ e^{-2t} \end{bmatrix} = \begin{bmatrix} 2te^{-t} + e^{-2t} \\ -2e^{-t} + 3te^{-t} + 2e^{-2t} \\ -e^{-t} + te^{-t} + e^{-2t} \end{bmatrix}$$

Hence:

$$e^{At} = b_0 I + b_1 A + b_2 A^2 = \begin{bmatrix} b_0 & 0 & 0 \\ 0 & b_0 & 0 \\ 0 & 0 & b_0 \end{bmatrix} + \begin{bmatrix} b_1 & -2b_1 & -b_1 \\ 2b_1 & -3b_1 & -b_1 \\ 2b_1 & -2b_1 & -2b_1 \end{bmatrix} + \begin{bmatrix} -5b_2 & 6b_2 & 3b_2 \\ -6b_2 & 7b_2 & 3b_2 \\ -6b_2 & 6b_2 & 4b_2 \end{bmatrix}$$

$$= \begin{bmatrix} b_0 + b_1 - 5b_2 & -2b_1 + 6b_2 & -b_1 + 3b_2 \\ 2b_1 - 6b_2 & b_0 - 3b_1 + 7b_2 & -b_1 + 3b_2 \\ 2b_1 - 6b_2 & -2b_1 + 6b_2 & b_0 - 2b_1 + 4b_2 \end{bmatrix}$$

$$= \begin{bmatrix} 3e^{-t} - 2e^{-2t} & -2e^{-t} + 2e^{-2t} & -e^{-t} + e^{-2t} \\ 2e^{-t} - 2e^{-2t} & -e^{-t} + 2e^{-2t} & -e^{-t} + e^{-2t} \\ 2e^{-t} - 2e^{-2t} & -2e^{-t} + 2e^{-2t} & e^{-2t} \end{bmatrix}$$

This yields:

$$e^{At} = \begin{bmatrix} 3e^{-t} - 2e^{-2t} & -2e^{-t} + 2e^{-2t} & -e^{-t} + e^{-2t} \\ 2e^{-t} - 2e^{-2t} & -e^{-t} + 2e^{-2t} & -e^{-t} + e^{-2t} \\ 2e^{-t} - 2e^{-2t} & -2e^{-t} + 2e^{-2t} & e^{-2t} \end{bmatrix}$$

Example 5.11

Find $f(A) = \ln(A)$ if:
$$A = \begin{bmatrix} 4.5 & -2.5 \\ 1.5 & 0.5 \end{bmatrix}$$

Solution: The eigenvalues of A are:
$$|\lambda I - A| = \begin{vmatrix} \lambda - 4.5 & 2.5 \\ -1.5 & \lambda - 0.5 \end{vmatrix} = (\lambda - 2)(\lambda - 3) = 0$$

$\lambda_1 = 2$ and $\lambda_2 = 3$.
Therefore:

$$\begin{bmatrix} 1 & \lambda_1 \\ 1 & \lambda_2 \end{bmatrix} \begin{bmatrix} b_0 \\ b_1 \end{bmatrix} = \begin{bmatrix} f(\lambda_1) \\ f(\lambda_2) \end{bmatrix} \rightarrow \begin{bmatrix} 1 & 2 \\ 1 & 3 \end{bmatrix} \begin{bmatrix} b_0 \\ b_1 \end{bmatrix} = \begin{bmatrix} \ln 2 \\ \ln 3 \end{bmatrix} \rightarrow \begin{bmatrix} b_0 \\ b_1 \end{bmatrix} = \begin{bmatrix} 3\ln 2 - 2\ln 3 \\ -\ln 2 + \ln 3 \end{bmatrix}$$

$$\ln(A) = b_0 I + b_1 A = \begin{bmatrix} b_0 & 0 \\ 0 & b_0 \end{bmatrix} + \begin{bmatrix} 4.5 b_1 & -2.5 b_1 \\ 1.5 b_1 & 0.5 b_1 \end{bmatrix} = \begin{bmatrix} b_0 + 4.5 b_1 & -2.5 b_1 \\ 1.5 b_1 & b_0 + 0.5 b_1 \end{bmatrix}$$

$$\ln(A) = \begin{bmatrix} -1.5\ln 2 + 2.5\ln 3 & 2.5\ln 2 - 2.5\ln 3 \\ -1.5\ln 2 + 1.5\ln 3 & 2.5\ln 2 - 1.5\ln 3 \end{bmatrix}$$

The MATLAB® command to compute $B = f(A)$ is $B =$ **funm(A,@name)**, where "name" stands for the name of the function. For example, if $f(A) = \ln(A)$, then the command to find $B = f(A)$ is $B =$ **funm(A,@log)**. If $f(A) = \cos(\omega_0 A)$, the command would be $B =$ **funm(ω_0A,@cos)**.

Example 5.12

Let the 2×2 matrix A be:
$$A = \begin{bmatrix} 1 & -1 \\ 2 & 4 \end{bmatrix}$$

Utilize Cayley–Hamilton technique to find the following functions and use MATLAB to check the answer in each case.

a. e^A
b. $\ln(A)$
c. A^k
d. $\cos\left(\dfrac{\pi}{2} A\right)$
e. $\sin\left(\dfrac{\pi}{2} A\right)$
f. $\cos^2\left(\dfrac{\pi}{2} A\right) + \sin^2\left(\dfrac{\pi}{2} A\right)$

Solution: The eigenvalues of A are:
$$|\lambda I - A| = \begin{vmatrix} \lambda - 1 & 1 \\ -2 & \lambda - 4 \end{vmatrix} = (\lambda - 2)(\lambda - 3) = 0$$

$\lambda_1 = 2$ and $\lambda_2 = 3$.

Therefore:

$$\begin{bmatrix} 1 & \lambda_1 \\ 1 & \lambda_2 \end{bmatrix} \begin{bmatrix} b_0 \\ b_1 \end{bmatrix} = \begin{bmatrix} f(\lambda_1) \\ f(\lambda_2) \end{bmatrix} \rightarrow \begin{bmatrix} 1 & 2 \\ 1 & 3 \end{bmatrix} \begin{bmatrix} b_0 \\ b_1 \end{bmatrix} = \begin{bmatrix} f(2) \\ f(3) \end{bmatrix} \rightarrow \begin{bmatrix} b_0 \\ b_1 \end{bmatrix} = \begin{bmatrix} 3f(2) - 2f(3) \\ -f(2) + f(3) \end{bmatrix}$$

The function $f(A)$ is:

$$f(A) = b_0 I + b_1 A = \begin{bmatrix} b_0 & 0 \\ 0 & b_0 \end{bmatrix} + \begin{bmatrix} b_1 & -b_1 \\ 2b_1 & 4b_1 \end{bmatrix} = \begin{bmatrix} b_0 + b_1 & -b_1 \\ 2b_1 & b_0 + 4b_1 \end{bmatrix}$$

or

$$f(A) = \begin{bmatrix} 2f(2) - f(3) & f(2) - f(3) \\ -2f(2) + 2f(3) & -f(2) + 2f(3) \end{bmatrix}$$

a. e^A

$$e^A = \begin{bmatrix} 2e^2 - e^3 & e^2 - e^3 \\ -2e^2 + 2e^3 & -e^2 + 2e^3 \end{bmatrix} = \begin{bmatrix} -5.3074 & -12.6965 \\ 25.3930 & 32.782 \end{bmatrix}$$

MATLAB: expm(A) or funm(A,@exp).

b. $\ln(A)$

$$\ln(A) = \begin{bmatrix} 2\ln(2) - \ln(3) & \ln(2) - \ln(3) \\ -2\ln(2) + 2\ln(3) & -\ln(2) + 2\ln(3) \end{bmatrix} = \begin{bmatrix} 0.2877 & -0.4055 \\ 0.8109 & 1.5041 \end{bmatrix}$$

MATLAB: logm(A) or funm(A,@log).

c. A^k

$$A^k = \begin{bmatrix} 2 \times 2^k - 3^k & 2^k - 3^k \\ -2 \times 2^k + 2 \times 3^k & -2^k + 2 \times 3^k \end{bmatrix}$$

MATLAB: for $k=4$ the command to find $B=A^4$ is B=A^4.

d. $\cos\left(\frac{\pi}{2}A\right)$

$$\cos\left(\frac{\pi}{2}A\right) = \begin{bmatrix} 2\cos\left(2\frac{\pi}{2}\right) - \cos\left(3\frac{\pi}{2}\right) & \cos\left(2\frac{\pi}{2}\right) - \cos\left(3\frac{\pi}{2}\right) \\ -2\cos\left(2\frac{\pi}{2}\right) + 2\cos\left(3\frac{\pi}{2}\right) & -\cos\left(2\frac{\pi}{2}\right) + 2\cos\left(3\frac{\pi}{2}\right) \end{bmatrix} = \begin{bmatrix} -2 & -1 \\ 2 & 1 \end{bmatrix}$$

MATLAB: funm(pi*A/2,@cos).

e. $\sin\left(\frac{\pi}{2}A\right)$

$$\sin\left(\frac{\pi}{2}A\right) = \begin{bmatrix} 2\sin\left(2\frac{\pi}{2}\right) - \cos\left(3\frac{\pi}{2}\right) & \sin\left(2\frac{\pi}{2}\right) - \sin\left(3\frac{\pi}{2}\right) \\ -2\sin\left(2\frac{\pi}{2}\right) + 2\cos\left(3\frac{\pi}{2}\right) & -\sin\left(2\frac{\pi}{2}\right) + 2\sin\left(3\frac{\pi}{2}\right) \end{bmatrix} = \begin{bmatrix} 1 & 1 \\ -2 & -2 \end{bmatrix}$$

MATLAB: funm(pi*A/2,@sin).

Matrix Polynomials and Functions of Square Matrices

f. $\cos^2\left(\frac{\pi}{2}A\right) + \sin^2\left(\frac{\pi}{2}A\right)$

$$\cos^2\left(\frac{\pi}{2}A\right) + \sin^2\left(\frac{\pi}{2}A\right) = \begin{bmatrix} -2 & -1 \\ 2 & 1 \end{bmatrix}\begin{bmatrix} -2 & -1 \\ 2 & 1 \end{bmatrix} + \begin{bmatrix} 1 & 1 \\ -2 & -2 \end{bmatrix}\begin{bmatrix} 1 & 1 \\ -2 & -2 \end{bmatrix} = \begin{bmatrix} 1 & 0 \\ 0 & 1 \end{bmatrix}$$

MATLAB: (funm(pi*A/2,@cos))^2+(funm(pi*A/2,@sin))^2.

5.4.3 Modal Matrix Technique

Assume that the $n \times n$ matrix A has n distinct eigenvalues and let M be the modal matrix of A, then:

$$A = M\Lambda M^{-1} \tag{5.47}$$

Let the Taylor series expansion of the function $f(A)$ be:

$$f(A) = \sum_{k=0}^{\infty} \alpha_k A^k \tag{5.48}$$

Now substitute Equation 5.48 into Equation 5.47.

$$f(A) = \sum_{k=0}^{\infty} \alpha_k (M\Lambda M^{-1})^k = \sum_{k=0}^{\infty} \alpha_k M\Lambda^k M^{-1} = M \sum_{k=0}^{\infty} \alpha_k \Lambda^k M^{-1} \tag{5.49}$$

But

$$\sum_{k=0}^{\infty} \alpha_k \Lambda^k = \begin{bmatrix} \sum_{k=0}^{\infty} \alpha_k \lambda_1^k & 0 & \cdots & 0 & 0 \\ 0 & \sum_{k=0}^{\infty} \alpha_k \lambda_2^k & \cdots & 0 & 0 \\ \vdots & \vdots & \ddots & \vdots & \vdots \\ 0 & 0 & \cdots & \sum_{k=0}^{\infty} \alpha_k \lambda_{n-1}^k & 0 \\ 0 & 0 & \cdots & 0 & \sum_{k=0}^{\infty} \alpha_k \lambda_n^k \end{bmatrix}$$

$$= \begin{bmatrix} f(\lambda_1) & 0 & \cdots & 0 & 0 \\ 0 & f(\lambda_2) & \cdots & 0 & 0 \\ \vdots & \vdots & \ddots & \vdots & \vdots \\ 0 & 0 & \cdots & f(\lambda_{n-1}) & 0 \\ 0 & 0 & \cdots & 0 & f(\lambda_n) \end{bmatrix} \tag{5.50}$$

Hence:

$$f(A) = M \begin{bmatrix} f(\lambda_1) & 0 & \cdots & 0 & 0 \\ 0 & f(\lambda_2) & & 0 & 0 \\ \vdots & \vdots & \ddots & \vdots & \vdots \\ 0 & 0 & \cdots & f(\lambda_{n-1}) & 0 \\ 0 & 0 & \cdots & 0 & f(\lambda_n) \end{bmatrix} M^{-1} \quad (5.51)$$

Example 5.13

Find $f(A) = A^k$ if:

$$A = \begin{bmatrix} 0.5 & 0.1 \\ 0.1 & 0.5 \end{bmatrix}$$

Solution: The eigenvalues and eigenvectors of A are:

$$|\lambda I - A| = \begin{vmatrix} \lambda - 0.5 & -0.1 \\ -0.1 & \lambda - 0.5 \end{vmatrix} = (\lambda - 0.4)(\lambda - 0.6) = 0$$

$\lambda_1 = 0.4$ and $\lambda_2 = 0.6$
The corresponding eigenvectors are $x_1 = \begin{bmatrix} 1 \\ -1 \end{bmatrix}$ and $x_2 = \begin{bmatrix} 1 \\ 1 \end{bmatrix}$
Therefore:

$$f(A) = Mf(\Lambda)M^{-1} = \begin{bmatrix} 1 & 1 \\ -1 & 1 \end{bmatrix} \begin{bmatrix} (0.4)^k & 0 \\ 0 & (0.6)^k \end{bmatrix} \begin{bmatrix} 1 & 1 \\ -1 & 1 \end{bmatrix}^{-1} = \frac{1}{2} \begin{bmatrix} (0.4)^k & (0.6)^k \\ -(0.4)^k & (0.6)^k \end{bmatrix} \begin{bmatrix} 1 & -1 \\ 1 & 1 \end{bmatrix}$$

$$= \begin{bmatrix} 0.5(0.4)^k + 0.5(0.6)^k & -0.5(0.4)^k + 0.5(0.6)^k \\ -0.5(0.4)^k + 0.5(0.6)^k & 0.5(0.4)^k + 0.5(0.6)^k \end{bmatrix}$$

Hence:

$$A^k = \begin{bmatrix} 0.5(0.4)^k + 0.5(0.6)^k & -0.5(0.4)^k + 0.5(0.6)^k \\ -0.5(0.4)^k + 0.5(0.6)^k & 0.5(0.4)^k + 0.5(0.6)^k \end{bmatrix}$$

Note that $A^0 = I$ and $A^1 = A$.

5.4.4 Special Matrix Functions

Certain functions of matrices have applications in linear system theory, control and communication. Among them are two functions frequently used in modern control system design. These are e^{At} and A^k.

5.4.4.1 Matrix Exponential Function e^{At}

The matrix exponential function e^{At} is a very important function with applications in analysis and design of continuous time control systems. It is defined by the infinite series:

$$e^{At} = I + At + \frac{A^2 t^2}{2!} + \frac{A^3 t^3}{3!} + \cdots + \frac{A^n t^n}{n!} + \cdots \quad (5.52)$$

This series is a convergent series for all values of t and can be computed using different techniques, such as Cayley–Hamilton or Sylvester's technique. The derivative of this function with respect to t is:

$$\begin{aligned} \frac{de^{At}}{dt} &= A + \frac{A^2 t}{1!} + \frac{A^3 t^2}{2!} + \cdots + \frac{A^n t^{n-1}}{(n-1)!} + \cdots \\ &= A(I + At + \frac{A^2 t^2}{2!} + \frac{A^3 t^3}{3!} + \cdots) \\ &= A e^{At} = e^{At} A \end{aligned} \quad (5.53)$$

The integral of the exponential function e^{At} is computed by:

$$\int_0^t e^{A\tau} d\tau = A^{-1}(e^{At} - I) \quad (5.54)$$

The above closed form solution is valid if and only if matrix A is nonsingular. If matrix A is singular, there is no simple closed form solution.

Example 5.14

Find e^{At}, $\dfrac{de^{At}}{dt}$ and $\displaystyle\int_0^t e^{A\tau} d\tau$ if:

a. $A = \begin{bmatrix} 0 & 1 \\ -2 & -3 \end{bmatrix}$

b. $A = \begin{bmatrix} 0 & 1 \\ 0 & -1 \end{bmatrix}$

Solution: (a) First we find e^{At} using the modal matrix technique. The eigenvalues of matrix A are:

$$|\lambda I - A| = \begin{vmatrix} \lambda & -1 \\ 2 & \lambda + 3 \end{vmatrix} = (\lambda + 1)(\lambda + 2) = 0 \quad \rightarrow \quad \lambda_1 = -1, \ \lambda_2 = -2$$

The corresponding eigenvectors are:
$Ax_1 = \lambda_1 x_1$, therefore $x_1 = [1 \ -1]^T$
$Ax_2 = \lambda_2 x_2$, therefore $x_2 = [1 \ -2]^T$

Then:

$$e^{At} = M\Lambda(t)M^{-1} = \begin{bmatrix} 1 & 1 \\ -1 & -2 \end{bmatrix} \begin{bmatrix} e^{-t} & 0 \\ 0 & e^{-2t} \end{bmatrix} \begin{bmatrix} 1 & 1 \\ -1 & -2 \end{bmatrix} = \begin{bmatrix} 2e^{-t} - e^{-2t} & e^{-t} - e^{-2t} \\ -2e^{-t} + 2e^{-2t} & -e^{-t} + 2e^{-2t} \end{bmatrix}$$

The derivative of e^{At} is:

$$\frac{de^{At}}{dt} = Ae^{At} = \begin{bmatrix} 0 & 1 \\ -2 & -3 \end{bmatrix} \begin{bmatrix} 2e^{-t} - e^{-2t} & e^{-t} - e^{-2t} \\ -2e^{-t} + 2e^{-2t} & -e^{-t} + 2e^{-2t} \end{bmatrix} = \begin{bmatrix} -2e^{-t} + 2e^{-2t} & -e^{-t} + 2e^{-2t} \\ 2e^{-t} - 4e^{-2t} & e^{-t} - 4e^{-2t} \end{bmatrix}$$

Since matrix A is nonsingular (both eigenvalues are nonzero), then:

$$\int_0^t e^{A\tau}d\tau = A^{-1}(e^{At} - I) = \begin{bmatrix} 0 & 1 \\ -2 & -3 \end{bmatrix}^{-1} \left(\begin{bmatrix} 2e^{-t} - e^{-2t} & e^{-t} - e^{-2t} \\ -2e^{-t} + 2e^{-2t} & -e^{-t} + 2e^{-2t} \end{bmatrix} - \begin{bmatrix} 1 & 0 \\ 0 & 1 \end{bmatrix} \right)$$

$$= \begin{bmatrix} -1.5 & -0.5 \\ 1 & 0 \end{bmatrix} \begin{bmatrix} 2e^{-t} - e^{-2t} - 1 & e^{-t} - e^{-2t} \\ -2e^{-t} + 2e^{-2t} & -e^{-t} + 2e^{-2t} - 1 \end{bmatrix}$$

$$= \begin{bmatrix} -2e^{-t} + 0.5e^{-2t} + 1.5 & -e^{-t} - 0.5e^{-2t} \\ 2e^{-t} - e^{-2t} - 1 & e^{-t} - e^{-2t} \end{bmatrix}$$

b. The eigenvalues of matrix A are:

$$|\lambda I - A| = \begin{vmatrix} \lambda & 1 \\ 0 & \lambda + 1 \end{vmatrix} = \lambda(\lambda + 1) = 0 \quad \rightarrow \quad \lambda_1 = 0, \lambda_2 = -1$$

The corresponding eigenvectors are:
$Ax_1 = \lambda_1 x_1$, therefore $x_1 = [1 \ 0]^T$
$Ax_2 = \lambda_2 x_2$, therefore $x_2 = [1 \ -1]^T$
Then:

$$e^{At} = M\Lambda(t)M^{-1} = \begin{bmatrix} 1 & 1 \\ 0 & -1 \end{bmatrix} \begin{bmatrix} 1 & 0 \\ 0 & e^{-t} \end{bmatrix} \begin{bmatrix} 1 & 1 \\ 0 & -1 \end{bmatrix} = \begin{bmatrix} 1 & 1 - e^{-t} \\ 0 & e^{-t} \end{bmatrix}$$

The derivative of e^{At} is:

$$\frac{de^{At}}{dt} = Ae^{At} = \begin{bmatrix} 0 & 1 \\ 0 & -1 \end{bmatrix} \begin{bmatrix} 1 & 1 - e^{-t} \\ 0 & e^{-t} \end{bmatrix} = \begin{bmatrix} 0 & e^{-t} \\ 0 & -e^{-t} \end{bmatrix}$$

Matrix A is singular, therefore Equation 5.54 is not valid. Direct integration results in

$$\int_0^t e^{A\tau}d\tau = \begin{bmatrix} \int_0^t d\tau & \int_0^t (1 - e^{-\tau})d\tau \\ 0 & \int_0^t e^{-\tau}d\tau \end{bmatrix} = \begin{bmatrix} t & t - 1 + e^{-t} \\ 0 & 1 - e^{-t} \end{bmatrix}$$

Matrix Polynomials and Functions of Square Matrices 249

5.4.4.2 Matrix Function A^k

Matrix function A^k is also useful in the analysis and design of discrete-time control systems. It is defined as:

$$A^k = \overbrace{A \times A \times \cdots \times A}^{k \text{ times}} \tag{5.55}$$

It can be computed using techniques such as Cayley–Hamilton or Sylvester's theorem.

Example 5.15

For the matrix A given by:

$$A = \begin{bmatrix} 0.75 & -1 & -0.5 \\ -0.75 & 1 & 1.5 \\ 0.25 & -0.5 & 0 \end{bmatrix}$$

a. Find A^k.
b. $\lim_{k \to \infty} A^k$.

Solution: (a) The eigenvalues and eigenvectors of A are:

$$\lambda_1 = 1 \quad x_1 = \begin{bmatrix} 2 \\ -1 \\ 1 \end{bmatrix}$$

$$\lambda_2 = 0.5 \quad x_2 = \begin{bmatrix} 2 \\ 0 \\ 1 \end{bmatrix}$$

$$\lambda_3 = 0.25 \quad x_3 = \begin{bmatrix} 3 \\ 1 \\ 1 \end{bmatrix}$$

Using the modal matrix approach, we have:

$$A^k = M\Lambda^k M^{-1} = \begin{bmatrix} 2 & 2 & 3 \\ -1 & 0 & 1 \\ 1 & 1 & 1 \end{bmatrix} \begin{bmatrix} 1^k & 0 & 0 \\ 0 & 0.5^k & 0 \\ 0 & 0 & 0.25^k \end{bmatrix} \begin{bmatrix} 2 & 2 & 3 \\ -1 & 0 & 1 \\ 1 & 1 & 1 \end{bmatrix}^{-1}$$

$$= \begin{bmatrix} 2 & 2 & 3 \\ -1 & 0 & 1 \\ 1 & 1 & 1 \end{bmatrix} \begin{bmatrix} 1^k & 0 & 0 \\ 0 & 0.5^k & 0 \\ 0 & 0 & 0.25^k \end{bmatrix} \begin{bmatrix} 1 & -1 & -2 \\ -2 & 1 & 5 \\ 1 & 0 & -2 \end{bmatrix}$$

$$= \begin{bmatrix} 2 & 2 & 3 \\ -1 & 0 & 1 \\ 1 & 1 & 1 \end{bmatrix} \begin{bmatrix} 1 & -1 & -2 \\ -2 \times 0.5^k & 0.5^k & 5 \times 0.5^k \\ 0.25^k & 0 & -2 \times 0.25^k \end{bmatrix}$$

Therefore:

$$A^k = \begin{bmatrix} 2-4\times 0.5^k + 3\times 0.25^k & -2+2\times 0.5^k & -4+10\times 0.5^k - 6\times 0.25^k \\ -1+0.25^k & 1 & 2-2\times 0.5^k \\ 1-2\times 0.5^k + 0.25^k & -1+0.5^k & -2+5\times 0.5^k - 2\times 0.25^k \end{bmatrix}$$

b. $\lim_{k\to\infty} A^k = \begin{bmatrix} 2 & -2 & -4 \\ -1 & 1 & 2 \\ 1 & -1 & -2 \end{bmatrix}$

5.5 The State Space Modeling of Linear Continuous-time Systems

Conventional control system design is rooted in transfer function and frequency response analysis. Modern control theory, on the other hand, is based on the state space formulation of the system. One advantage of modern control theory over the conventional approach is in its applicability to both single-input single-output (SISO) and multiple-input multiple-output (MIMO) systems. Another advantage of the state space formulation is that it is a unified approach which can be applied to both linear as well as nonlinear dynamic systems. In this section, we present: (i) the fundamentals of system modeling in state space, (ii) the solutions of state equations, and (iii) the determination of controllability and observability of linear time invariant (LTI) systems for both continuous as well as discrete-time systems.

5.5.1 Concept of States

Definition: The state of a continuous-time dynamic system is the minimum number of variables called states variables such that the knowledge of these variables at time $t=t_0$, together with input for $t \geq t_0$, uniquely determines the behavior of the system for $t \geq t_0$. If n variables are needed, then these n variables are considered components of a n-dimensional vector x called the state vector. The n-dimensional space whose coordinates are the states of the system is called the state space. The state of a system at time t is a point in the n-dimensional state space.

5.5.2 State Equations of Continuous Time Systems

The state space equations describing a multiple-input-multiple-output (MIMO) continuous-time dynamic system have a general form given by:

$$\frac{dx(t)}{dt} = f(x(t), u(t), t) \tag{5.56}$$

$$y(t) = h(x(t), u(t), t) \tag{5.57}$$

where $x(t)=[x_1(t) \;\; x_2(t) \;\; \cdots \;\; x_n(t)]^T$ is $n\times 1$ state vector, $u(t)$ is the $M\times 1$ input vector and $y(t)$ is the $P\times 1$ output vector. The vector functions f and h are nonlinear functions of states and input vectors given by:

$$f(x(t),u(t),t) = \begin{bmatrix} f_1(x_1(t),x_2(t),\cdots x_n(t),u_1(t),u_2(t),\cdots,u_M(t),t) \\ f_2(x_1(t),x_2(t),\cdots x_n(t),u_1(t),u_2(t),\cdots,u_M(t),t) \\ \vdots \\ f_n(x_1(t),x_2(t),\cdots x_n(t),u_1(t),u_2(t),\cdots,u_M(t),t) \end{bmatrix} \quad (5.58)$$

and

$$h(x(t),u(t),t) = \begin{bmatrix} h_1(x_1(t),x_2(t),\cdots x_n(t),u_1(t),u_2(t),\cdots,u_M(t),t) \\ h_2(x_1(t),x_2(t),\cdots x_n(t),u_1(t),u_2(t),\cdots,u_M(t),t) \\ \vdots \\ h_P(x_1(t),x_2(t),\cdots x_n(t),u_1(t),u_2(t),\cdots,u_M(t),t) \end{bmatrix} \quad (5.59)$$

The state equations given by Equations 5.57 and 5.58 are applicable to both linear and nonlinear, time invariant (TI) and time varying systems. If the system is linear and time invariant, then the state and output equations can be written as:

$$\frac{dx(t)}{dt} = Ax(t) + Bu(t) \quad (5.60)$$

$$y(t) = Cx(t) + Du(t) \quad (5.61)$$

where A, B, C and D are $n\times n$, $n\times M$, $P\times n$ and $P\times M$ constant matrices, respectively. The block diagram of a continuous LTI control system in state-space form is shown in Figure 5.1.

In the next section, we show how the state equations can be derived for electrical and mechanical systems.

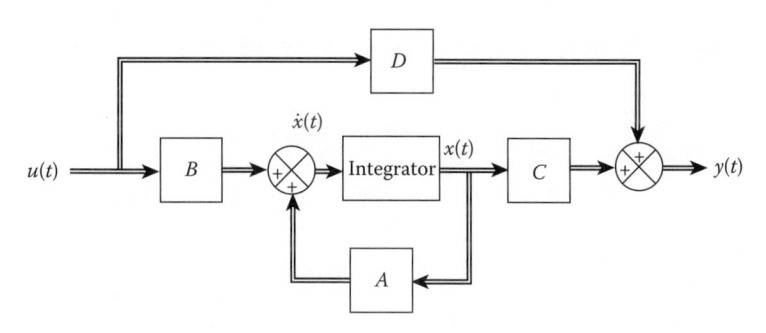

FIGURE 5.1
Block diagram of a linear continuous time invariant system in state space.

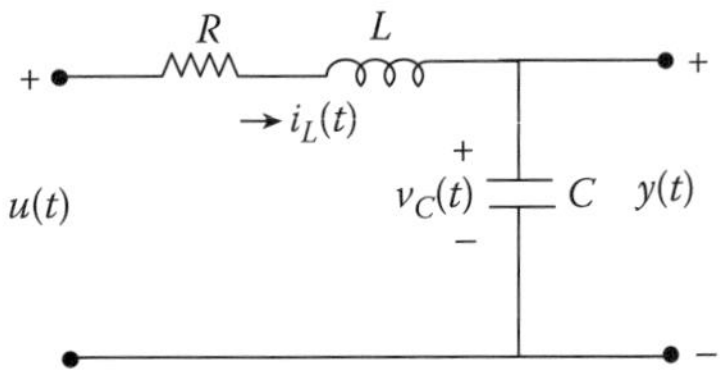

FIGURE 5.2
Series RLC circuit.

This introduction effectively creates a simple filter.

Example 5.16

Consider the RLC circuit shown in Figure 5.2 with the input u(t) and output y(t).

To determine the output of this system, one needs to know the input, the initial charge on the capacitor and the initial current through the inductor. Therefore, the system has two states. The first state is defined as the voltage across the capacitor and the second state as the current through the inductor. These are denoted as follows:

$$x_1(t) = v_c(t)$$
$$x_2(t) = i_L(t)$$
(5.62)

Differentiating both sides of Equation 5.62 yields:

$$\dot{x}_1(t) = \frac{dv_c(t)}{dt} = \frac{i_L(t)}{C} = \frac{1}{C}x_2(t)$$

$$\dot{x}_2(t) = \frac{di_L(t)}{dt} = \frac{v_L(t)}{L} = \frac{1}{L}[u(t) - v_R(t) - v_C(t)] = \frac{1}{L}[u(t) - Rx_2(t) - x_1(t)]$$
(5.63)

$$= \frac{u(t)}{L} - \frac{R}{L}x_2(t) - \frac{1}{L}x_1(t)$$

The output equation is:

$$y(t) = x_1(t)$$
(5.64)

The above equations can be written in matrix form as:

$$\begin{bmatrix} \dot{x}_1(t) \\ \dot{x}_2(t) \end{bmatrix} = \begin{bmatrix} \frac{1}{C} & 0 \\ -\frac{1}{L} & -\frac{R}{L} \end{bmatrix} \begin{bmatrix} x_1(t) \\ x_2(t) \end{bmatrix} + \begin{bmatrix} 0 \\ \frac{1}{L} \end{bmatrix} u(t)$$

(5.65)

$$y(t) = \begin{bmatrix} 1 & 0 \end{bmatrix} \begin{bmatrix} x_1(t) \\ x_2(t) \end{bmatrix}$$

Therefore:

$$A = \begin{bmatrix} \frac{1}{C} & 0 \\ -\frac{1}{L} & -\frac{R}{L} \end{bmatrix}, \quad B = \begin{bmatrix} 0 \\ \frac{1}{L} \end{bmatrix}, \quad C = \begin{bmatrix} 1 & 0 \end{bmatrix}, \quad \text{and} \quad D = 0 \quad (5.66)$$

As can be seen from this simple example, the number of states needed to model an electric circuit in state space form is equal to the total number of independent energy storage elements in the circuit. In the circuit shown in Figure 5.2, there are two elements that can store energy: the capacitor and the inductor. Hence, two states are needed. In the next example, we consider a mechanical system with two degrees of freedom.

Example 5.17

Consider the spring-mass system with friction shown in Figure 5.3.
Applying Newton's law, we have:

$$m\frac{d^2x(t)}{dt^2} = u(t) - kx(t) - c\frac{dx(t)}{dt} \quad (5.67)$$

Defining the two states of the system to be the mass displacement $x(t)$ and its velocity $\dot{x}(t)$ yields:

$$\begin{aligned} x_1(t) &= x(t) \\ x_2(t) &= \dot{x}(t) \end{aligned} \quad (5.68)$$

Differentiating both sides of Equations 5.68 results in:

$$\dot{x}_1(t) = \frac{dx(t)}{dt} = x_2(t)$$

$$\dot{x}_2(t) = \frac{d^2x(t)}{dt^2} = -\frac{k}{m}x(t) - \frac{c}{m}\frac{dx(t)}{dt} + \frac{1}{m}u(t) = -\frac{k}{m}x_1(t) - \frac{c}{m}x_2(t) + \frac{1}{m}u(t) \quad (5.69)$$

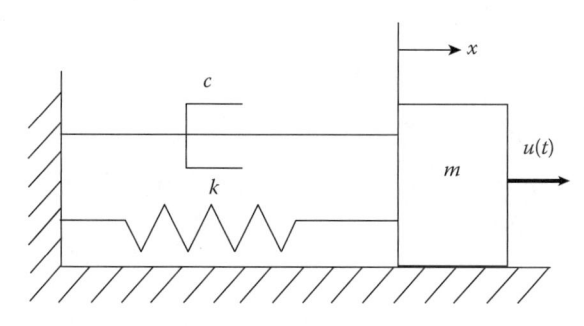

FIGURE 5.3
Mass-spring system with friction.

The above equations can be written in matrix form as:

$$\begin{bmatrix} \dot{x}_1(t) \\ \dot{x}_2(t) \end{bmatrix} = \begin{bmatrix} 0 & 1 \\ -\dfrac{k}{m} & -\dfrac{c}{m} \end{bmatrix} \begin{bmatrix} x_1(t) \\ x_2(t) \end{bmatrix} + \begin{bmatrix} 0 \\ \dfrac{1}{m} \end{bmatrix} u(t) \tag{5.70}$$

5.5.3 State Space Representation of Continuous LTI Systems

To obtain the state equations, consider a SISO system described by the nth order constant coefficients differential equation with input $u(t)$ and output $y(t)$:

$$\frac{dy^n(t)}{dt^n} + a_1 \frac{dy^{n-1}(t)}{dt^{n-1}} + \cdots + a_n y(t) = u(t) \tag{5.71}$$

The states of the above dynamic system can be defined as:

$$x_1(t) = y(t)$$

$$x_2(t) = \frac{dy(t)}{dt}$$

$$x_3(t) = \frac{d^2 y(t)}{dt^2} \tag{5.72}$$

$$\vdots$$

$$x_n(t) = \frac{dy^{n-1}(t)}{dt^{n-1}}$$

Differentiating the above equations with respect to t results in the state equations below.

$$\dot{x}_1(t) = \frac{dy(t)}{dt} = x_2(t)$$

$$\dot{x}_2(t) = \frac{dy^2(t)}{dt^2} = x_3(t)$$

$$\dot{x}_3(t) = \frac{d^3 y(t)}{dt^3} = x_4(t) \tag{5.73}$$

$$\vdots$$

$$\dot{x}_n(t) = \frac{dy^n(t)}{dt^n} = u(t) - a_1 \frac{dy^{n-1}(t)}{dt^{n-1}} - a_2 \frac{dy^{n-2}(t)}{dt^{n-2}} - \cdots - a_n y(t)$$

$$= u(t) - a_1 x_{n-1}(t) - a_2 x_{n-2}(t) - \cdots - a_n x_1(t)$$

Matrix Polynomials and Functions of Square Matrices 255

The above equation can be put in matrix form as:

$$\begin{bmatrix} \dot{x}_1(t) \\ \dot{x}_2(t) \\ \vdots \\ \dot{x}_{n-1}(t) \\ \dot{x}_n(t) \end{bmatrix} = \begin{bmatrix} 0 & 1 & 0 & \cdots & 0 \\ 0 & 0 & 1 & \cdots & 0 \\ \vdots & \vdots & \vdots & \vdots & \vdots \\ 0 & 0 & 0 & \cdots & 1 \\ -a_n & -a_{n-1} & -a_{n-2} & \cdots & -a_1 \end{bmatrix} \begin{bmatrix} x_1(t) \\ x_2(t) \\ \vdots \\ x_{n-1}(t) \\ x_n(t) \end{bmatrix} + \begin{bmatrix} 0 \\ 0 \\ \vdots \\ 0 \\ 1 \end{bmatrix} u(t) \quad (5.74)$$

$$y(t) = x_1(t) = \begin{bmatrix} 1 & 0 & \cdots & 0 \end{bmatrix} \begin{bmatrix} x_1(t) \\ x_2(t) \\ \vdots \\ x_{n-1}(t) \\ x_n(t) \end{bmatrix} \quad (5.75)$$

In general, if the differential equation describing the system is given by:

$$\frac{dy^n(t)}{dt^n} + a_1 \frac{dy^{n-1}(t)}{dt^{n-1}} + \cdots + a_n y(t) = b_0 \frac{du^n(t)}{dt^n} + b_1 \frac{du^{n-1}(t)}{dt^{n-1}} + \cdots + b_n u(t) \quad (5.76)$$

then the state equations are:

$$\begin{bmatrix} \dot{x}_1(t) \\ \dot{x}_2(t) \\ \vdots \\ \dot{x}_{n-1}(t) \\ \dot{x}_n(t) \end{bmatrix} = \begin{bmatrix} 0 & 1 & 0 & \cdots & 0 \\ 0 & 0 & 1 & \cdots & 0 \\ \vdots & \vdots & \vdots & \vdots & \vdots \\ 0 & 0 & 0 & \cdots & 1 \\ -a_n & -a_{n-1} & -a_{n-2} & \cdots & -a_1 \end{bmatrix} \begin{bmatrix} x_1(t) \\ x_2(t) \\ \vdots \\ x_{n-1}(t) \\ x_n(t) \end{bmatrix} + \begin{bmatrix} 0 \\ 0 \\ \vdots \\ 0 \\ 1 \end{bmatrix} u(t) \quad (5.77)$$

and the output equation is:

$$y(t) = \begin{bmatrix} b_n - a_n b_0 & b_{n-1} - a_{n-1} b_0 & b_{n-2} - a_{n-2} b_0 & \cdots & b_1 - a_1 b_0 \end{bmatrix} \begin{bmatrix} x_1(t) \\ x_2(t) \\ \vdots \\ x_{n-1}(t) \\ x_n(t) \end{bmatrix} + b_0 u(t) \quad (5.78)$$

The above realization is known as the controllable canonical form. Another realization known as the observable canonical form is given by the following state equations:

$$\begin{bmatrix} \dot{x}_1(t) \\ \dot{x}_2(t) \\ \vdots \\ \dot{x}_{n-1}(t) \\ \dot{x}_n(t) \end{bmatrix} = \begin{bmatrix} 0 & 0 & \cdots & 0 & -a_n \\ 1 & 0 & \cdots & 0 & -a_{n-1} \\ 0 & 1 & \cdots & 0 & a_{n-3} \\ \vdots & \vdots & \vdots & \vdots & \vdots \\ 0 & 0 & \cdots & 1 & -a_1 \end{bmatrix} \begin{bmatrix} x_1(t) \\ x_2(t) \\ \vdots \\ x_{n-1}(t) \\ x_n(t) \end{bmatrix} + \begin{bmatrix} b_n - a_n b_0 \\ b_{n-1} - a_{n-1} b_0 \\ b_{n-2} - a_{n-2} b_0 \\ \vdots \\ b_1 - a_1 b_0 \end{bmatrix} u(t) \quad (5.79)$$

with the output equation

$$y(t) = \begin{bmatrix} 0 & 0 & 0 & \cdots & 1 \end{bmatrix} \begin{bmatrix} x_1(t) \\ x_2(t) \\ \vdots \\ x_{n-1}(t) \\ x_n(t) \end{bmatrix} + b_0 u(t) \qquad (5.80)$$

The MATLAB® commands to find the state space realization of a dynamic system in controllable canonical form are:

$$\text{num} = \begin{bmatrix} b_0 & b_1 & \cdots & b_n \end{bmatrix};$$

$$\text{den} = \begin{bmatrix} 1 & a_1 & \cdots & a_n \end{bmatrix};$$

$$[A, \ B, \ C, \ D,] = \text{tf2ss(num,den)};$$

Example 5.18

Consider the following system:

$$\frac{dy^3(t)}{dt^3} + 4\frac{dy^2(t)}{dt^2} + \frac{dy(t)}{dt} + 8y = \frac{du^2(t)}{dt^2} + 8\frac{du(t)}{dt} + 9u(t)$$

Obtain a state space representation of this system in:
a. Controllable canonical form.
b. Observable canonical form.

Solution:
a. Controllable canonical form:

$$\begin{bmatrix} \dot{x}_1(t) \\ \dot{x}_2(t) \\ \dot{x}_3(t) \end{bmatrix} = \begin{bmatrix} 0 & 1 & 0 \\ 0 & 0 & 1 \\ -8 & -1 & -4 \end{bmatrix} \begin{bmatrix} x_1(t) \\ x_2(t) \\ x_3(t) \end{bmatrix} + \begin{bmatrix} 0 \\ 0 \\ 1 \end{bmatrix} u(t), \text{ and } y(t) = \begin{bmatrix} 9 & 8 & 1 \end{bmatrix} \begin{bmatrix} x_1(t) \\ x_2(t) \\ x_3(t) \end{bmatrix}$$

b. Observable canonical form:

$$\begin{bmatrix} \dot{x}_1(t) \\ \dot{x}_2(t) \\ \dot{x}_3(t) \end{bmatrix} = \begin{bmatrix} 0 & 0 & -8 \\ 1 & 0 & -1 \\ 0 & 1 & -4 \end{bmatrix} \begin{bmatrix} x_1(t) \\ x_2(t) \\ x_3(t) \end{bmatrix} + \begin{bmatrix} 9 \\ 8 \\ 1 \end{bmatrix} u(t), \text{ and } y(t) = \begin{bmatrix} 0 & 0 & 1 \end{bmatrix} \begin{bmatrix} x_1(t) \\ x_2(t) \\ x_3(t) \end{bmatrix}$$

5.5.4 Solution of Continuous-time State Space Equations

Consider a LTI system described by the state space equations:

$$\dot{x}(t) = Ax(t) + Bu(t) \qquad (5.81)$$

$$y(t) = Cx(t) + Du(t) \qquad (5.82)$$

Matrix Polynomials and Functions of Square Matrices

To obtain the output of this system for a given input and an initial state, we first solve the state equations given by Equation 5.81. The solution is then substituted into algebraic Equation 5.82 in order to find the output $y(t)$. The solution to the state equation has two parts: a homogenous solution and a particular solution. We first consider the homogenous solution.

5.5.5 Solution of Homogenous State Equations and State Transition Matrix

The homogenous solution represents the solution of the state equations to an arbitrary initial condition with zero input. The homogenous state equation is given by:

$$\dot{x}(t) = Ax(t) \tag{5.83}$$

Assuming that the initial state is $x(0)$, then:

$$x(t) = e^{At}x(0) \tag{5.84}$$

where e^{At} is the matrix exponential function. To verify that this is the solution to the homogenous equation, we need to show that it satisfies the initial condition as well as the differential equation. The initial condition is satisfied since:

$$x(0) = e^{A0}x(0) = Ix(0) = x(0) \tag{5.85}$$

Differentiating both sides of Equation 5.83, we have:

$$\dot{x}(t) = \frac{de^{At}x(0)}{dt} = Ae^{At}x(0) = Ax(t) \tag{5.86}$$

Therefore, the homogenous part of the solution is $x(t) = e^{At}x(0)$. The matrix exponential e^{At} is called the state transition matrix and is denoted by $\phi(t)$. The state transition matrix $\phi(t)$ is an $n \times n$ matrix and is the solution to the homogenous equation:

$$\frac{d\phi(t)}{dt} = A\phi(t)$$
$$\phi(0) = I \tag{5.87}$$

The state transition matrix is given by:

$$\phi(t) = e^{At} \tag{5.88}$$

5.5.6 Properties of State Transition Matrix

State transition matrix $\phi(t) = e^{At}$ has the following properties:

a. $\phi(0) = I$
b. $\phi(-t) = \phi^{-1}(t)$
c. $\phi(t_1 + t_2) = \phi(t_1)\phi(t_2) = \phi(t_2)\phi(t_1)$
d. $\dfrac{d\phi(t)}{dt} = A\phi(t) = \phi(t)A$

The proofs are left as an exercise.

5.5.7 Computing State Transition Matrix

There are several methods to compute the state transition matrix e^{At} as described before. Here we discuss a new approach based on the Laplace transform (for a brief review of the Laplace transforms, see Appendix A). Taking the Laplace-transform from both sides of Equation 5.88, we have

$$s\Phi(s) - \phi(0) = A\Phi(s) \qquad (5.89)$$

Since $\phi(0) = I$, then:

$$(sI - A)\Phi(s) = I \qquad (5.90)$$

and

$$\Phi(s) = (sI - A)^{-1} \qquad (5.91)$$

Therefore:

$$\phi(t) = L^{-1}[(sI - A)^{-1}] \qquad (5.92)$$

where L^{-1} stands for the inverse Laplace transform.

Example 5.19

Find the state-transition matrix of the following system using the Laplace transform approach:

$$\dot{x}(t) = \begin{bmatrix} 0 & 1 \\ -3 & -4 \end{bmatrix} x(t)$$

Using the Laplace transform, we have:

$$\Phi(s) = (sI - A)^{-1} = \begin{bmatrix} s & -1 \\ 3 & s+4 \end{bmatrix}^{-1} = \frac{1}{s^2 + 4s + 3} \begin{bmatrix} s+4 & 1 \\ -3 & s \end{bmatrix}$$

or

$$\Phi(s) = \begin{bmatrix} \dfrac{s+4}{(s+1)(s+3)} & \dfrac{1}{(s+1)(s+3)} \\ \dfrac{-3}{(s+1)(s+3)} & \dfrac{s}{(s+1)(s+3)} \end{bmatrix}$$

Using the partial fraction expansion, the above matrix can be transformed as follows:

$$\Phi(s) = \begin{bmatrix} \dfrac{1.5}{s+1} - \dfrac{0.5}{s+3} & \dfrac{0.5}{s+1} - \dfrac{0.5}{s+3} \\ -\dfrac{1.5}{s+1} + \dfrac{1.5}{s+3} & -\dfrac{0.5}{s+1} + \dfrac{1.5}{s+3} \end{bmatrix}$$

Hence:

$$\phi(t) = L^{-1}[\Phi(s)] = \begin{bmatrix} 1.5e^{-t} - 0.5e^{-3t} & 0.5e^{-t} - 0.5e^{-3t} \\ -1.5e^{-t} + 1.5e^{-3t} & -0.5e^{-t} + 1.5e^{-3t} \end{bmatrix}$$

5.5.8 Complete Solution of State Equations

The complete solution of the state equations given by Equation 5.81 is the sum of the homogenous and particular solution and is given by:

$$x(t) = \phi(t)x(0) + \int_0^t \phi(t-\tau)Bu(\tau)d\tau \qquad (5.93)$$

The first term $\phi(t)x(0)$ is referred to as the zero input response (ZIR) and the second term $\int_0^t \phi(t-\tau)Bu(\tau)d\tau$ as the zero state response. To verify that this is the total solution, we need to show that it satisfies the state equation and the initial condition. The initial condition is satisfied since:

$$x(0) = \phi(0)x(0) + \int_0^0 \phi(0-\tau)Bu(\tau)d\tau = I \times x(0) + 0 = x(0) \qquad (5.94)$$

To show that it satisfies the state equation, differentiate both sides of Equation 5.93 with respect to t:

$$\dot{x}(t) = \dot{\phi}(t)x(0) + \frac{d}{dt}\int_0^t \phi(t-\tau)Bu(\tau)d\tau = A\phi(t)x(0) + \phi(t-t)Bu(t)$$

$$+ \int_0^t \dot{\phi}(t-\tau)Bu(\tau)d\tau$$

$$= A\phi(t)x(0) + \phi(0)Bu(t) + \int_0^t A\phi(t-\tau)Bu(\tau)d\tau = A\phi(t)x(0) + Bu(t)$$

$$+ A\int_0^t \phi(t-\tau)Bu(\tau)d\tau$$

$$= A[\phi(t)x(0) + \int_0^t \phi(t-\tau)Bu(\tau)d\tau] + Bu(t) = Ax(t) + Bu(t) \qquad (5.95)$$

The first term in the total solution, that is $\phi(t)x(0)$, is the homogenous solution or the zero input response. The second term is the particular solution, which is also refered to as the ZSP.

Example 5.20

Find the output of the dynamic system described by the state equation

$$\dot{x}(t) = \begin{bmatrix} 0 & 1 \\ -3 & -4 \end{bmatrix} x(t) + \begin{bmatrix} 1 \\ 0 \end{bmatrix} u(t)$$

$$y(t) = \begin{bmatrix} 1 & -1 \end{bmatrix} x(t)$$

with the input and initial conditions given by $u(t) = \begin{cases} 1 & t > 0 \\ 0 & t < 0 \end{cases}$ and $x(0) = \begin{bmatrix} 1 \\ -1 \end{bmatrix}$.

Solution: The ZIR is given by:

$$x_{ZIR}(t) = \varphi(t)x(0) = \begin{bmatrix} 1.5e^{-t} - 0.5e^{-3t} & 0.5e^{-t} - 0.5e^{-3t} \\ -1.5e^{-t} + 1.5e^{-3t} & -0.5e^{-t} + 1.5e^{-3t} \end{bmatrix}\begin{bmatrix} 1 \\ -1 \end{bmatrix} = \begin{bmatrix} e^{-t} \\ -e^{-t} \end{bmatrix}$$

The ZSR or the particular solution is:

$$x_{ZSR}(t) = \int_0^t \varphi(t-\tau)Bu(\tau)d\tau$$

$$= \int_0^t \begin{bmatrix} 1.5e^{-t+\tau} - 0.5e^{-3t+3\tau} & 0.5e^{-t+\tau} - 0.5e^{-3t+3\tau} \\ -1.5e^{-t+\tau} + 1.5e^{-3t+3\tau} & -0.5e^{-t+\tau} + 1.5e^{-3t+3\tau} \end{bmatrix}\begin{bmatrix}1\\0\end{bmatrix}d\tau$$

$$= \int_0^t \begin{bmatrix} 1.5e^{-t+\tau} - 0.5e^{-3t+3\tau} \\ -1.5e^{-t+\tau} + 1.5e^{-3t+3\tau} \end{bmatrix}d\tau = \begin{bmatrix} 1.5e^{-t}\int_0^t e^\tau d\tau - 0.5e^{-3t}\int_0^t e^{3\tau}d\tau \\ -1.5e^{-t}\int_0^t e^\tau d\tau + 1.5e^{-3t}\int_0^t e^{3\tau}d\tau \end{bmatrix}$$

$$= \begin{bmatrix} 1.5e^{-t}(e^t-1) - \dfrac{0.5}{3}e^{-3t}(e^{3t}-1) \\ -1.5e^{-t}(e^t-1) + \dfrac{1.5}{3}e^{-3t}(e^{3t}-1) \end{bmatrix} = \begin{bmatrix} \dfrac{4}{3} - \dfrac{3}{2}e^{-t} + \dfrac{1}{6}e^{-3t} \\ -1 + \dfrac{3}{2}e^{-t} - \dfrac{1}{2}e^{-3t} \end{bmatrix}$$

Therefore, the complete solution is:

$$x(t) = x_{ZIR}(t) + x_{ZSR}(t) = \begin{bmatrix} e^{-t} \\ -e^{-t} \end{bmatrix} + \begin{bmatrix} \dfrac{4}{3} - \dfrac{3}{2}e^{-t} + \dfrac{1}{6}e^{-3t} \\ -1 + \dfrac{3}{2}e^{-t} - \dfrac{1}{2}e^{-3t} \end{bmatrix} = \begin{bmatrix} \dfrac{4}{3} - \dfrac{1}{2}e^{-t} + \dfrac{1}{6}e^{-3t} \\ -1 + \dfrac{1}{2}e^{-t} - \dfrac{1}{2}e^{-3t} \end{bmatrix}$$

The output $y(t)$ is:

$$y(t) = \begin{bmatrix}1 & -1\end{bmatrix}x(t) = \begin{bmatrix}1 & -1\end{bmatrix}\begin{bmatrix} \dfrac{4}{3} - \dfrac{1}{2}e^{-t} + \dfrac{1}{6}e^{-3t} \\ -1 + \dfrac{1}{2}e^{-t} - \dfrac{1}{2}e^{-3t} \end{bmatrix} = \dfrac{7}{3} - e^{-t} + \dfrac{2}{3}e^{-3t}$$

Example 5.21

Use MATLAB to compute and plot the states of the system described by the state equations:

$$\dot{x}(t) = \begin{bmatrix} 0 & 1 & 0 \\ 0 & 0 & 1 \\ -1 & -2 & -3 \end{bmatrix}x(t) + \begin{bmatrix}0\\0\\1\end{bmatrix}u(t)$$

with the input and initial conditions given by $u(t) = \begin{cases}1 & t>0 \\ 0 & t<0\end{cases}$ and $x(0) = \begin{bmatrix}-1\\1\\0\end{bmatrix}$.

The following MATLAB code (Table 5.1) computes the total response of the system to a unit step function with the given initial conditions. The plots of the three states are shown in Figure 5.4.

TABLE 5.1

MATLAB® Code for Example 5.21

```
% Step response with nonzero
  initial condition
A = [0 1 0;0 0 1; -1 -2 -3];
B = [0 0 1]';
C = eye(3);
D = 0;
sys = ss(A,B,C,D);
t = 0:0.005:15;
x0 = [-1 1 0]';
xzir = initial(sys,x0,t);
xzsr = step(sys,t);
x = xzir+xzsr;
plot(t,x)
grid on
xlabel('t')
ylabel('x_1(t),x_2(t),x_3(t)')
text(2.2,1.3,'x_1(t)')
text(2.3,0.6,'x_2(t)')
text(7.0,0.2,'x_3(t)')
```

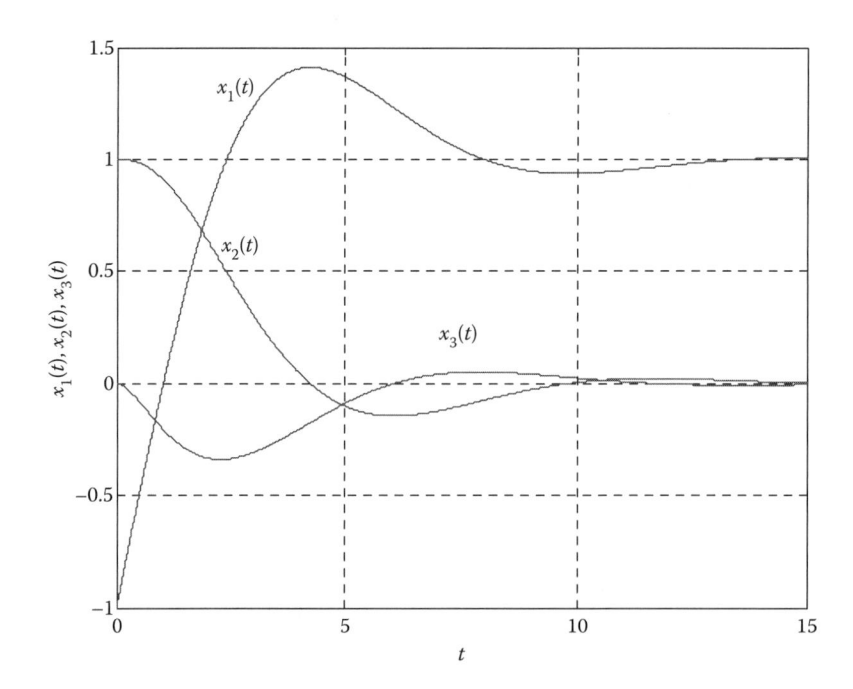

FIGURE 5.4
Plots of the three states of the system versus time.

5.6 State Space Representation of Discrete-time Systems

5.6.1 Definition of States

Similar to continuous time systems, The state of a discrete-time dynamic system is the minimum number of variables called states variables such that the knowledge of these variables at time $k=k_0$, together with input for $k \geq k_0$, uniquely determines the behavior of the system for $k \geq k_0$. If n variables are needed, then these n variables are considered components of a n dimensional vector x called the state vector. The n dimensional space whose coordinates are states of the system is called the state space. The state of a system at time n is a point in the state space.

5.6.2 State Equations

The general expressions for the state equation of a MIMO discrete-time system are given by:

$$x(k+1) = f(x(k), u(k), k) \tag{5.96}$$

$$y(k) = h(x(k), u(k), k) \tag{5.97}$$

where $x(k) = [x_1(k) \ x_2(k) \ \ldots \ x_n(k)]^T$ is $n \times 1$ state vector, $u(k)$ is the $M \times 1$ input vector and $y(n)$ is the $P \times 1$ output vector. The functions f and h are:

$$f(x(k), u(k), k) = \begin{bmatrix} f_1(x_1(k), x_2(k), \cdots x_n(k), u_1(k), u_2(k), \cdots, u_M(k), k) \\ f_2(x_1(k), x_2(k), \cdots x_n(k), u_1(k), u_2(k), \cdots, u_M(k), k) \\ \vdots \\ f_n(x_1(k), x_2(k), \cdots x_n(k), u_1(k), u_2(k), \cdots, u_M(k), k) \end{bmatrix} \tag{5.98}$$

$$h(x(k), u(k), k) = \begin{bmatrix} h_1(x_1(k), x_2(k), \cdots x_n(k), u_1(k), u_2(k), \cdots, u_M(k), k) \\ h_2(x_1(k), x_2(k), \cdots x_n(k), u_1(k), u_2(k), \cdots, u_M(k), k) \\ \vdots \\ h_P(x_1(k), x_2(k), \cdots x_n(k), u_1(k), u_2(k), \cdots, u_M(k), k) \end{bmatrix} \tag{5.99}$$

If the system is LTI, then the state equations can be written as:

$$x(k+1) = Ax(k) + Bu(k) \tag{5.100}$$

$$y(k) = Cx(k) + Du(k) \tag{5.101}$$

where A, B, C and D are $n \times n$, $n \times M$, $P \times n$ and $P \times M$ constant matrices, respectively. The block diagram of a LTI control system in state-space form is shown in Figure 5.5.

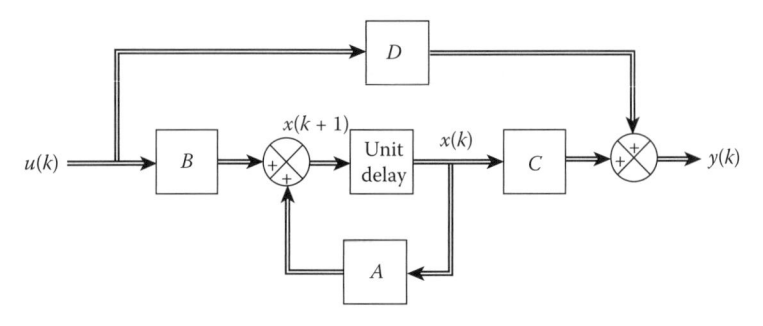

FIGURE 5.5
Block diagram of a LTI discrete system in state space.

5.6.3 State Space Representation of Discrete-time LTI Systems

Consider the LTI discrete-time system described by the difference equation:

$$y(k) = -\sum_{i=1}^{n} a_i y(k-i) + \sum_{i=0}^{n} b_i u(k-i) \tag{5.102}$$

Similar to the continuous case, there are two realizations: the controllable canonical form and the observable canonical form. The state equations in the controllable canonical form are given by:

$$\begin{bmatrix} x_1(k+1) \\ x_2(k+1) \\ \vdots \\ x_{n-1}(k+1) \\ x_n(k+1) \end{bmatrix} = \begin{bmatrix} 0 & 1 & 0 & \cdots & 0 \\ 0 & 0 & 1 & \cdots & 0 \\ \vdots & \vdots & \vdots & \vdots & \vdots \\ 0 & 0 & 0 & \cdots & 1 \\ -a_n & -a_{n-1} & -a_{n-2} & \cdots & -a_1 \end{bmatrix} \begin{bmatrix} x_1(k) \\ x_2(k) \\ \vdots \\ x_{n-1}(k) \\ x_n(k) \end{bmatrix} + \begin{bmatrix} 0 \\ 0 \\ \vdots \\ 0 \\ 1 \end{bmatrix} u(k) \tag{5.103}$$

with the output equation:

$$y(k) = \begin{bmatrix} b_n - a_n b_0 & b_{n-1} - a_{n-1} b_0 & b_{n-2} - a_{n-2} b_0 & \cdots & b_1 - a_1 b_0 \end{bmatrix} \begin{bmatrix} x_1(k) \\ x_2(k) \\ \vdots \\ x_{n-1}(k) \\ x_n(k) \end{bmatrix} + b_0 u(k) \tag{5.104}$$

The observable canonical form is given by:

$$\begin{bmatrix} x_1(k+1) \\ x_2(k+1) \\ \vdots \\ x_{n-1}(k+1) \\ x_n(k+1) \end{bmatrix} = \begin{bmatrix} 0 & 0 & \cdots & 0 & -a_n \\ 1 & 0 & \cdots & 0 & -a_{n-1} \\ 0 & 1 & \cdots & 0 & a_{n-3} \\ \vdots & \vdots & & \vdots & \vdots \\ 0 & 0 & \cdots & 1 & -a_1 \end{bmatrix} \begin{bmatrix} x_1(k) \\ x_2(k) \\ \vdots \\ x_{n-1}(k) \\ x_n(k) \end{bmatrix} + \begin{bmatrix} b_n - a_n b_0 \\ b_{n-1} - a_{n-1} b_0 \\ b_{n-2} - a_{n-2} b_0 \\ \vdots \\ b_1 - a_1 b_0 \end{bmatrix} u(k) \tag{5.105}$$

with the output equation:

$$y(k) = \begin{bmatrix} 0 & 0 & 0 & \cdots & 1 \end{bmatrix} \begin{bmatrix} x_1(k) \\ x_2(k) \\ \vdots \\ x_{n-1}(k) \\ x_n(k) \end{bmatrix} + b_0 u(k) \quad (5.106)$$

Example 5.22

Consider the following SISO discrete-time LTI system:

$$y(k) - 0.75y(k-1) + 0.125y(k-2) = u(k) + 3u(k-1) - 4u(k-2)$$

Obtain a state space representation of this system in
a. Controllable canonical form.
b. Observable canonical form.

Solution: The state equations in controllable canonical form are:

$$\begin{bmatrix} x_1(k+1) \\ x_1(k+1) \end{bmatrix} = \begin{bmatrix} 0 & 1 \\ -0.125 & 0.75 \end{bmatrix} \begin{bmatrix} x_1(k) \\ x_1(k) \end{bmatrix} + \begin{bmatrix} 0 \\ 1 \end{bmatrix} u(k),$$

and

$$y(k) = \begin{bmatrix} -4.25 & 2.25 \end{bmatrix} \begin{bmatrix} x_1(k) \\ x_2(k) \end{bmatrix} + u(k)$$

The observable canonical form is:

$$\begin{bmatrix} x_1(k+1) \\ x_1(k+1) \end{bmatrix} = \begin{bmatrix} 0 & -0.125 \\ 1 & 0.75 \end{bmatrix} \begin{bmatrix} x_1(k) \\ x_1(k) \end{bmatrix} + \begin{bmatrix} -4.25 \\ 2.25 \end{bmatrix} u(k), \text{ and } y(k) = \begin{bmatrix} 0 & 1 \end{bmatrix} \begin{bmatrix} x_1(k) \\ x_2(k) \end{bmatrix} + u(k)$$

5.6.4 Solution of Discrete-time State Equations

Consider a LTI discrete-time system described by the state space and output equations given below:

$$x(k+1) = Ax(k) + Bu(k) \quad (5.107)$$

$$y(k) = Cx(k) + Du(k) \quad (5.108)$$

To obtain the output of this system for a given input and an initial state, we first solve the state Equation 5.107. The solution is then substituted into the algebraic Equation 5.108. The solution to the state equation has two parts: a homogenous solution and a particular solution. We first consider the homogenous solution.

5.6.4.1 Solution of Homogenous State Equation and State Transition Matrix

The homogenous solution is the solution of the state equations to an arbitrary initial condition with zero input. The homogenous state equation is given by:

$$x(k+1) = Ax(k) \tag{5.109}$$

Assuming that the initial state is $x(0)$, we have:

$$x(1) = Ax(0)$$

$$x(2) = Ax(1) = A \times Ax(0) = A^2 x(0)$$

$$x(3) = Ax(2) = A \times A^2 x(0) = A3x(0) \tag{5.110}$$

$$\vdots$$

$$x(k) = Ax(k-1) = A \times A^{k-1} x(0) = A^k x(0)$$

Therefore, in matrix form, the solution to the homogenous equation is:

$$x(k) = A^k x(0) \tag{5.111}$$

Therefore, the homogenous part of the solution is $x(k) = A^k x(0)$. The matrix exponential A^k is called the state transition matrix and is denoted by $\phi(k)$. The state transition matrix $\phi(k)$ is an $n \times n$ matrix and is the solution to the homogenous equation:

$$\phi(k+1) = A\phi(k)$$

$$\phi(0) = I \tag{5.112}$$

The state transition matrix is given by:

$$\phi(k) = A^k \tag{5.113}$$

5.6.4.2 Properties of State Transition Matrix

State transition matrix $\phi(k) = A^k$ has the following properties:

a. $\phi(0) = I$
b. $\phi(-k) = \phi^{-1}(k)$
c. $\phi(k_1 + n) = \phi(k) \phi(n) = \phi(n) \phi(k)$
d. $\phi(k+1) = A\phi(k) = \phi(k)A$

The proofs are left as an exercise.

5.6.4.3 Computing the State Transition Matrix

Similar to the continuous case, the discrete state transition matrix can be computed using Cayley–Hamilton technique, modal matrix, or Sylvester theorem as discussed in Section 5.2. Here we discuss an approach based on the Z-transform (for a brief review of the Z-transforms, see Appendix B). Taking the z-transform from both sides of Equation 5.109, we have:

$$z\Phi(z) - z\phi(0) = A\Phi(z) \tag{5.114}$$

Since $\phi(0) = I$, then:

$$(zI - A)\Phi(z) = zI \tag{5.115}$$

and

$$\Phi(z) = z(zI - A)^{-1} \tag{5.116}$$

Therefore:

$$\phi(k) = Z^{-1}[z(zI - A)^{-1}] \tag{5.117}$$

Example 5.23

Find the state-transition matrix of the following system:

$$\begin{bmatrix} x_1(k+1) \\ x_2(k+1) \end{bmatrix} = \begin{bmatrix} 1 & -0.25 \\ 1.5 & -0.25 \end{bmatrix} \begin{bmatrix} x_1(k) \\ x_2(k) \end{bmatrix}$$

Solution:

$$\Phi(z) = z(zI - A)^{-1} = z \begin{bmatrix} z-1 & 0.25 \\ -1.5 & z+0.25 \end{bmatrix}^{-1} = \frac{z}{z^2 - 0.75z + 0.125} \begin{bmatrix} z+0.25 & -0.25 \\ 1.5 & z-1 \end{bmatrix}$$

or

$$\Phi(z) = \begin{bmatrix} \dfrac{z+0.25}{(z-0.5)(z-0.25)} & -\dfrac{0.25}{(z-0.5)(z-0.25)} \\ \dfrac{1.5}{(z-0.5)(z-0.25)} & \dfrac{z-1}{(z-0.5)(z-0.25)} \end{bmatrix}$$

Using partial fraction expansion, we have:

$$\Phi(z) = z \begin{bmatrix} \dfrac{3}{z-0.5} + \dfrac{-2}{z-0.25} & \dfrac{-1}{z-0.5} + \dfrac{1}{z-0.25} \\ \dfrac{6}{z-0.5} + \dfrac{-6}{z-0.25} & \dfrac{-2}{z-0.5} + \dfrac{3}{z-0.25} \end{bmatrix}$$

$$= \begin{bmatrix} \dfrac{3z}{z-0.5} + \dfrac{-2z}{z-0.25} & \dfrac{-z}{z-0.5} + \dfrac{z}{z-0.25} \\ \dfrac{6z}{z-0.5} + \dfrac{-6z}{z-0.25} & \dfrac{-2z}{z-0.5} + \dfrac{3z}{z-0.25} \end{bmatrix}$$

Hence:

$$\phi(k) = Z^{-1}[\Phi(z)] = \begin{bmatrix} 3(0.5)^k - 2(0.25)^k & -(0.5)^k + (0.25)^k \\ 6(0.5)^k - 6(0.25)^k & -2(0.5)^k + 3(0.25)^k \end{bmatrix}$$

5.6.4.4 Complete Solution of the State Equations

Assuming that the initial state is $x(0)$, we have:

$x(1) = Ax(0) + Bu(0)$

$x(2) = Ax(1) + Bu(1) = A^2x(0) + ABu(0) + Bu(1)$

$x(3) = Ax(2) + Bu(2) = A^3x(0) + A^2Bu(0) + ABu(1) + Bu(2)$ (5.118)

$\vdots$

$x(k) = Ax(k-1) + Bu(k-1) = A^k x(0) + A^{k-1}Bu(0) + A^{k-2}Bu(1)$
$+ \cdots + Bu(k-1)$

Therefore:

$$x(k) = A^k x(0) + \sum_{i=0}^{k-1} A^{k-1-i} Bu(i) \qquad (5.119)$$

In terms of the state-transition matrix, the total solution is given by:

$$x(k) = \phi(k)x(0) + \sum_{i=0}^{k-1} \phi(k-1-i)Bu(i) \qquad (5.120)$$

The total solution consists of two terms. The first term $\phi(k)x(0)$ is referred to as the ZIR and the second term $\sum_{i=0}^{k-1} \phi(k-1-i)Bu(i)$ is called the ZSR. The ZIR is the response of the system to initial conditions only and the ZSR is the response due to the input with zero initial conditions. In the following example, we compute the ZSR and ZIR of a second order system.

Example 5.24

Find the output of the following dynamic system if $u(k) = \begin{cases} 1 & k \geq 0 \\ 0 & k < 0 \end{cases}$ and $x(0) = \begin{bmatrix} 1 \\ -1 \end{bmatrix}$:

$$x(k+1) = \begin{bmatrix} 1 & -0.25 \\ 1.5 & -0.25 \end{bmatrix} x(k) + \begin{bmatrix} 1 \\ -1 \end{bmatrix} u(k)$$

$$y(k) = \begin{bmatrix} 1 & 1 \end{bmatrix} x(k)$$

Matrix Polynomials and Functions of Square Matrices

First we find the ZIR:

$$\phi(k)x(0) = \begin{bmatrix} 3(0.5)^k - 2(0.25)^k & -(0.5)^k + (0.25)^k \\ 6(0.5)^k - 6(0.25)^k & -2(0.5)^k + 3(0.25)^k \end{bmatrix}\begin{bmatrix} 1 \\ -1 \end{bmatrix} = \begin{bmatrix} 4(0.5)^k - 3(0.25)^k \\ 8(0.5)^k - 9(0.25)^k \end{bmatrix}$$

Next we find the ZSR:

$$\sum_{i=0}^{k-1}\phi(k-1-i)Bu(i) = \sum_{i=0}^{k-1}\begin{bmatrix} \phi_{11}(k-1-i) & \phi_{12}(k-1-i) \\ \phi_{21}(k-1-i) & \phi_{22}(k-1-i) \end{bmatrix}\begin{bmatrix} 1 \\ -1 \end{bmatrix}$$

$$= \sum_{i=0}^{k-1}\begin{bmatrix} \phi_{11}(k-1-i) - \phi_{12}(k-1-i) \\ \phi_{21}(k-1-i) - \phi_{22}(k-1-i) \end{bmatrix}$$

$$= \sum_{i=0}^{k-1}\begin{bmatrix} 4(0.5)^{k-1-i} - 3(0.25)^{k-1-i} \\ 8(0.5)^{k-1-i} - 9(0.25)^{k-1-i} \end{bmatrix}$$

or

$$\sum_{i=0}^{k-1}\phi(k-1-i)Bu(i) = \begin{bmatrix} 4(0.5)^{k-1}\sum_{i=0}^{k-1}2^i - 3(0.25)^{k-1}\sum_{i=0}^{k-1}4^i \\ 8(0.5)^{k-1}\sum_{i=0}^{k-1}2^i - 9(0.25)^{k-1}\sum_{i=0}^{k-1}4^i \end{bmatrix}$$

$$= \begin{bmatrix} 4(0.5)^{k-1}\dfrac{1-2^k}{1-2} - 3(0.25)^{k-1}\dfrac{1-4^k}{1-4} \\ 8(0.5)^{k-1}\dfrac{1-2^k}{1-2} - 3(0.25)^{k-1}\dfrac{1-4^k}{1-4} \end{bmatrix}$$

$$= \begin{bmatrix} -4(0.5)^{k-1} + 4 + (0.25)^{k-1} \\ -8(0.5)^{k-1} + 12 + (0.25)^{k-1} \end{bmatrix} = \begin{bmatrix} 4 - 8(0.5)^k + 4(0.25)^k \\ 4 - 16(0.5)^k + 12(0.25)^k \end{bmatrix}$$

The total response is the sum of the zero-state and the ZIRs. Therefore:

$$x(k) = \phi(k)x(0) + \sum_{i=0}^{k-1}\phi(k-1-i)Bu(i) = \begin{bmatrix} 4(0.5)^k - 3(0.25)^k \\ 8(0.5)^k - 9(0.25)^k \end{bmatrix} + \begin{bmatrix} 4 - 8(0.5)^k + 4(0.25)^k \\ 4 - 16(0.5)^k + 12(0.25)^k \end{bmatrix}$$

$$= \begin{bmatrix} 4 - 4(0.5)^k + (0.25)^k \\ 4 - 8(0.5)^k + 3(0.25)^k \end{bmatrix}$$

The output signal is:

$$y(k) = \begin{bmatrix} 1 & 1 \end{bmatrix}x(k) = \begin{bmatrix} 1 & 1 \end{bmatrix}\begin{bmatrix} 4 - 4(0.5)^k + (0.25)^k \\ 4 - 8(0.5)^k + 3(0.25)^k \end{bmatrix} = 8 - 12(0.5)^k + 4(0.25)^k$$

5.7 Controllability of LTI Systems

In general, a system is controllable if there exists an input that can transfer the states of the system from an arbitrary initial state to a final state in a finite number of steps. In this section, we examine the controllability of LTI discrete-time systems. Since the results obtained are identical for continuous and discrete systems, we only consider the controllability of discrete-time LTI systems.

5.7.1 Definition of Controllability

A discrete-time LTI system is said to be controllable at time k_0 if there exists an input $u(k)$ for $k \geq k_0$ that can transfer the system from any initial state $x(k_0)$ to the origin in a finite number of steps.

5.7.2 Controllability Condition

Theorem 5.4: A LTI discrete-time system given by:

$$x(k+1) = Ax(k) + Bu(k) \qquad (5.121)$$

is completely state controllable if the controllability matrix Q is full rank, where:

$$Q = \begin{bmatrix} B & AB & A^2B & \cdots & A^{n-1}B \end{bmatrix} \qquad (5.122)$$

Proof: Let the initial state be $x(k_0)$ at time k_0 and the input for time $k \geq k_0$ be $u(k_0), u(k_0+1), \ldots, u(k_0+n-1)$. Then the state of the system at time $k = n + k_0$ is given by:

$$x(n+k_0) = A^n x(k_0) + \sum_{i=k_0}^{n+k_0-1} A^{n+k_0-1-i} Bu(i) \qquad (5.123)$$

To drive the system to the origin, we must have:

$$A^n x(k_0) + A^{n-1} Bu(k_0) + A^{n-2} Bu(k_0+1) + \cdots + Bu(n+k_0-1) = 0 \qquad (5.124)$$

or

$$\begin{bmatrix} B & AB & A^2B & \cdots & A^{n-1}B \end{bmatrix} \begin{bmatrix} u(n+k_0-1) \\ u(n+k_0-2) \\ \vdots \\ u(k_0+1) \\ u(k_0) \end{bmatrix} = -A^n x(k_0) \qquad (5.125)$$

Matrix Polynomials and Functions of Square Matrices 271

Equation 5.125 has a solution if matrix $Q=[B \quad AB \quad A^2B \quad \cdots \quad A^{n-1}B]$ is full rank.

The MATLAB® command for computation of controllability matrix Q is $Q=\mathbf{ctrb}(A,B)$.

Example 5.25

Consider the following system:

$$x(k+1) = \begin{bmatrix} 1 & -0.26 \\ 1.4 & -0.2 \end{bmatrix} x(k) + \begin{bmatrix} -1 \\ 1 \end{bmatrix} u(k)$$

The controllability matrix Q is:

$$Q = \begin{bmatrix} B & AB \end{bmatrix} = \begin{bmatrix} -1 & -1.26 \\ 1 & -1.6 \end{bmatrix}$$

Since $\det(Q)=2.86 \neq 0$, Q is full rank and the system is completely state controllable.

Example 5.26

Consider the system

$$x(k+1) = \begin{bmatrix} 0.5 & 1 & 1 \\ 0 & 0.7 & 1 \\ 0 & 1 & 0.9 \end{bmatrix} x(k) + \begin{bmatrix} 1 \\ -1 \\ 1 \end{bmatrix} u(k)$$

The controllability matrix Q is:

$$Q = \begin{bmatrix} B & AB & A^2B \end{bmatrix} = \begin{bmatrix} 1 & 0.5 & 0.45 \\ -1 & 0.3 & 0.11 \\ 1 & 0.1 & 0.21 \end{bmatrix}$$

Since $\det(Q)=-0.144 \neq 0$, Q is full rank and the system is completely state controllable.

Example 5.27

Consider the system given by:

$$x(k+1) = \begin{bmatrix} 1 & 0 & 0 \\ 0 & 1 & 0 \\ 0 & 0 & 1 \end{bmatrix} x(k) + \begin{bmatrix} 1 & 0 \\ 0 & 1 \\ -1 & 2 \end{bmatrix} \begin{bmatrix} u_1(k) \\ u_2(k) \end{bmatrix}$$

The controllability matrix Q is:

$$Q = \begin{bmatrix} B & AB & A^2B \end{bmatrix} = \begin{bmatrix} 1 & 0 & 1 & 0 & 1 & 0 \\ 0 & 1 & 0 & 1 & 0 & 1 \\ -1 & 2 & -1 & 2 & -1 & 2 \end{bmatrix}$$

Since $\text{rank}(Q)=2$, Q is not full rank and the system is not completely state controllable.

5.8 Observability of LTI Systems

In general, a system is completely state observable if the states of the systems can be found from the knowledge of the inputs and outputs of the system. Since the system and its inputs are assumed to be known, the only unknown in determining the states of the system is the initial conditions. Therefore, if we can derive the initial conditions, then the system is completely state observable. In this section, we examine the observability of LTI systems. Since the results obtained are identical for continuous and discrete systems, we only consider the observability of discrete-time LTI systems.

5.8.1 Definition of Observability

A discrete-time LTI system is said to be completely state observable if the states of the system can be estimated from the knowledge of the inputs and outputs of the system.

5.8.2 Observability Condition

Theorem 5.5: A linear time-invariant discrete-time system given by:

$$x(k+1) = Ax(k) + Bu(k)$$
$$y(k) = Cx(k) + Du(k) \tag{5.126}$$

is completely state observable if the observability matrix P is full rank, where:

$$P = \begin{bmatrix} C \\ CA \\ CA^2 \\ \vdots \\ CA^{n-1} \end{bmatrix} \tag{5.127}$$

Proof: Let the initial state be $x(0)$ at time $k=0$ and the input and output for time $k \geq 0$ be $u(k)$, $y(k)$, respectively. Then:

$$y(k) = Cx(k) + Du(k) = CA^k x(0) + C\sum_{i=0}^{k-1} A^{k-1-i} Bu(i) + Du(k) \tag{5.128}$$

Equation 5.128 can be written in matrix form as:

Matrix Polynomials and Functions of Square Matrices

$$\begin{bmatrix} y(0) \\ y(1) \\ y(2) \\ \vdots \\ y(n-1) \end{bmatrix} = \begin{bmatrix} C \\ CA \\ CA^2 \\ \vdots \\ CA^{n-1} \end{bmatrix} x(0) + C \begin{bmatrix} 0 \\ Bu(0) \\ ABu(1) + Bu(0) \\ \vdots \\ \sum_{i=0}^{n-2} A^{n-2-i} Bu(i) \end{bmatrix} + D \begin{bmatrix} u(0) \\ u(1) \\ u(2) \\ \vdots \\ u(n-1) \end{bmatrix} \quad (5.129)$$

or

$$\begin{bmatrix} C \\ CA \\ CA^2 \\ \vdots \\ CA^{n-1} \end{bmatrix} x(0) = \begin{bmatrix} y(0) \\ y(1) \\ y(2) \\ \vdots \\ y(n-1) \end{bmatrix} - C \begin{bmatrix} 0 \\ Bu(0) \\ ABu(1) + Bu(0) \\ \vdots \\ \sum_{i=0}^{n-2} A^{n-2-i} Bu(i) \end{bmatrix} - D \begin{bmatrix} u(0) \\ u(1) \\ u(2) \\ \vdots \\ u(n-1) \end{bmatrix} \quad (5.130)$$

Equation 5.130 can be used to solve for $x(0)$ if and only if matrix P is full rank. The MATLAB® command for the computation of observability matrix P is $P=\text{obsv}(A,C)$.

Example 5.28

Consider the following system:

$$x(k+1) = \begin{bmatrix} 1 & -0.25 \\ 1.5 & -0.25 \end{bmatrix} x(k) + \begin{bmatrix} -1 \\ 1 \end{bmatrix} u(k)$$

$$y(k) = \begin{bmatrix} 1 & 0 \end{bmatrix} x(k)$$

The observability matrix P is:

$$P = \begin{bmatrix} C \\ CA \end{bmatrix} = \begin{bmatrix} 1 & 0 \\ 1 & -0.25 \end{bmatrix}$$

Since $\det(P) = -0.25 \neq 0$, P is full rank and the system is completely state observable.

Example 5.29

Consider the system:

$$x(k+1) = \begin{bmatrix} 0.5 & 1 & 0 \\ 1 & 0.75 & 1 \\ 2 & -1 & 0.8 \end{bmatrix} x(k) + \begin{bmatrix} 1 \\ -1 \\ 2 \end{bmatrix} u(k)$$

$$y(k) = \begin{bmatrix} 1 & 0 & -2 \end{bmatrix} x(k) + 3u(k)$$

The observability matrix P is:

$$P = \begin{bmatrix} C \\ CA \\ CA^2 \end{bmatrix} = \begin{bmatrix} 1 & 0 & -2 \\ -3.5 & 3 & -1.6 \\ -1.95 & 0.35 & 1.72 \end{bmatrix}$$

Since det(P)=−3.53≠0, P is full rank and the system is completely state observable.

Example 5.30

Consider the MIMO system:

$$x(k+1) = \begin{bmatrix} 1 & 0 & 0 \\ 0 & 1 & 0 \\ 0 & 0 & 1 \end{bmatrix} x(k) + \begin{bmatrix} 1 & 0 \\ 0 & 1 \\ -1 & 2 \end{bmatrix} \begin{bmatrix} u_1(k) \\ u_2(k) \end{bmatrix}$$

$$y(k) = \begin{bmatrix} -1 & 0 & 1 \\ 1 & 1 & 0 \end{bmatrix} x(k)$$

The observability matrix P is:

$$P = \begin{bmatrix} C \\ CA \\ CA^2 \end{bmatrix} = \begin{bmatrix} -1 & 0 & 1 \\ 1 & 1 & 0 \\ -1 & 0 & 1 \\ 1 & 1 & 0 \\ -1 & 0 & 1 \\ 1 & 1 & 0 \end{bmatrix}$$

Since rank (Q)=2, matrix Q is not full rank and the system is not completely state observable.

Example 5.31

Consider the MIMO system

$$x(k+1) = \begin{bmatrix} 0.24 & 0 & 0.025 & 0 \\ 0 & 0.24 & 0 & 0.025 \\ -0.525 & 0.85 & 0.74 & 0.0625 \\ 0.3 & 0.575 & 0.375 & -0.26 \end{bmatrix} x(k) + \begin{bmatrix} 1 & 0 \\ 0 & 1 \\ -1 & 2 \\ 1 & 1 \end{bmatrix} \begin{bmatrix} u_1(k) \\ u_2(k) \end{bmatrix}$$

$$y(k) = \begin{bmatrix} 1 & 0 & 0 & 1 \\ 0 & 1 & 0 & 0 \end{bmatrix} x(k)$$

Use MATLAB to check the controllability and observability of the system and display the results.

Matrix Polynomials and Functions of Square Matrices

Solution: The MATLAB function file (Table 5.2) determines if a system is observable and/or controllable. The inputs to the function file are *A*, *B*, and *C* and the outputs are the ranks of the controllability and observability matrices. For the above example, the code to call the function file and the results are shown in Tables 5.3 and 5.4, respectively.

TABLE 5.2

MATLAB Function File to Test Controllability and Observability of a System

```
function [rc,ro] = testco(A,B,C)
rc = rank(ctrb(A,B));
ro = rank(obsv(A,C));
[n,n] = size(A);
if rc==n
 disp('System is controllable')
else
 disp('System is not completely state
 controllable')
end
if ro==n
 disp('System is observable')
else
 disp('System is not completely state
 observable')
end
```

TABLE 5.3

MATLAB Code for Example 5.31

```
A = [0.24 0 0.025 0;0 0.24 0 0.025
-0.525 0.85 0.74 0.0625;0.3 0.575
 0.375-0.26];
B = [1 0;0 1; -1 2;1 1];
C = [1 0 0 1;0 1 10 0];
[rc,ro] = testco(A,B,C)
```

TABLE 5.4

Results of Executing the MATLAB Code Given in Table 5.3

```
Output:
r_c = 4
r_0 = 4
System is controllable
System is observable
```

Problems

PROBLEM 5.1

Consider the following square matrix $A = \begin{bmatrix} 2 & -1 \\ 0 & 3 \end{bmatrix}$.

a. Compute A^k and e^{At} using the Cayley–Hamilton theorem.
b. Find the square roots of the matrix, i.e., find all matrices B such that $B^2=A$ using the Cayley–Hamilton theorem method.

PROBLEM 5.2

Let $A = \begin{bmatrix} -1 & 2 & -1 \\ -0 & 1 & 3 \\ -1 & 2 & 5 \end{bmatrix}$.

a. Find $P(\lambda)$, the characteristic polynomial of A.
b. Find eigenvalues of A.
c. Show that $P(A)=0$.

PROBLEM 5.3

Consider the following square matrix $A = \begin{bmatrix} 1 & 1 & 0 \\ 0 & 0 & 1 \\ 0 & 0 & 1 \end{bmatrix}$.

a. Compute A^{10} and A^{103} using the Cayley–Hamilton theorem.
b. Compute e^{At} using the Cayley–Hamilton theorem.

PROBLEM 5.4

Assume that the following matrix is nonsingular:

$$C = \begin{bmatrix} a & 1 & 0 \\ 0 & a & 0 \\ 0 & 0 & b \end{bmatrix}$$

Compute the natural logarithm of the matrix C; that is, identify the matrix $B=\ln(C)$ that satisfies the equation $C=e^B$. Is the assumption of nonsingularity of C really needed for the solution of this problem?

PROBLEM 5.5

Given the 3×3 matrix:

$$A = \begin{bmatrix} 0 & 1 & 0 \\ 0 & 0 & 1 \\ 2 & -1 & 4 \end{bmatrix}$$

Express the matrix polynomial $f(A)=A^4+2A^3+A^2-A+3I$ as a linear combination of the matrices A^2, A and I only.

Matrix Polynomials and Functions of Square Matrices

PROBLEM 5.6
Let the 2×2 matrix A be defined as:

$$A = \begin{bmatrix} 0 & 1 \\ 1 & 0 \end{bmatrix}$$

Compute A^{-20} and e^{At}.

PROBLEM 5.7
Let the 3×3 matrix A be defined as:

$$A = \begin{bmatrix} 0 & 1 & 0 \\ 1 & 0 & 1 \\ 0 & 1 & 0 \end{bmatrix}$$

Compute A^{2004} and e^{At}.

PROBLEM 5.8
Show that functions of matrix A commute. That is:

$$f(A)\,g(A) = g(A)\,f(A)$$

for any two functions $f(.)$ and $g(.)$ and any square matrix A.

PROBLEM 5.9

Let $B = \begin{bmatrix} 3 & 1 \\ 1 & 3 \end{bmatrix}$, and $C = \begin{bmatrix} 2 & 1 \\ 1 & 2 \end{bmatrix}$, and $A = BC$.

a. Show that $BC = CB$.
b. Find $\ln(B)$, $\ln(C)$, and $\ln(A)$.
c. Is $\ln(A) = \ln(B) + \ln(C)$? Can this be generalized?

PROBLEM 5.10
Find the continuous state transition matrix corresponding to:

$$A = \begin{bmatrix} \sigma_0 & \omega_0 \\ -\omega_0 & \sigma_0 \end{bmatrix}$$

PROBLEM 5.11
Show that the discrete state transition matrix $\phi(k) = A^k$ satisfies the following properties:

a. $\phi(0) = I$
b. $\phi(-k) = A^{-k} = (A^k)^{-1} = \phi^{-1}(k)$
c. $\phi(k_1 + k_2) = A^{k_1 + k_2} = A^{k_1} A^{k_2} = \phi(k_1)\phi(k_2)$
d. $\phi(k_3 - k_2)\,\phi(k_2 - k_1) = \phi(k_3 - k_1)$

PROBLEM 5.12
Show that the continuous state transition matrix $\phi(t)=e^{At}$ satisfies the following properties:

a. $\phi(0)=I$
b. $\phi(-t)=e^{-At}=(e^{At})^{-1}=\phi^{-1}(t)$
c. $\phi(t_1+t_2)=\phi(t_1)\phi(t_2)$
d. $\phi(t_3-t_2)\phi(t_2-t_1)=\phi(t_3-t_1)$

PROBLEM 5.13
Find the state space representation of the following dynamical systems:

a. $\dfrac{dy^3(t)}{dt^3}+4\dfrac{dy^2(t)}{dt^2}+4\dfrac{dy(t)}{dt}+3y(t)=u(t)$

b. $y(k+2)+0.98y(k+1)+0.8y(k)=4u(k)$

PROBLEM 5.14
Find the state space representation of the following dynamical systems in controllable canonical form:

a. $\dfrac{dy^3(t)}{dt^3}+3\dfrac{dy^2(t)}{dt^2}+4\dfrac{dy(t)}{dt}+2y(t)=7u(t)$

b. $3\dfrac{dy^3(t)}{dt^3}+\dfrac{dy^2(t)}{dt^2}+\dfrac{dy(t)}{dt}+2y(t)=6\dfrac{du(t)}{dt}+9u(t)$

PROBLEM 5.15
Find the state space representation of the following dynamical systems in observable canonical form:

a. $\dfrac{dy^3(t)}{dt^3}+3\dfrac{dy^2(t)}{dt^2}+4\dfrac{dy(t)}{dt}+2y(t)=3\dfrac{du^3(t)}{dt^3}+8\dfrac{du^2(t)}{dt^2}+\dfrac{du(t)}{dt}+4u(t)$

b. $\dfrac{dy^3(t)}{dt^3}+\dfrac{dy^2(t)}{dt^2}+\dfrac{dy(t)}{dt}+2y(t)=6\dfrac{du(t)}{dt}+27u(t)$

PROBLEM 5.16
Consider the discrete-time system:

$$x_1(k+1)=x_1(k)-\dfrac{1}{3}x_2(k)$$

$$x_2(k+1)=3x_1(k)-x_2(k)$$

$$x_3(k+1)=4x_1(k)-\dfrac{4}{3}x_2(k)-\dfrac{1}{3}x_3(k)$$

$$y(k)=x_3(k)$$

a. Find the state transition matrix A^k.
b. Find $y(k)$ if $x(0)=[1\ \ 1\ \ 1]^T$.
c. Find the initial condition $x(0)$ if $y(k)=4(-1/3)^k$ for $k\geq 0$.

Matrix Polynomials and Functions of Square Matrices 279

PROBLEM 5.17
Consider the discrete-time system:

$$x(k+1) = \begin{bmatrix} 0 & 1 & 0 & 0 \\ 0 & 0 & 1 & 0 \\ 0 & 0 & 0 & 1 \\ 0 & 0 & 0 & 0 \end{bmatrix} x(k) + \begin{bmatrix} 1 \\ 0 \\ 0 \\ 0 \end{bmatrix} u(k)$$

$$y(k) = \begin{bmatrix} 1 & 0 & 0 & 0 \end{bmatrix} x(k)$$

a. Find the state transition matrix A^k.
b. Find $y(k)$ if $x(0) = \begin{bmatrix} 1 & 1 & 1 & 1 \end{bmatrix}^T$ and $u(k) = 0$.
c. Find $y(k)$ if $x(0) = \begin{bmatrix} 1 & 1 & 1 & 1 \end{bmatrix}^T$ and $u(k) = 1$ for $k \geq 0$.

PROBLEM 5.18
Is the following system state controllable? Is it state observable?

$$\begin{bmatrix} x_1(k+1) \\ x_2(k+1) \\ x_3(k+1) \end{bmatrix} = \begin{bmatrix} 0 & 1 & 0 \\ 1 & 0 & 1 \\ -1 & -2 & -3 \end{bmatrix} \begin{bmatrix} x_1(k) \\ x_2(k) \\ x_3(k) \end{bmatrix} + \begin{bmatrix} 1 & 0 \\ 0 & 1 \\ -1 & -2 \end{bmatrix} \begin{bmatrix} u_1(k) \\ u_2(k) \end{bmatrix}$$

$$\begin{bmatrix} y_1(k) \\ y_2(k) \end{bmatrix} = \begin{bmatrix} 1 & -1 & 2 \\ 0 & 1 & 1 \end{bmatrix} \begin{bmatrix} x_1(k) \\ x_2(k) \\ x_3(k) \end{bmatrix}$$

PROBLEM 5.19 (MATLAB)
Consider the dynamic system

$$x(k+1) = Ax(k)$$
$$y(k) = Cx(k)$$

where all the computations are performed modulo 2 arithmetic. Let

$$A = \begin{bmatrix} 1 & 1 & 0 & 0 \\ 1 & 1 & 1 & 1 \\ 0 & 0 & 0 & 1 \\ 0 & 1 & 1 & 1 \end{bmatrix}, \text{ and } C = \begin{bmatrix} 1 & 0 & 0 & 0 \end{bmatrix}$$

a. Find $y(k)$ for $0 \leq k \leq 15$. Assume initial conditions to be $x(0) = \begin{bmatrix} 1 & 0 & 0 & 0 \end{bmatrix}^T$.
b. Find A^{16}.
c. Use the result of part (b) to show that $y(k)$ is periodic. What is the period of $y(k)$?

PROBLEM 5.20 (MATLAB)
Consider the nonlinear system described by:

$$\dot{x}_1 = -x_1(x_1^2 + x_2^2 - 1) + x_2$$
$$\dot{x}_2 = -x_1 - x_2(x_1^2 + x_2^2 - 1)$$

a. Find the equilibrium points of the system.
b. Using MATLAB, plot the trajectory of the motion of the system in the x_1-x_2 plane for different initial conditions. Is this system stable?

PROBLEM 5.21 (MATLAB)

Consider the nonlinear system described by:

$$\dot{x}_1 = x_1(x_1^2 + x_2^2 - 1) - x_2$$

$$\dot{x}_2 = x_1 + x_2(x_1^2 + x_2^2 - 1)$$

a. Find the equilibrium points of the system.
b. Using MATLAB, plot the trajectory of the motion of the system in the x_1-x_2 plane for different initial conditions. Is this system stable?

PROBLEM 5.22 (MATLAB)

A continuous time system can be converted to a discrete-time system using the following transformation.

$$\dot{x}(t) = Ax(t) + Bu(t) \qquad x(k+1) = \tilde{A}x(k) + \tilde{B}u(k)$$

$$\Rightarrow$$

$$y(t) = Cx(t) + Du(t) \qquad y(k) = \tilde{C}x(k) + \tilde{D}u(k)$$

where $\tilde{A} = e^{AT}$, $\tilde{B} = (\int_0^T e^{A\tau}d\tau)B$, $\tilde{C} = C$, $\tilde{D} = D$, and T=sampling period.

a. Write a general MATLAB function file that accepts A, B, C, D and T as inputs and produces $\tilde{A}$, $\tilde{B}$, $\tilde{C}$, and $\tilde{D}$ as outputs.
b. Use the code developed in part (a) to convert the following continuous time systems into discrete systems. Assume the sampling period $T=1$ sec.

(i) $\dot{x}(t) = \begin{bmatrix} 0 & 1 \\ 0 & 0 \end{bmatrix} x(t) + \begin{bmatrix} 0 \\ 1 \end{bmatrix} u(t)$

$y(t) = \begin{bmatrix} 1 & 0 \end{bmatrix} x(t)$

(ii) $\dot{x}(t) = \begin{bmatrix} 0 & 1 & 0 \\ 0 & 0 & 1 \\ -1 & -2 & -3 \end{bmatrix} x(t) + \begin{bmatrix} 1 \\ 2 \\ -1 \end{bmatrix} u(t)$

$y(t) = \begin{bmatrix} 1 & 0 & -1 \end{bmatrix} x(t) + 2u(t)$

PROBLEM 5.23 (MATLAB)

a. Find the transformation matrix P that transforms matrix A into a Jordan canonical form.

$$A = \begin{bmatrix} -2 & 1 & 0 & 3 \\ 0 & -3 & 1 & 1 \\ 0 & 0 & -2 & 1 \\ 0 & 0 & -1 & -4 \end{bmatrix}$$

b. Repeat part (a) for the matrix B given by:

$$B = \frac{1}{2}\begin{bmatrix} 0 & 0 & 2 & 0 \\ 0 & 0 & 2 & 2 \\ 0 & -2 & 2 & 0 \\ 2 & 2 & -4 & 2 \end{bmatrix} \begin{bmatrix} 3 & 1 & 0 & 0 \\ 0 & 3 & 0 & 0 \\ 0 & 0 & -2 & 0 \\ 0 & 0 & 0 & -3 \end{bmatrix} \begin{bmatrix} 2 & -1 & 1 & 1 \\ 1 & 0 & -1 & 0 \\ 1 & 0 & 0 & 0 \\ -1 & 1 & 0 & 0 \end{bmatrix}$$

6

Introduction to Optimization

6.1 Introduction

The term "optimization" refers to finding the best solution to a given problem. Optimization is used in many different disciplines. In real life, we are always trying to find an optimal approach to perform the day to day tasks that we all encounter. In engineering, every problem we tackle is in one way or another an optimization problem. This often means finding the maximum or minimum of a cost function subject to certain constraints and in accordance with a performance metric. In this chapter, we introduce several approaches for constrained and unconstrained optimization problems with emphasis on engineering applications.

6.2 Stationary Points of Functions of Several Variables

Let $f(x)=f(x_1,x_2,\ldots,x_n)$ be a continuous function of n variables, $x=(x_1,x_2,\ldots,x_n)$. The stationary points (e.g. maxima, minima, and saddle-point) of $f(x_1,x_2,\ldots,x_n)=f(x)$ are the solutions to the following n simultaneous equations:

$$\frac{\partial f(x)}{\partial x_1}=0$$

$$\frac{\partial f(x)}{\partial x_2}=0$$

$$\vdots$$

$$\frac{\partial f(x)}{\partial x_n}=0$$

(6.1)

This can be written in compact form as:

$$\nabla f(x)=0 \qquad (6.2)$$

Where $\nabla f(x)$ is the gradient vector defined as:

$$\nabla f(x) = \frac{\partial f}{\partial x} = \begin{bmatrix} \dfrac{\partial f}{\partial x_1} \\ \dfrac{\partial f}{\partial x_2} \\ \vdots \\ \dfrac{\partial f}{\partial x_n} \end{bmatrix} \qquad (6.3)$$

The stationary points of $f(x)$ are the points where:

$$\Delta f = \frac{\partial f}{\partial x_1}\Delta x_1 + \frac{\partial f}{\partial x_2}\Delta x_2 + \cdots + \frac{\partial f}{\partial x_n}\Delta x_n = 0 \qquad (6.4)$$

for arbitrary differential values of $\Delta x_1, \Delta x_2, \ldots, \Delta x_n$. Thus, it is imperative that the following equations hold at any point for which Equation 6.4 is true:

$$\nabla f(x) = \begin{bmatrix} \dfrac{\partial f}{\partial x_1} \\ \dfrac{\partial f}{\partial x_2} \\ \vdots \\ \dfrac{\partial f}{\partial x_n} \end{bmatrix} = \begin{bmatrix} 0 \\ 0 \\ \vdots \\ 0 \end{bmatrix} = 0 \qquad (6.5)$$

Equation 6.5 provides n simultaneous equations with n unknown variables $x_1, x_2, \ldots, x_n$ whose solutions will provide the "stationary points" of $f(x)$. Note that $\nabla f(x) = 0$ at any stationary point of the function. The question of whether these stationary points are a maxima, minima, or "saddle point" may be determined from the nature of the problem or may require further differentiation of the function $f(x)$.

Example 6.1

Find the stationary points of the function of two variables:

$$f(x) = f(x_1, x_2) = x_1^2 + 2x_1x_2 - x_2^3 + 5x_2 + 8 \qquad (6.6)$$

Solution: The stationary points can be found by computing the gradient of $f(x)$ with respect to x, setting it equal to zero and solving for vector x. This is shown below:

Introduction to Optimization

$$\nabla f(x) = \begin{bmatrix} \frac{\partial f}{\partial x_1} \\ \frac{\partial f}{\partial x_2} \end{bmatrix} = \begin{bmatrix} 2x_1 + 2x_2 \\ 2x_1 - 3x_2^2 + 5 \end{bmatrix} = \begin{bmatrix} 0 \\ 0 \end{bmatrix} \rightarrow \begin{cases} x_1 + x_2 = 0 \\ 2x_1 - 3x_2^2 + 5 = 0 \end{cases} \quad (6.7)$$

From which we obtain two stationary points:

$$x_a^* = \begin{bmatrix} x_1 \\ x_2 \end{bmatrix} = \begin{bmatrix} -1 \\ 1 \end{bmatrix} \text{ and } x_b^* = \begin{bmatrix} x_1 \\ x_2 \end{bmatrix} = \begin{bmatrix} \frac{5}{3} \\ -\frac{5}{3} \end{bmatrix} \quad (6.8)$$

To determine the corresponding types of the stationary points of f(x), we employ the second derivative of the function. For functions of several variables, the Hessian matrix is defined as the generalization of the second derivative.

6.2.1 Hessian Matrix

The Hessian matrix of $f(x)$, where $x \in R^n$ is an $n \times n$ matrix defined as:

$$\nabla^2 f(x) = \begin{bmatrix} \frac{\partial^2 f(x)}{\partial x_1^2} & \frac{\partial^2 f(x)}{\partial x_1 \partial x_2} & \cdots & \frac{\partial^2 f(x)}{\partial x_1 \partial x_n} \\ \frac{\partial^2 f(x)}{\partial x_2 \partial x_1} & \frac{\partial^2 f(x)}{\partial x_2^2} & \cdots & \frac{\partial^2 f(x)}{\partial x_2 \partial x_n} \\ \vdots & \vdots & & \vdots \\ \frac{\partial^2 f(x)}{\partial x_n \partial x_1} & \frac{\partial^2 f(x)}{\partial x_n \partial x_2} & \cdots & \frac{\partial^2 f(x)}{\partial x_n^2} \end{bmatrix} \quad (6.9)$$

Note that the Hessian matrix is symmetric since $\frac{\partial^2 f(x)}{\partial x_n \partial x_m} = \frac{\partial^2 f(x)}{\partial x_m \partial x_n}$.

Theorem 6.1: Let $x \in R^n$ and $f(x)$ be a mapping from $R^n \rightarrow R^1$, and let x^* be a point in R^n such that $\nabla f(x^*) = 0$, then:

(a) x^* is a local minimum of $f(x)$ if $\nabla^2 f(x^*)$ is a positive definite matrix.

(b) x^* is a local maximum of $f(x)$ if $\nabla^2 f(x^*)$ is a negative definite matrix.

(c) x^* is a saddle point of $f(x)$ if $\nabla^2 f(x^*)$ is an indefinite matrix.

Proof: We prove the first part of the theorem. Using Taylor series to expand $f(x^* + \varepsilon)$ about $\varepsilon = 0$, we have:

$$f(x^* + \varepsilon) = f(x^*) + [\nabla f(x^*)]^T \varepsilon + 0.5 \varepsilon^T \nabla^2 f(x^*) \varepsilon + \text{hot} \quad (6.10)$$

where hot represents higher order terms which are the terms of the order of $O(\|\varepsilon\|^2)$. Since $\nabla f(x^*)=0$, Equation 6.10 can be written as:

$$f(x^* + \varepsilon) - f(x^*) = 0.5\varepsilon^T \nabla^2 f(x^*)\varepsilon + O(\|\varepsilon\|^2) \qquad (6.11)$$

If $\nabla^2 f(x^*)$ is a positive definite matrix, then for any $\varepsilon \in R^n$ there is a positive number β such that:

$$\varepsilon^T \nabla^2 f(x^*)\varepsilon \geq \beta\|\varepsilon\|^2 \qquad (6.12)$$

Using this inequality in Equation 6.11, we have:

$$f(x^* + \varepsilon) - f(x^*) \geq 0.5\beta\|\varepsilon\|^2 + O(\|\varepsilon\|^2) \qquad (6.13)$$

For small ε, the first term is the dominant term and hence $f(x^*+\varepsilon)-f(x^*) \geq 0$. Therefore, x^* is a local minimum of $f(x)$. Parts (b) and (c) could be proved similarly. If the Hessian is positive on negative semidefinite, x^* can be local minimum, maximum or a saddle point. The test is inconclusive.

Example 6.2

Find the nature of the stationary points of the function:

$$f(x_1, x_2) = x_1^2 + 2x_1x_2 - x_2^3 + 5x_2 + 8$$

Solution: The two stationary points of $f(x)$ are (see Example 6.1):

$$x_a^* = \begin{bmatrix} x_1 \\ x_2 \end{bmatrix} = \begin{bmatrix} -1 \\ 1 \end{bmatrix} \text{ and } x_b^* = \begin{bmatrix} x_1 \\ x_2 \end{bmatrix} = \begin{bmatrix} \frac{5}{3} \\ -\frac{5}{3} \end{bmatrix}$$

The Hessian matrix computed at x_a^* is:

$$\nabla^2 f(x_a^*) = \begin{bmatrix} \frac{\partial^2 f(x)}{\partial x_1^2} & \frac{\partial^2 f(x)}{\partial x_1 \partial x_2} \\ \frac{\partial^2 f(x)}{\partial x_2 \partial x_1} & \frac{\partial^2 f(x)}{\partial x_2^2} \end{bmatrix} = \begin{bmatrix} 2 & 2 \\ 2 & -6x_2 \end{bmatrix} = \begin{bmatrix} 2 & 2 \\ 2 & -6 \end{bmatrix}$$

This is an indefinite matrix since the two eigenvalues of $\nabla^2 f(x_a^*)$ are 2.4721 which is greater than zero, and -6.4721 which is less than zero. Hence, the first stationary point is a saddle point. The Hessian matrix at the second stationary point is:

$$\nabla^2 f(x_b^*) = \begin{bmatrix} 2 & 2 \\ 2 & -6x_2 \end{bmatrix} = \begin{bmatrix} 2 & 2 \\ 2 & 10 \end{bmatrix}$$

which is a positive definite matrix since the two eigenvalues of $\nabla^2 f(x_b^*)$ are 1.5279 and 10.4721. Therefore, the second point is a global minimum of $f(x)$.

6.3 Least-Square (LS) Technique

Let x be an unknown vector in the vector space V which it defined as C^n and let y be a vector in subspace S of vector space C^n. Furthermore, assume that x and y are related through a linear transformation given by:

$$y = Ax + e \tag{6.14}$$

where A is a transformation from V to S and e is the model error vector. We wish to estimate x while minimizing the norm of the error defined by:

$$E = \|y - Ax\|^2 = (y - Ax)^H(y - Ax) \tag{6.15}$$

Expanding the above inner product, we have:

$$E = \|y\|^2 + x^H A^H A x - 2 y^H A x \tag{6.16}$$

To minimize E with respect to x, we set the gradient of E with respect to x equal to zero, that is:

$$\nabla_x E = \frac{\partial E}{\partial x} = 2A^H A x - 2 A^H y = 0 \tag{6.17}$$

Solving for x, we obtain:

$$x_{LS} = (A^H A)^{-1} A^H y \tag{6.18}$$

The resulting LS error is:

$$E = \|y\|^2 + x_{LS}^H A^H A x_{LS} - 2 y^H A x_{LS} = \|y\|^2 - y^H A (A^H A)^{-1} A^H y$$

$$= y^H [I - A(A^H A)^{-1} A^H] y \tag{6.19}$$

The matrix $(A^H A)^{-1} A^H$ is called the pseudo inverse of A. The MATLAB® command to compute LS solution is $x = A \backslash y$.

Example 6.3

Find the LS solution to the following set of linear equations:

$$\begin{bmatrix} 1 & 3 \\ 2 & 4 \\ 1 & -1 \\ -2 & 3 \end{bmatrix} \begin{bmatrix} x_1 \\ x_2 \end{bmatrix} = \begin{bmatrix} 4.1 \\ 5.8 \\ 0.1 \\ 1.12 \end{bmatrix}$$

Solution: The LS solution is computed as shown in Equation 6.18 as follows:

$$x_{LS} = (A^H A)^{-1} A^H y = \left(\begin{bmatrix} 1 & 2 & 1 & -2 \\ 3 & 4 & -1 & 3 \end{bmatrix} \begin{bmatrix} 1 & 3 \\ 2 & 4 \\ 1 & -1 \\ -2 & 3 \end{bmatrix} \right)^{-1} \begin{bmatrix} 1 & 2 & 1 & -2 \\ 3 & 4 & -1 & 3 \end{bmatrix} \begin{bmatrix} 4.1 \\ 5.8 \\ 0.1 \\ 1.12 \end{bmatrix}$$

$$x_{LS} = \begin{bmatrix} 0.9568 \\ 0.9981 \end{bmatrix}$$

The resulting mean square error (MSE) is:

$$MSE = E = y^H [I - A(A^H A)^{-1} A^H] y = 0.0549$$

6.3.1 LS Computation Using QR Factorization

Another technique to compute the LS solution is through the use of QR factorization. In this approach, matrix A is first decomposed into its QR form as follows:

$$A = QR \tag{6.20}$$

Once the decomposition is accomplished, matrix A in Equation 6.18 is then replaced by its QR decomposition. Hence, the LS solution given by Equation 6.18 will become:

$$x_{LS} = (R^H Q^H Q R)^{-1} R^H Q^H y = (R^H R)^{-1} R^H Q^H y$$
$$= R^{-1} (R^H)^{-1} R^H Q^H y = R^{-1} Q^H y \tag{6.21}$$

The above simplification is made possible since $Q^H Q = I$ as was shown in Chapter 4. Equation 6.21 can be expressed in an alternate form as:

$$R x_{LS} = Q^H y \tag{6.22}$$

Since R is a triangular matrix, the set of linear equations given by Equation 6.22 can be solved recursively without matrix inversion by utilizing back substitution.

Example 6.4

Find the LS solution of the following equations using QR factorization.

$$\begin{bmatrix} 1 & 3 \\ 2 & 4 \\ 1 & -1 \\ -2 & 3 \end{bmatrix} \begin{bmatrix} x_1 \\ x_2 \end{bmatrix} = \begin{bmatrix} 4.1 \\ 5.8 \\ 0.1 \\ 1.12 \end{bmatrix}$$

Introduction to Optimization

Solution: The QR factorization of A is:

$$A = QR = \begin{bmatrix} 0.3162 & 0.4499 \\ 0.6325 & 0.5537 \\ 0.3162 & -0.2422 \\ -0.6325 & 0.6575 \end{bmatrix} \begin{bmatrix} 3.1623 & 1.2649 \\ 0 & 5.7793 \end{bmatrix}$$

The LS solution is (see Equation 6.22):

$$Rx_{LS} = Q^H y$$

$$\begin{bmatrix} 3.1623 & 1.2649 \\ 0 & 5.7793 \end{bmatrix} \begin{bmatrix} x_1 \\ x_2 \end{bmatrix} = \begin{bmatrix} 0.3162 & 0.4499 \\ 0.6325 & 0.5537 \\ 0.3162 & -0.2422 \\ -0.6325 & 0.6575 \end{bmatrix}^T \begin{bmatrix} 4.1 \\ 5.8 \\ 0.1 \\ 1.12 \end{bmatrix}$$

Simplifying the above equation results in:

$$\begin{bmatrix} 3.1623 & 1.2649 \\ 0 & 5.7793 \end{bmatrix} \begin{bmatrix} x_1 \\ x_2 \end{bmatrix} = \begin{bmatrix} 4.288 \\ 5.7682 \end{bmatrix}$$

Therefore:

$$x_2 = \frac{5.7682}{5.7793} = 0.9981$$

$$x_1 = \frac{4.288 - 1.2649 x_2}{3.1623} = \frac{4.288 - 1.2649 \times 0.9981}{3.1623} = 0.9568$$

6.3.2 LS Computation Using Singular Value Decomposition (SVD)

If the matrix $A^H A$ is ill conditioned, the computation of its inverse becomes numerically unstable. In this case, matrix A is first expressed in terms of its SVD which is in turn utilized to find the LS solution. Using SVD, we can write:

$$A = \sum_{i=1}^{r} \sigma_i u_i v_i^H \quad (6.23)$$

where r is the rank of the $n \times m$ matrix A, u_i and v_i are unit norm pairwise orthogonal vectors. The LS solution is:

$$x_{LS} = (A^H A)^{-1} A^H y = \left(\sum_{i=1}^{r} \sigma_i v_i u_i^H \sum_{j=1}^{r} \sigma_j u_j v_j^H \right)^{-1} \sum_{k=1}^{r} \sigma_k v_k u_k^H y$$

$$= \left(\sum_{i=1}^{r} \sum_{j=1}^{r} \sigma_i \sigma_j v_i u_i^H u_j v_j^H \right)^{-1} \sum_{k=1}^{r} \sigma_k v_k u_k^H y \quad (6.24)$$

Since $u_i u_j^H = \delta_{ij}$, then:

$$x_{LS} = \left(\sum_{i=1}^{r} \sigma_i^2 v_i v_i^H\right)^{-1} \sum_{k=1}^{r} \sigma_k v_k u_k^H y$$

$$= \sum_{i=1}^{r} \sigma_i^{-2} v_i v_i^H \sum_{k=1}^{r} \sigma_k v_k u_k^H y = \sum_{i=1}^{r} \sum_{k=1}^{r} \sigma_i^{-2} \sigma_k v_i v_i^H v_k u_k^H y \quad (6.25)$$

Also $v_i v_k^H = \delta_{ik}$, therefore:

$$x_{LS} = \sum_{i=1}^{r} \sum_{k=1}^{r} \sigma_i^{-2} \sigma_k v_i v_i^H v_k u_k^H y = \sum_{i=1}^{r} \sigma_i^{-1} v_i u_i^H y \quad (6.26)$$

If A is ill-conditioned, some of its singular values will be very small positive numbers, so we approximate the summation in Equation 6.26 by using the first $k<r$ dominant singular values. Therefore:

$$x_{LS} = \sum_{i=1}^{k} \sigma_i^{-1} v_i u_i^H y \quad (6.27)$$

Example 6.5

Assume that we would like to estimate x_1, x_2 and x_3 using the following five noisy measurements:

$$y_1 = x_1 + 6x_2 + 11x_3 + e_1$$
$$y_2 = 3x_1 + 8x_2 + 13x_3 + e_2$$
$$y_3 = 2x_1 + 7x_2 + 12x_3 + e_3$$
$$y_4 = 2x_1 + 4.5x_2 + 7x_3 + e_4$$
$$y_5 = -5x_1 - 10x_2 - 15x_3 + e_5$$

Find the LS estimate of x_1, x_2 and x_3 if the measurement vector y is:

$$y = [y_1 \quad y_2 \quad y_3 \quad y_4 \quad y_5]^T = [22 \quad 25 \quad 21 \quad 15 \quad -36]^T$$

Solution: The matrix A is:

$$A = \begin{bmatrix} 1 & 6 & 11 \\ 3 & 8 & 13 \\ 2 & 7 & 12 \\ 2 & 4.5 & 7 \\ -5 & -10 & -15 \end{bmatrix}$$

Introduction to Optimization

The matrix $A^T A$ is ill conditioned. Hence, the direct computation of the LS solution using Equation 6.18 yields inaccurate results, since the computation of $(A^T A)^{-1}$ is numerically unstable. As a result, an SVD of A is first employed in order to facilitate the computation of the pseudo inverse prior to obtaining the LS solution. This will allow for a more accurate solution. To this effect, the SVD of A is: $\sigma_1^2 = 31.8515$, $\sigma_2^2 = 2.3993$ and $\sigma_3^2 = 0$ with corresponding vectors

$$u_1 = \begin{bmatrix} -0.3914 \\ -0.4884 \\ -0.4399 \\ -0.2684 \\ 0.5854 \end{bmatrix}, u_2 = \begin{bmatrix} -0.6684 \\ -0.0132 \\ -0.3408 \\ 0.1572 \\ -0.6420 \end{bmatrix}, u_3 = \begin{bmatrix} -0.0544 \\ 0.8218 \\ -0.4192 \\ -0.3371 \\ 0.1797 \end{bmatrix}$$

and

$$v_1 = \begin{bmatrix} -0.1947 \\ -0.5148 \\ -0.8349 \end{bmatrix}, v_2 = \begin{bmatrix} 0.8919 \\ 0.2614 \\ -0.3691 \end{bmatrix}, v_3 = \begin{bmatrix} -0.4082 \\ 0.8165 \\ 0.4082 \end{bmatrix}$$

We ignore the small singular values (i.e. $\sigma_3^2 = 0$) and consider the two largest ones only for the computation of the pseudo inverse as shown in Equation 6.27. As a result, the least square solution becomes:

$$x_{LS} = \sum_{i=1}^{2} \sigma_i^{-1} v_i u_i^T y = (\sigma_1^{-1} v_1 u_1^T + \sigma_2^{-1} v_2 u_2^T) y$$

$$= \begin{bmatrix} -0.2467 & -0.0019 & -0.1243 & 0.0602 & -0.2428 \\ -0.0667 & 0.0065 & -0.0301 & 0.0215 & -0.0796 \\ 0.1133 & 0.0148 & 0.0641 & -0.0172 & 0.0837 \end{bmatrix} \begin{bmatrix} 22 \\ 25 \\ 21 \\ 15 \\ -36 \end{bmatrix} = \begin{bmatrix} 1.5585 \\ 1.2495 \\ 0.9404 \end{bmatrix}$$

6.3.3 Weighted Least Square (WLS)

Assume that the variances of the different components of the error model e in Equation 6.14 are not equal to each other. In other words, some of the components of the measurement vector y are noisier than others. In this or similar situations, the different components of the error vector in the LS technique are weighted differently yielding a more accurate estimation. That is, the objective function to be minimized is chosen to be:

$$E = (y - Ax)^H W(y - Ax) \tag{6.28}$$

where W is a positive definite Hermitian weighting matrix. Setting the gradient of E with respect to x equal to zero gives:

$$\nabla_x E = \frac{\partial E}{\partial x} = 2A^H W A x - 2A^H W y = 0 \tag{6.29}$$

Solving for x, we obtain:

$$x_{WLS} = (A^H W A)^{-1} A^H W y \qquad (6.30)$$

The best choice for W is the inverse of the covariance matrix of the measurement noise e. That is:

$$W = R^{-1} \qquad (6.31)$$

Where $R = E(ee^H)$ is the covariance matrix of e. It is assumed that the noise is zero mean hence the covariance matrix is equal to the correlation matrix.

Example 6.6

Consider the following set of measurements:

$$y_1 = 2x + e_1$$
$$y_2 = 3x + e_2$$
$$y_3 = x + e_3$$

where x is the scalar unknown variable to be estimated. Assume that the measurement errors e_1, e_2, and e_3 are zero mean independent Gaussian random variables with variances $\sigma_1^2 = 2$, $\sigma_2^2 = 1$, and $\sigma_3^2 = 4$, respectively. Find x_{LS} and x_{WLS} if the measurement vector y is:

$$y = [4.2 \quad 7 \quad 4]^T$$

For the weighted least square, use $W = R^{-1}$, where R is the covariance matrix of the measurement noise bearing in mind that the mean of the noise is assumed to be zero.

Solution:
a. LS solution

$$y = \begin{bmatrix} 2 \\ 3 \\ 1 \end{bmatrix} x + e$$

$$x_{LS} = (A^H A)^{-1} A^H y = \left\{ [2 \quad 3 \quad 1] \begin{bmatrix} 2 \\ 3 \\ 1 \end{bmatrix} \right\}^{-1} [2 \quad 3 \quad 1] \begin{bmatrix} 4.2 \\ 7 \\ 4 \end{bmatrix} = 2.3857$$

b. WLS solution with $W = R^{-1}$

$$W = R^{-1} = \begin{bmatrix} \sigma_1^2 & 0 & 0 \\ 0 & \sigma_2^2 & 0 \\ 0 & 0 & \sigma_3^2 \end{bmatrix}^{-1} = \begin{bmatrix} 0.5 & 0 & 0 \\ 0 & 1 & 0 \\ 0 & 0 & 0.25 \end{bmatrix}$$

Introduction to Optimization

$$x_{LS} = (A^H W A)^{-1} A^H W y$$

$$= \left\{ \begin{bmatrix} 2 & 3 & 1 \end{bmatrix} \begin{bmatrix} 0.5 & 0 & 0 \\ 0 & 1 & 0 \\ 0 & 0 & 0.25 \end{bmatrix} \begin{bmatrix} 2 \\ 3 \\ 1 \end{bmatrix} \right\}^{-1} \begin{bmatrix} 2 & 3 & 1 \end{bmatrix} \begin{bmatrix} 0.5 & 0 & 0 \\ 0 & 1 & 0 \\ 0 & 0 & 0.25 \end{bmatrix} \begin{bmatrix} 4.2 \\ 7 \\ 4 \end{bmatrix} = 2.3289$$

6.3.4 LS Curve Fitting

As an application of LS, consider the following curve fitting problem. Assume that we have a set of data points (x_i, y_i) $i=1,2,\ldots,N$ in a two dimensional space and would like to fit a polynomial of degree m to the data. We first consider fitting a straight line $y = ax + b$ to the data using the LS approach. This can be written in matrix form as:

$$\begin{bmatrix} y_1 \\ y_2 \\ \vdots \\ y_N \end{bmatrix} = \begin{bmatrix} x_1 & 1 \\ x_2 & 1 \\ \vdots & \vdots \\ x_N & 1 \end{bmatrix} \begin{bmatrix} a \\ b \end{bmatrix} + \begin{bmatrix} e_1 \\ e_2 \\ \vdots \\ e_N \end{bmatrix} \qquad (6.32)$$

Solving for a and b using LS yields:

$$\begin{bmatrix} a \\ b \end{bmatrix} = \left\{ \begin{bmatrix} x_1 & x_2 & \cdots & x_N \\ 1 & 1 & \cdots & 1 \end{bmatrix} \begin{bmatrix} x_1 & 1 \\ x_2 & 1 \\ \vdots & \vdots \\ x_N & 1 \end{bmatrix} \right\}^{-1} \begin{bmatrix} x_1 & x_2 & \cdots & x_N \\ 1 & 1 & \cdots & 1 \end{bmatrix} \begin{bmatrix} y_1 \\ y_2 \\ \vdots \\ y_N \end{bmatrix}$$

$$= \begin{bmatrix} \sum_{i=1}^{N} x_i^2 & \sum_{i=1}^{N} x_i \\ \sum_{i=1}^{N} x_i & N \end{bmatrix}^{-1} \begin{bmatrix} \sum_{i=1}^{N} x_i y_i \\ \sum_{i=1}^{N} y_i \end{bmatrix} \qquad (6.33)$$

or

$$a = \frac{N \sum_{i=1}^{N} x_i y_i - \sum_{i=1}^{N} x_i \sum_{i=1}^{N} y_i}{N \sum_{i=1}^{N} x_i^2 - \left(\sum_{i=1}^{N} x_i \right)^2} \qquad (6.34)$$

$$b = \frac{-\sum_{i=1}^{N} x_i \sum_{i=1}^{N} x_i y_i + \sum_{i=1}^{N} x_i^2 \sum_{i=1}^{N} y_i}{N \sum_{i=1}^{N} x_i^2 - \left(\sum_{i=1}^{N} x_i\right)^2} \quad (6.35)$$

In general, the above formulation can also be used to fit a polynomial of degree m to the data (x_i, y_i) $i = 1, 2, \ldots, N$. Let the fitting polynomial be:

$$P(x) = P_m x^m + P_{m-1} x^{m-1} + \cdots + P_1 x + P_0 \quad (6.36)$$

Then:

$$\begin{bmatrix} y_1 \\ y_2 \\ \vdots \\ y_N \end{bmatrix} = \begin{bmatrix} x_1^m & x_1^{m-1} & \cdots & x_1 & 1 \\ x_2^m & x_2^{m-1} & \cdots & x_2 & 1 \\ x_3^m & x_3^{m-1} & \cdots & x_3 & 1 \\ \vdots & \vdots & & \vdots & \vdots \\ x_N^m & x_N^{m-1} & \cdots & x_N & 1 \end{bmatrix} \begin{bmatrix} P_m \\ P_{m-1} \\ \vdots \\ P_0 \end{bmatrix} + \begin{bmatrix} e_1 \\ e_2 \\ \vdots \\ e_N \end{bmatrix} \quad (6.37)$$

Thus, the LS solution is:

$$P = (A^H A)^{-1} A^H y \quad (6.38)$$

where

$$A = \begin{bmatrix} x_1^m & x_1^{m-1} & \cdots & x_1 & 1 \\ x_2^m & x_2^{m-1} & \cdots & x_2 & 1 \\ x_3^m & x_3^{m-1} & \cdots & x_3 & 1 \\ \vdots & \vdots & & \vdots & \vdots \\ x_N^m & x_N^{m-1} & \cdots & x_N & 1 \end{bmatrix}, \quad y = \begin{bmatrix} y_1 \\ y_2 \\ \vdots \\ y_N \end{bmatrix}, \text{ and } P = \begin{bmatrix} P_m \\ P_{m-1} \\ \vdots \\ P_0 \end{bmatrix} \quad (6.39)$$

The MATLAB® command to fit a polynomial of degree m to a data set stored in the x and y vectors is $P = \mathbf{polyfit}(x, y, m)$. Vector P contains the coefficients of the resulting polynomial. For linear curve fitting, the parameter $m = 1$. The above approach is illustrated by Example 6.7.

Example 6.7

For the data given in Table 6.1, find the best linear fit in the mean square error sense.

TABLE 6.1

X and Y data

x_i	0.059	0.347	0.833	1.216	2.117	2.461	4.879	5.361	5.501	5.612
y_i	2.685	2.029	4.792	5.721	6.088	9.114	13.947	13.686	14.330	14.400

Introduction to Optimization

Solution:

$$\begin{bmatrix} a \\ b \end{bmatrix} = \begin{bmatrix} \sum_{i=1}^{N} x_i^2 & \sum_{i=1}^{N} x_i \\ \sum_{i=1}^{N} x_i & N \end{bmatrix}^{-1} \begin{bmatrix} \sum_{i=1}^{N} x_i y_i \\ \sum_{i=1}^{N} y_i \end{bmatrix} = \begin{bmatrix} 127.164 & 28.39 \\ 28.39 & 10 \end{bmatrix}^{-1} \begin{bmatrix} 348.236 \\ 86.793 \end{bmatrix} = \begin{bmatrix} 2.1868 \\ 2.4709 \end{bmatrix}$$

Therefore, the linear fit is:

$$y = 2.1868x + 2.4709$$

The plot of data and the linear fit are shown in Figure 6.1.

6.3.5 Applications of LS Technique

6.3.5.1 One Dimensional Wiener Filter

An engineering application of LS is the Wiener filter. The Wiener filter is a useful signal and image processing technique for signal denoising and reconstruction in both one dimensional and two dimensional applications. It can be used for deblurring or noise removal. Here, we first derive the one

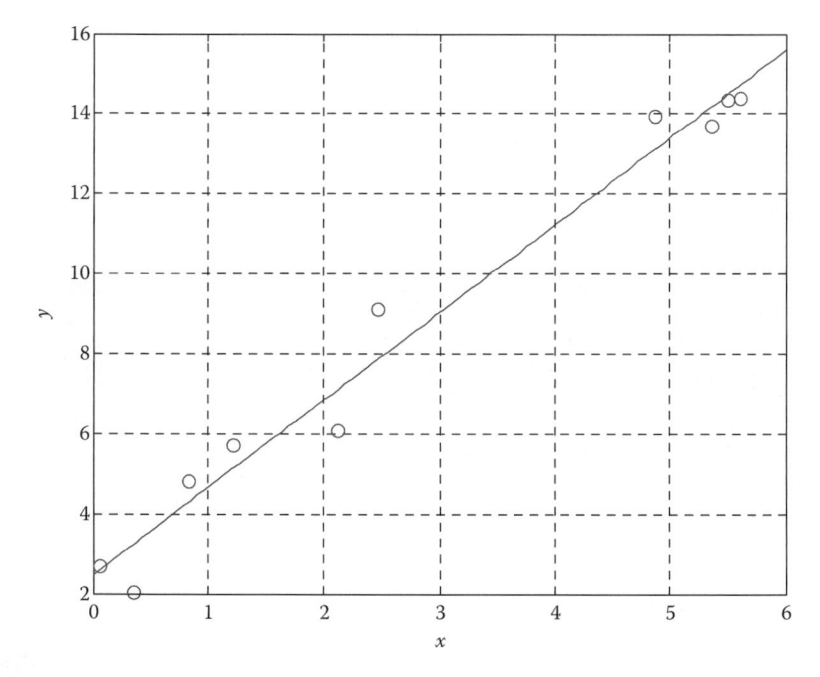

FIGURE 6.1
Linear curve fitting example.

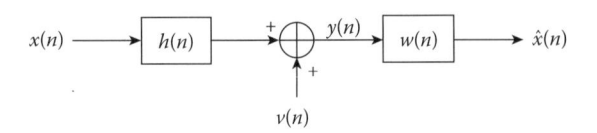

FIGURE 6.2
1-d Wiener filter.

dimensional Wiener filter and then extend the results to the two dimensional case.

Consider the following linear model for the measurement of signal $x(n)$:

$$y(n) = x(n) * h(n) + v(n) \qquad (6.40)$$

In this model, $x(n)$ is the signal, $y(n)$ is the noisy measurement, $h(n)$ is the filter impulse response, and $v(n)$ is the zero mean white additive noise with variance σ^2. The goal is to derive an estimate for $x(n)$ in the minimum mean square error sense. The overall process is illustrated with the block diagram shown in Figure 6.2.

Using the matrix formulation of the discrete convolution (see Chapter 1), Equation 6.40 can be written in matrix form as:

$$y = Hx_e + V \qquad (6.41)$$

where:

$$y = [y(0) \quad y(1) \quad y(2) \quad \cdots \quad y(M+L-2) \quad y(M+L-1) \quad 0 \quad \cdots \quad 0]^T \qquad (6.42)$$

$$x_e = [x(0) \quad x(1) \quad \cdots \quad x(M-1) \quad 0 \quad 0 \quad \cdots \quad 0]^T \qquad (6.43)$$

$$V = [v(0) \quad v(1) \quad v(2) \quad \cdots \quad v(M+L-2) \quad v(M+L-1) \quad 0 \quad \cdots \quad 0]^T \qquad (6.44)$$

$$h_e = [h(0) \quad h(1) \quad \cdots \quad h(L-1) \quad 0 \quad 0 \quad \cdots \quad 0]^T \qquad (6.45)$$

$$H = \begin{bmatrix} h_e(0) & h_e(N-1) & h_e(N-2) & \cdots & h_e(2) & h_e(1) \\ h_e(1) & h_e(0) & h_e(N-1) & h_e(N-2) & \vdots & h_e(2) \\ h_e(2) & h_e(1) & h_e(0) & h_e(N-1) & \vdots & h_e(3) \\ h_e(3) & h_e(2) & h_e(1) & h_e(0) & \vdots & h_e(4) \\ \vdots & \vdots & \vdots & \vdots & \cdots & \vdots \\ h_e(N-1) & h_e(N-2) & h_e(N-3) & h_e(N-4) & \cdots & h_e(0) \end{bmatrix} \qquad (6.46)$$

Using LS (see Equation 6.18), the solution to Equation 6.41 is given by:

$$\hat{x}_e = (H^T H)^{-1} H^T y \qquad (6.47)$$

Introduction to Optimization

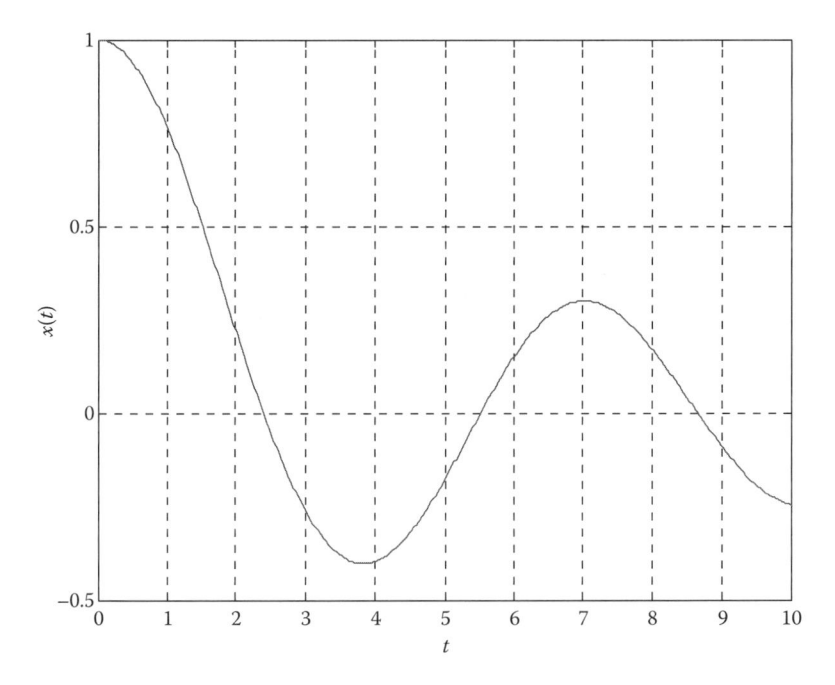

FIGURE 6.3
Continuous signal $x(t)$.

Example 6.8

Consider the continuous signal $x(t)$ defined for $t = [0, 10]$ as shown in Figure 6.3.

This signal is sampled in the interval $t = [0, 10]$ using a sampling interval of $T = 0.025$ yielding $10/0.025 = 400$ samples. The sampled signal $x(n)$ is convolved with a linear filter that possesses the impulse response $h = [1\ 2\ 4\ 3\ -2\ 1]$. The output of the filter is corrupted by zero mean additive white Gaussian noise with variance equal to 0.1 to generate the corrupted output signal $y(n)$. In order to recover an estimate for $x(n)$ using LS, the corrupted signal $y(n)$ is passed through the LS filter implemented using Equation 6.46 as discussed earlier yielding the deconvolved signal shown in Figure 6.4. As can be seen, the result is not smooth due to the presence of measurement noise.

To compensate for the noise and obtain a more accurate solution, an additional term is added to the cost function. This additional term is in the form of $x_e^T Q x_e$. This term penalizes any deviation from a smooth solution. The matrix Q is the smoothing matrix and is selected to minimize the second derivative of the deconvolved signal. Hence, the updated cost function is given by:

$$J = \|y - Hx_e\|^2 + \beta x_e^T Q x_e \qquad (6.48)$$

The positive factor β controls the smoothness of the solution. If $\beta = 0$, the solution tends to fit the noisy data yielding a nonsmooth estimate. On the other hand,

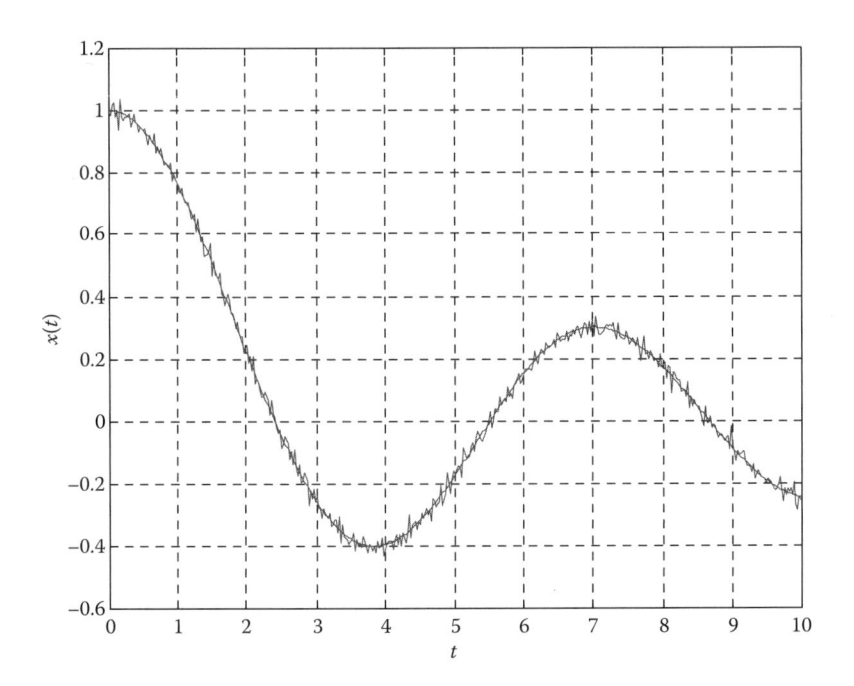

FIGURE 6.4
(See color insert following page 174.) Deconvolved signal in presence of noise.

if β is chosen to be very large, then the solution is smooth but the mean square error is large. Setting the gradient of J with respect to x_e equal to zero yields:

$$\frac{\partial J}{\partial x_e} = 2H^T(y - Hx_e) + 2\beta Q x_e = 0 \quad (6.49)$$

Therefore:

$$\hat{x}_e = (H^T H + \beta Q)^{-1} H^T y \quad (6.50)$$

The smoothing operator matrix Q is a positive definite matrix that is chosen to satisfy some smoothing criteria.

6.3.5.2 Choice of Q Matrix and Scale Factor β

To choose the smoothing operator matrix Q, we need some a priori knowledge regarding the solution. Suppose that our a priori knowledge is that the solution is close to a constant; then, a reasonable cost function associated with the smoothness of the solution would be:

$$J = \int \left(\frac{dx(t)}{dt}\right)^2 dt \approx \sum_n (x(n+1) - x(n))^2 \quad (6.51)$$

This can be written in matrix form as:

$$J = (Bx_e)^T Bx_e = x_e^T B^T B x_e = x_e^T Q x_e \quad (6.52)$$

Introduction to Optimization

where B is the (N×N) matrix given by:

$$B = \begin{pmatrix} 1 & -1 & 0 & 0 & 0 & 0 & 0 & \cdots & 0 \\ 0 & 1 & -1 & 0 & 0 & 0 & 0 & \cdots & 0 \\ \vdots & \vdots & \vdots & \ddots & & & & & \vdots \\ 0 & \cdots & 0 & 0 & 0 & 0 & 1 & -1 & 0 \\ 0 & \cdots & 0 & 0 & 0 & 0 & 0 & 1 & -1 \end{pmatrix} \quad (6.53)$$

and Q is the (N×N) matrix defined as $Q = B^T B$. If our a priori knowledge is that the solution is close to a linear function, then a reasonable cost function associated with the smoothness of the solution should penalize any deviation from a linear solution. Therefore, the second derivative must be close to zero. If we assume that a linear function is a better approximation, then the cost function to be minimized can be written as:

$$J = \int \left(\frac{d^2 x(t)}{dt^2} \right)^2 dt \approx \sum_n (x(n+2) - 2x(n+1) + x(n))^2 \quad (6.54)$$

This implies that:

$$B = \begin{pmatrix} 1 & -2 & 1 & 0 & 0 & 0 & 0 & \cdots & 0 \\ 0 & 1 & -2 & 1 & 0 & 0 & 0 & \cdots & 0 \\ \vdots & \vdots & \vdots & \ddots & & & & & \vdots \\ 0 & \cdots & 0 & 0 & 0 & 1 & -2 & 1 & 0 \\ 0 & \cdots & 0 & 0 & 0 & 0 & 1 & -2 & 1 \end{pmatrix} \quad (6.55)$$

and

$$Q = B^T B \quad (6.56)$$

The constant β is chosen by trial and error. An initial estimate for β would be:

$$\beta = \frac{\text{Trace}(H^T H)}{\text{Trace}(Q)} \quad (6.57)$$

If this value of β does not provide a satisfactory smooth estimate, we need to change it until we get an appropriate solution.

Example 6.9

Consider the deconvolution problem described in Example 6.8. The solution obtained using the second derivative smoothing operator given by Equation 6.55 and 6.56 is shown in Figure 6.5. As can be seen, the solution is very smooth yielding a more accurate estimate.

The 1-d matrix form of the Wiener filter solution given by Equation 6.50 can also be implemented in the frequency domain using the Fourier transform. The resulting filter is:

$$W(\omega) = \frac{H^*(j\omega)}{|H(j\omega)|^2 + \beta Q(j\omega)} \quad (6.58)$$

where $H(j\omega)$ is the frequency response of $h(n)$ and $Q(\omega)$ is the frequency response of the smoothing operator. This filter is typically implemented using the FFT algorithm.

6.3.5.3 Two Dimensional Wiener Filter

The 2-d Wiener filter is similar to the 1-d case. The block diagram of the image degradation model and the Wiener filter is shown in Figure 6.6.

The solution is similar to 1-d and the results are stated without proof. The filter in matrix form is given by:

$$\hat{f} = (H^T H + \beta Q)^{-1} H^T g \tag{6.59}$$

The 2-d Wiener filter in the frequency domain is given by:

$$W(\omega_1, \omega_2) = \frac{H^*(j\omega_1, j\omega_2)}{|H(j\omega_1, j\omega_2)|^2 + \beta Q(j\omega_1, j\omega_2)} \tag{6.60}$$

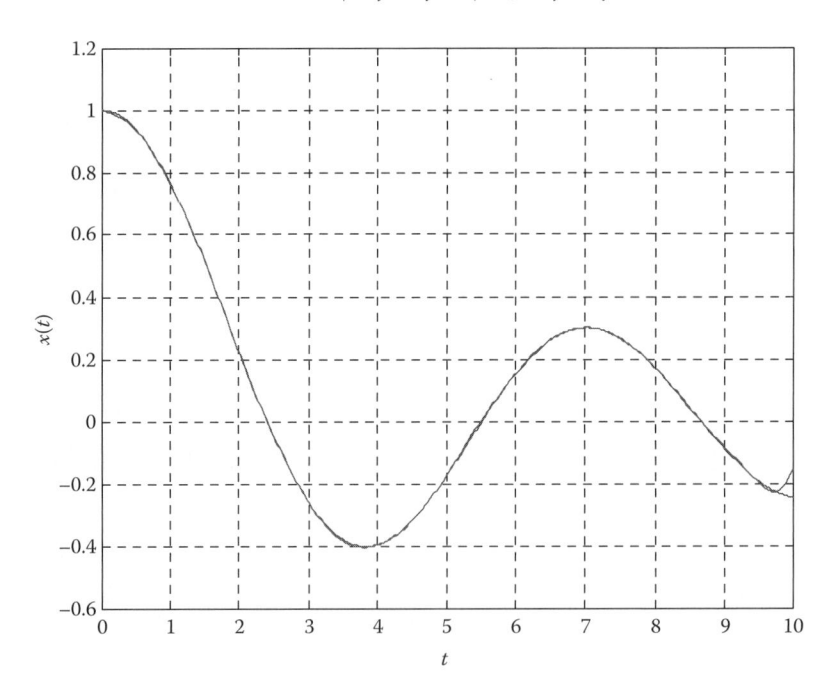

FIGURE 6.5
(See color insert following page 174.) Deconvolution with a smooth operator.

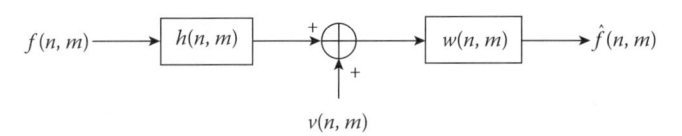

FIGURE 6.6
2-d Wiener filter.

Example 6.10

Consider the LENA image shown in Figure 6.7a and assume that the original has been blurred by a linear motion blur in both the horizontal and vertical directions and further corrupted by additive white zero mean Gaussian noise with variance of $\sigma^2 = 10^{-4}$ resulting in the image shown in Figure 6.7b. The image pixel values are normalized between zero and one. The Wiener filter is implemented in the frequency domain using the following equation (Note: $Q(j\omega_1, j\omega_2) = 1$:

$$W(j\omega_1, j\omega_2) = \frac{H^*(j\omega_1, j\omega_2)}{|H(j\omega_1, j\omega_2)|^2 + \beta} \qquad (6.61)$$

where the optimal value of the parameter $\beta = 0.022$ has been selected by trial and error yielding the reconstructed image shown in Figure 6.7c. The error image computed from the difference between the original image (see Figure 6.7a) and the reconstructed image (see Figure 6.7c) is shown in Figure 6.7d.

FIGURE 6.7
(a) Original image; (b) blurred and noisy image; (c) restored image; (d) error image.

6.4 Total Least-Squares (TLS)

Consider the solution to the problem $Ax=b$, where b is the measurement vector which is assumed to contain some amount of error. That is: $Ax=b+\Delta b$. The LS solution that minimizes $\|\Delta b\|$, as discussed earlier, is given by:

$$x_{Ls} = \arg\min_{x}\|Ax-b\| = (A^H A)^{-1} A^H b \tag{6.62}$$

In many practical engineering problems, the matrix A is a function of measured data and is also assumed to contain some amount of measurement noise. In these cases, a solution to the following problem is sought:

$$(A+\Delta A)x = b+\Delta b \tag{6.63}$$

Assume that matrix A is $m \times n$. To find the TLS solution to Equation 6.63, we proceed to write it in the form:

$$[A+\Delta A \quad b+\Delta b]\begin{bmatrix} x \\ -1 \end{bmatrix} = 0 \tag{6.64}$$

Let the $m \times (n+1)$ matrices H and ΔH be defined as $H=[A\ b]$ and $\Delta H=[\Delta A\ \Delta b]$. Then, Equation 6.64 is reduced to:

$$(H+\Delta H)\begin{bmatrix} x \\ -1 \end{bmatrix} = 0 \tag{6.65}$$

Therefore, the solution $\begin{bmatrix} x \\ -1 \end{bmatrix}$ is a vector in the null space of $H+\Delta H$. Since the solution is a nontrivial solution, the matrix $H+\Delta H$ must be rank deficient. Hence, TLS finds ΔH with the smallest norm that makes $H+\Delta H$ rank deficient. Let the SVD of H be:

$$H = \sum_{i=1}^{n+1} \sigma_i u_i v_i^H \tag{6.66}$$

where:

$$\sigma_1 \geq \sigma_2 \geq \cdots \geq \sigma_n \geq \sigma_{n+1} > 0 \tag{6.67}$$

Then:

$$H+\Delta H = \sum_{i=1}^{n} \sigma_i u_i v_i^H + \sigma_{n+1} u_{n+1} v_{n+1}^H - \sigma_{n+1} u_{n+1} v_{n+1}^H \tag{6.68}$$

Therefore:

$$\Delta H = -\sigma_{n+1} u_{n+1} v_{n+1}^H \tag{6.69}$$

Introduction to Optimization

and

$$x_{TLS} = -\frac{v_{n+1}(1:n)}{v_{n+1}(n+1)} \quad (6.70)$$

Example 6.11

Fit a straight line to the following noisy data using TLS.

$$x = \begin{bmatrix} 0 \\ 1 \\ 1.9 \\ 2.9 \end{bmatrix} \quad y = \begin{bmatrix} 2 \\ 2.8 \\ 4.8 \\ 7.1 \end{bmatrix}$$

Solution: In order to solve the problem using TLS, we begin by defining the following matrices:

$$A = [x \quad 1] = \begin{bmatrix} 0 & 1 \\ 1 & 1 \\ 1.9 & 1 \\ 2.9 & 1 \end{bmatrix}, b = y = \begin{bmatrix} 2 \\ 2.8 \\ 4.8 \\ 7.1 \end{bmatrix}, \text{ and } H = [A \quad b] = \begin{bmatrix} 0 & 1 & 2 \\ 1 & 1 & 2.8 \\ 1.9 & 1 & 4.8 \\ 2.9 & 1 & 7.1 \end{bmatrix}$$

The SVD of matrix H is given by:

$$H = \sum_{i=1}^{3} \sigma_i u_i v_i^H$$

where $\sigma_1 = 10.0489$, $\sigma_2 = 1.1107$, $\sigma_3 = 0.382$ and

$$v_1 = \begin{bmatrix} -0.3524 \\ -0.1792 \\ -0.9189 \end{bmatrix}, v_2 = \begin{bmatrix} -0.6389 \\ 0.7634 \\ 0.0954 \end{bmatrix}, v_3 = \begin{bmatrix} -0.6843 \\ -0.6206 \\ 0.3828 \end{bmatrix}$$

$$u_1 = \begin{bmatrix} -0.2007 \\ -0.3088 \\ -0.5232 \\ -0.7685 \end{bmatrix}, u_2 = \begin{bmatrix} 0.8591 \\ 0.3527 \\ 0.0068 \\ -0.3708 \end{bmatrix}, u_3 = \begin{bmatrix} 0.4703 \\ -0.7563 \\ -0.2707 \\ 0.3654 \end{bmatrix}$$

The TLS is given by:

$$x_{TLS} = -\frac{v_3(1:2)}{v_3(3)} = -\frac{\begin{bmatrix} -0.6843 \\ -0.6206 \end{bmatrix}}{0.3828} = \begin{bmatrix} 1.7877 \\ 1.6213 \end{bmatrix}$$

and

$$[\Delta A \quad \Delta b] = \sigma_{n+1} u_{n+1} v_{n+1}^T = \sigma_3 u_3 v_3^T = -0.3082 \begin{bmatrix} 0.4703 \\ -0.7563 \\ -0.2707 \\ 0.3654 \end{bmatrix} [-0.6843 \quad -0.6206 \quad 0.3828]$$

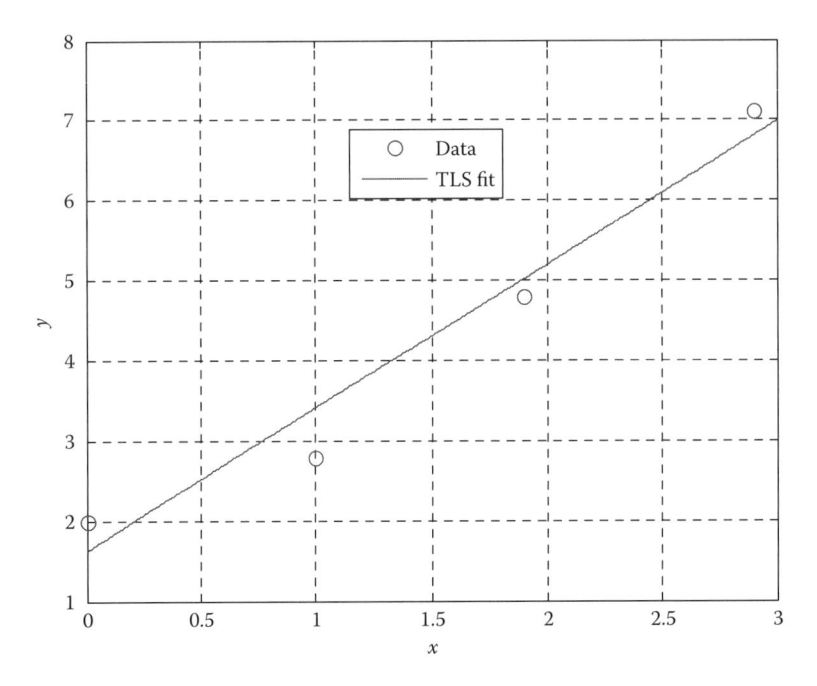

FIGURE 6.8
(See color insert following page 174.) TLS linear fit. The original data is shown by the red circles and the LS linear fit is illustrated by the blue line.

$$[\Delta A \quad \Delta b] = \begin{bmatrix} 0.0092 & 0.0900 & -0.0555 \\ -0.1595 & -0.1447 & -0.0892 \\ -0.0571 & -0.0518 & 0.0319 \\ 0.0771 & 0.0699 & -0.0431 \end{bmatrix}$$

Note that $(A+\Delta A) x_{TLS} = b + \Delta b$. The plots of the data and the TLS fit are shown in Figure 6.8.

6.5 Eigen Filters

Assume that a zero mean stochastic process $y(n)$ is observed in the presence of zero mean white Gaussian noise $v(n)$. Then, the observed signal $x(n)$ is given by:

$$x(n) = y(n) + v(n) \tag{6.71}$$

The signal plus noise is passed through an M-tap finite impulse response (FIR) filter with impulse response $h(n)$, as shown in Figure 6.9.

Introduction to Optimization

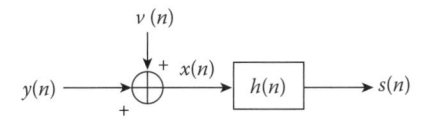

FIGURE 6.9
Eigen filter.

The output signal s(n) is given by:

$$s(n) = \sum_{i=0}^{M-1} h(i)x(n-i) = h^T x \tag{6.72}$$

where

$$h = [h(0) \quad h(1) \quad \ldots \quad h(M-1)]^T \text{ and } x = [x(n) \quad x(n-1) \quad \ldots \quad x(n-M+1)]^T$$

The output power due to the signal is:

$$P = E(s^2(n)) = E(s(n)s^T(n)) = E(h^T xx^T h) = h^T E(xx^T) h = h^T R h \tag{6.73}$$

where R is the $M \times M$ correlation matrix of $x(n)$. The covariance matrix of the additive white noise is $\sigma^2 I$. The output noise power is:

$$P_N = E(N^2(n)) = E(N(n)N^T(n)) = E(h^T vv^T h) = \sigma^2 h^T h \tag{6.74}$$

and the output signal to noise ratio SNR is given by:

$$\text{SNR} = \frac{h^T R h}{\sigma^2 h^T h} \tag{6.75}$$

The objective is to select the best filter h to maximize SNR. Since SNR is invariant to a scale factor σ^2, maximizing SNR is equivalent to maximizing:

$$\max_h \; h^T R h \tag{6.76}$$

$$\text{subject to the constraint } h^T h = 1 \tag{6.77}$$

Using Lagrange multiplier, we have:

$$F = h^T R h - \lambda(h^T h - 1) \tag{6.78}$$

Setting the gradient of F with respect to h equal to zero yields:

$$\frac{\partial F}{\partial h} = 2Rh - 2\lambda h = 0 \tag{6.79}$$

or

$$Rh = \lambda h \qquad (6.80)$$

Therefore, the solution h is an eigenvector of the signal covariance matrix R with λ being the corresponding eigenvalue. The maximum SNR is then written as:

$$\text{SNR}_{max} = \frac{h^T R h}{\sigma^2 h^T h} = \frac{\lambda}{\sigma^2} \qquad (6.81)$$

Hence, the optimal solution is the eigenvector corresponding to the maximum eigenvalue of signal covariance matrix R.

Example 6.12

Consider a sinusoidal signal $x(n)$ buried in white noise with variance σ^2. That is:

$$y(n) = x(n) + v(n) = A\cos(n\omega_0 + \theta) + v(n)$$

where θ is uniformly distributed between $-\pi$ and π. The autocorrelation of the signal $x(n)$ is:

$$r(k) = E(x(n)x(n+k)) = \frac{A^2}{2}\cos(k\omega_0)$$

Design a three-tap FIR filter to reduce the noise and improve SNR.

Solution: The 3×3 covariance matrix of the signal is given by:

$$R = E\begin{bmatrix} x^2(n) & x(n)x(n-1) & x(n)x(n-2) \\ x(n-1)x(n) & x^2(n-1) & x(n-1)x(n-2) \\ x(n-2)x(n) & x(n-2)x(n-1) & x^2(n-2) \end{bmatrix}$$

$$= \begin{bmatrix} r(0) & r(1) & r(2) \\ r(1) & r(0) & r(1) \\ r(2) & r(1) & r(0) \end{bmatrix} = \begin{bmatrix} \dfrac{A^2}{2} & \dfrac{A^2}{2}\cos\omega_0 & \dfrac{A^2}{2}\cos 2\omega_0 \\ \dfrac{A^2}{2}\cos\omega_0 & \dfrac{A^2}{2} & \dfrac{A^2}{2}\cos\omega_0 \\ \dfrac{A^2}{2}\cos 2\omega_0 & \dfrac{A^2}{2}\cos\omega_0 & \dfrac{A^2}{2} \end{bmatrix}$$

The characteristic polynomial of R is:

$$P(\lambda) = |\lambda I - R| = \lambda \left(\lambda - \frac{3A^2}{4} + \frac{A^2}{2}\left|\frac{1}{2} + \cos 2\omega_0\right| \right)\left(\lambda - \frac{3A^2}{4} - \frac{A^2}{2}\left|\frac{1}{2} + \cos 2\omega_0\right| \right)$$

The three eigenvalues are:

$$\lambda_1 = 0, \quad \lambda_2 = \frac{3}{4}A^2 - \frac{A^2}{2}\left|\frac{1}{2} + \cos 2\omega_0\right| \quad \text{and} \quad \lambda_3 = \frac{3}{4}A^2 + \frac{A^2}{2}\left|\frac{1}{2} + \cos 2\omega_0\right|.$$

Introduction to Optimization

The maximum eigenvalue is λ_3 and the corresponding normalized eigenvector is:

$$h = \frac{v}{\|v\|}$$

where

$$v = \left[r^2(1) - r(2)r(0) + r(2)\lambda_3 \quad r(1)r(2) - r(1)r(0) + r(1)\lambda_3 \quad (r(0) - \lambda_3)^2 - r^2(1) \right]^T$$

The maximum SNR is given by:

$$\text{SNR}_{\max} = \frac{\lambda_3}{\sigma^2} = \frac{A^2}{2\sigma^2}(1.5 + |0.5 + \cos(2\omega_0)|)$$

6.6 Stationary Points with Equality Constraints

We now consider optimization problems with equality constraints. The Lagrange multiplier technique is used to solve these types of problems.

6.6.1 Lagrange Multipliers

Let us assume that we have a function $f(x)$, $x \in R^n$ and that we wish to find a stationary point x subject to the following constraints:

$$g(x) = \begin{bmatrix} g_1(x) \\ g_2(x) \\ \vdots \\ g_m(x) \end{bmatrix} = \begin{bmatrix} 0 \\ 0 \\ \vdots \\ 0 \end{bmatrix} \tag{6.82}$$

where $m < n$. This implies that a point x is sought at which:

$$\Delta f = \frac{\partial f}{\partial x_1} \Delta x_1 + \frac{\partial f}{\partial x_2} \Delta x_2 + \cdots + \frac{\partial f}{\partial x_n} \Delta x_n = 0 \tag{6.83}$$

While, at the same point we have:

$$\Delta g_1 = \frac{\partial g_1}{\partial x_1} \Delta x_1 + \frac{\partial g_1}{\partial x_2} \Delta x_2 + \cdots + \frac{\partial g_1}{\partial x_n} \Delta x_n = 0 \tag{6.84}$$

$$\Delta g_2 = \frac{\partial g_2}{\partial x_1} \Delta x_1 + \frac{\partial g_2}{\partial x_2} \Delta x_2 + \cdots + \frac{\partial g_2}{\partial x_n} \Delta x_n = 0 \tag{6.85}$$

$$\vdots$$

$$\Delta g_m = \frac{\partial g_m}{\partial x_1} \Delta x_1 + \frac{\partial g_m}{\partial x_2} \Delta x_2 + \cdots + \frac{\partial g_m}{\partial x_n} \Delta x_n = 0 \tag{6.86}$$

It should be clear that, in this situation, the individual variations ($\Delta x_1, \Delta x_2, \ldots, \Delta x_n$) cannot be chosen arbitrarily, but must be chosen so as to maintain the validity of the constraints as expressed in Equation 6.82. Because of these constraints, $n-m$ of the Δx's may be chosen arbitrarily, but the remaining m are then determined from these chosen $n-m$ Δx's by using Equations 6.84 through 6.86. Suppose, however, that each of the equations in Equation 6.82 were multiplied by separate constants $\lambda_1, \lambda_2, \ldots, \lambda_m$ yielding:

$$\lambda_1 \Delta g_1 = \lambda_1 \frac{\partial g_1}{\partial x_1} \Delta x_1 + \lambda_1 \frac{\partial g_1}{\partial x_2} \Delta x_2 + \cdots + \lambda_1 \frac{\partial g_1}{\partial x_n} \Delta x_n = 0 \qquad (6.87)$$

$$\lambda_2 \Delta g_2 = \lambda_2 \frac{\partial g_2}{\partial x_1} \Delta x_1 + \lambda_2 \frac{\partial g_2}{\partial x_2} \Delta x_2 + \cdots + \lambda_2 \frac{\partial g_2}{\partial x_n} \Delta x_n = 0 \qquad (6.88)$$

$$\vdots$$

$$\lambda_m \Delta g_m = \lambda_m \frac{\partial g_m}{\partial x_1} \Delta x_1 + \lambda_m \frac{\partial g_m}{\partial x_2} \Delta x_2 + \cdots + \lambda_m \frac{\partial g_m}{\partial x_n} \Delta x_n = 0 \qquad (6.89)$$

The parameters $\lambda_1, \ldots, \lambda_m$ are known as Lagrange multipliers. The above equations yield:

$$\left(\frac{\partial f}{\partial x_1} + \lambda_1 \frac{\partial g_1}{\partial x_1} + \lambda_2 \frac{\partial g_2}{\partial x_1} + \cdots + \lambda_m \frac{\partial g_m}{\partial x_1} \right) \Delta x_1 + \left(\frac{\partial f}{\partial x_2} + \lambda_1 \frac{\partial g_1}{\partial x_2} + \lambda_2 \frac{\partial g_2}{\partial x_2} + \cdots + \lambda_m \frac{\partial g_m}{\partial x_2} \right)$$

$$\Delta x_2 + \cdots + \left(\frac{\partial f}{\partial x_m} + \lambda_1 \frac{\partial g_1}{\partial x_m} + \lambda_2 \frac{\partial g_2}{\partial x_m} + \cdots + \lambda_m \frac{\partial g_m}{\partial x_m} \right) \Delta x_n = 0 \qquad (6.90)$$

However, in Equation 6.90, even though m of the Δx's are dependent upon the others, the m multipliers $\lambda_1, \lambda_2, \ldots, \lambda_m$ could be chosen such that:

$$\frac{\partial f}{\partial x_1} + \lambda_1 \frac{\partial g_1}{\partial x_1} + \lambda_2 \frac{\partial g_2}{\partial x_1} + \cdots + \lambda_m \frac{\partial g_m}{\partial x_1} = 0$$

$$\frac{\partial f}{\partial x_2} + \lambda_1 \frac{\partial g_1}{\partial x_2} + \lambda_2 \frac{\partial g_2}{\partial x_2} + \cdots + \lambda_m \frac{\partial g_m}{\partial x_2} = 0$$

$$(6.91)$$

$$\vdots$$

$$\frac{\partial f}{\partial x_n} + \lambda_1 \frac{\partial g_1}{\partial x_n} + \lambda_2 \frac{\partial g_2}{\partial x_n} + \cdots + \lambda_m \frac{\partial g_m}{\partial x_n} = 0$$

Introduction to Optimization

These equations provide n equations in the unknown $x_1, x_2, \ldots, x_n$ and $\lambda_1, \lambda_2, \ldots, \lambda_m$. The original constraint equations:

$$g(x) = \begin{bmatrix} g_1(x) \\ g_2(x) \\ \vdots \\ g_m(x) \end{bmatrix} = \begin{bmatrix} 0 \\ 0 \\ \vdots \\ 0 \end{bmatrix} \qquad (6.92)$$

provide the remaining m equations to determine the n values of x and m values of λ. The resulting set of $x_1, x_2, \ldots, x_n$ provides the stationary point for the function $f(x)$.

Example 6.13

Let $f(x) = x_1^2 + x_2^2$. This is an upward-facing bowl with a true minimum at $x_1 = 0$ and $x_2 = 0$. Find the minimum value of $f(x)$ subject to the constraint:

$$g(x) = x_1 + x_2 - 2 = 0$$

Solution: We begin by forming the function:

$$H(x) = f(x) + \lambda g(x) = x_1^2 + x_2^2 + \lambda_1(x_1 + x_2 - 2)$$

Differentiating $H(x)$ with respect to x_1, x_2 and λ and setting these corresponding equations equal to zero yields:

$$\frac{\partial H}{\partial x_1} = 2x_1 + \lambda = 0$$

$$\frac{\partial H}{\partial x_2} = 2x_2 + \lambda = 0$$

$$\frac{\partial H}{\partial \lambda} = x_1 + x_2 - 2 = 0$$

Solving the above three equations for the three unknowns x_1, x_2 and λ yields:

$$x_1 = 1, \; x_2 = 1 \text{ and } \lambda = -2.$$

Example 6.14

Find the minimum of the function:

$$f(x) = x^T A x \qquad (6.93)$$

subject to the constraint:

$$C^T x = b \qquad (6.94)$$

where $x \in R^n$, A is an $n \times n$ symmetric matrix, C is an $n \times 1$ column vector and b is a scalar.

Solution: Using a Lagrange multiplier, we form:

$$F(x) = x^T A x + \lambda (C^T x - b) \tag{6.95}$$

The necessary conditions are:

$$\frac{\partial F(x)}{\partial x} = 0 \rightarrow 2Ax + \lambda C = 0 \tag{6.96}$$

$$\frac{\partial F(x)}{\partial \lambda} = 0 \rightarrow Cx^T - b = 0 \tag{6.97}$$

From Equating 6.96, we have:

$$x^* = -\frac{\lambda}{2} A^{-1} C \tag{6.98}$$

Substituting Equation 6.98 into Equation 6.97 results in the following:

$$-\frac{\lambda}{2} CC^T A^{-1} - b = 0 \rightarrow \lambda = -2(CC^T A^{-1})^{-1} b = -2A(CC^T)^{-1} b \tag{6.99}$$

Hence, the optimal solution is:

$$x^* = A^{-1} CA (CC^T)^{-1} b \tag{6.100}$$

6.6.2 Applications

6.6.2.1 Maximum Entropy Problem

Example 6.15

The entropy of a discrete memoryless random source S with K symbols and distribution $P = [p_1 \quad p_2 \quad \cdots \quad p_K]$ is given by:

$$H(S) = -\sum_{k=1}^{K} p_k \log_2 p_k$$

Find the distribution that maximizes the entropy. What is the maximum entropy in this case?

Solution: This is an optimization problem with one constraint:

$$\max_{p_k} H(S) = -\sum_{k=1}^{K} p_k \log_2 p_k$$

subject to:

$$\sum_{k=1}^{K} p_k = 1$$

Introduction to Optimization

Using a Lagrange multiplier, we form:

$$F(p_k) = H(S) + \lambda g(p_k) = -\sum_{k=1}^{K} p_k \log_2 p_k + \lambda\left(\sum_{k=1}^{K} p_k - 1\right)$$

Setting the gradient of F with respect to p_k and λ equal to zero yields:

$$\frac{\partial F(p_k)}{\partial p_k} = -\frac{1}{\ln 2}(1 + \ln(p_k)) + \lambda = 0$$

or

$$p_k = e^{\lambda \ln 2 - 1} = \text{constant}$$

Since $\sum_{k=1}^{K} p_k = 1$ (constraint), then:

$$p_k = \frac{1}{K} \quad k = 1, 2, \ldots, K$$

Therefore, maximum entropy is achieved when the source symbols are modeled by a uniform distribution. The maximum entropy is:

$$H(S)_{\max} = -\sum_{k=1}^{K} p_k \log_2 p_k = -\sum_{k=1}^{K} \frac{1}{K} \log_2 \frac{1}{K} = \log_2 K$$

Example 6.16

Find the probability distribution function $p(k)$ that maximizes the entropy of a nonnegative integer valued random source S subject to the constraint that the average value of S is a constant α. That is:

$$\max_{p_k} H(S) = -\sum_{k=0}^{\infty} p_k \log_2 p_k$$

subject to:

$$\sum_{k=0}^{\infty} k p_k = \alpha \quad \text{and} \quad \sum_{k=0}^{\infty} p_k = 1$$

Solution: Use Lagrange multipliers and form:

$$F(p_k) = H(S) + \lambda_1 g_1(p_k) + \lambda_2 g_2(p_k)$$

$$= -\sum_{k=0}^{\infty} p_k \log_2 p_k + \lambda_1\left(\sum_{k=0}^{\infty} k p_k - \alpha\right) + \lambda_2\left(\sum_{k=0}^{\infty} p_k - 1\right)$$

Setting the gradient of F with respect to p_k, λ_1, and λ_2 equal to zero gives:

$$\frac{\partial F(p_k)}{\partial p_k} = -\frac{1}{\ln 2}(1 + \ln(p_k)) + \lambda_1 k + \lambda_2 = 0$$

Therefore:

$$p_k = e^{\lambda_2 \ln 2 + k\lambda_1 \ln 2 - 1} = ab^k$$

Since $\sum_{k=0}^{\infty} p_k = 1$ (constraint), then:

$$\sum_{k=0}^{\infty} ab^k = a\frac{1}{1-b} = \frac{a}{1-b} = 1 \rightarrow a = (1-b) \quad (6.101)$$

Enforcing the second constraint results in the following:

$$\sum_{k=0}^{\infty} kp_k = \sum_{k=0}^{\infty} akb^k = a\sum_{k=0}^{\infty} kb^k = \frac{ab}{(1-b)^2} = \alpha \quad (6.102)$$

We can now utilize Equations 6.101 and 6.102 to solve for a and b yielding the following values:

$$a = \frac{1}{\alpha+1} \quad \text{and} \quad b = \frac{\alpha}{\alpha+1}$$

and the optimal distribution is thus:

$$p_k = \frac{1}{\alpha+1}\left(\frac{\alpha}{\alpha+1}\right)^k$$

Therefore, maximum entropy is achieved with an exponential distribution for the source symbols. The maximum entropy is:

$$H(S)_{max} = -\sum_{k=0}^{\infty} p_k \log_2 p_k = -\sum_{k=0}^{\infty} ab^k \log_2 ab^k = \sum_{k=0}^{\infty} ab^k (\log_2 a + k\log_2 b)$$

$$= a\log_2 a \sum_{k=0}^{\infty} b^k + a\log_2 b \sum_{k=0}^{\infty} kb^k = \frac{a\log_2 a}{1-b} + \frac{ab\log_2 b}{(1-b)^2}$$

$$= \frac{(1-b)\log_2 a}{1-b} + \frac{(1-b)b\log_2 b}{(1-b)^2}$$

$$= -\log_2(1+\alpha) + \alpha\log_2\frac{\alpha}{\alpha+1} = \alpha\log_2\alpha - (1+\alpha)\log_2(1+\alpha)$$

6.6.3 Design of Digital Finite Impulse Response (FIR) Filters

An example of a constrained optimization problem in digital signal processing is the design of digital filters. Consider the design of a linear phase FIR filter with an odd number of filter taps to approximate the ideal filter given by the desired frequency response shown in Figure 6.10. In this figure, ω_p is the passband frequency and ω_s is the stopband frequency.

Introduction to Optimization

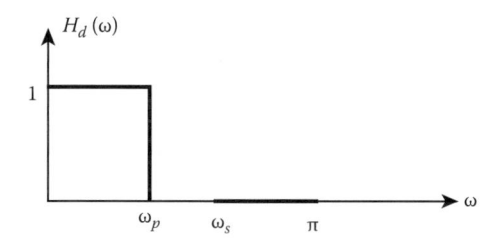

FIGURE 6.10
Ideal digital low pass filter.

Let the filter taps be represented by $h(0), h(1), \ldots, h(M-1)$, where M is an odd number. To have a linear phase filter, the filter taps must be symmetric with respect to the midpoint. That is:

$$h(k) = h(M-1-k), \quad k = 0, 1, \ldots, M-1 \tag{6.103}$$

The frequency response of this filter is given by:

$$H(e^{j\omega}) = \sum_{k=0}^{M-1} h(k)e^{-jk\omega} = h(0) + h(1)e^{-j\omega} + h(2)e^{-j2\omega} + \cdots + h(M-1)e^{-j(M-1)\omega}$$

$$= e^{-j\frac{M-1}{2}\omega} h(0)(e^{j\frac{M-1}{2}\omega} + e^{-j\frac{M-1}{2}\omega}) + e^{-j\frac{M-1}{2}\omega} h(1)(e^{j\frac{M-3}{2}\omega} + e^{-j\frac{M-3}{2}\omega})$$

$$+ \cdots + e^{-j\frac{M-1}{2}\omega} h\left(\frac{M-1}{2}\right)$$

$$= e^{-j\frac{M-1}{2}\omega} \left(2h(0)\cos\left(\frac{M-1}{2}\omega\right) + 2h(1)\cos\left(\frac{M-1}{2}\omega\right) + \cdots + h\left(\frac{M-1}{2}\right)\right)$$

$$= M(\omega)e^{j\theta(\omega)} \tag{6.104}$$

Therefore, the phase response is linear and is given by:

$$\angle H(e^{j\omega}) = -\frac{M-1}{2}\omega \tag{6.105}$$

and the magnitude of the frequency response is:

$$M(\omega) = 2h(0)\cos\left(\frac{M-1}{2}\omega\right) + 2h(1)\cos\left(\frac{M-1}{2}\omega\right) + \cdots + h\left(\frac{M-1}{2}\right) \tag{6.106}$$

Let us define the following quantities:

$N = (M-1)/2$, $x = [x(0) \ x(1) \ \ldots \ x(N)]^T = [h(N) \ 2h(N-1) \ \ldots \ 2h(0)]^T$ and $a(\omega) = [1 \ \cos(\omega) \ \cos(2\omega) \ \ldots \ \cos(N\omega)]$, then:

$$M(\omega) = \sum_{k=0}^{N} x(k)\cos(k\omega) = x^T a(\omega) \tag{6.107}$$

Assume that the DC gain of the filter is one, i.e. $M(0) = x^T a(0) = 1$ and let the cost function associated with the passband be defined as:

$$E_P = \frac{1}{\pi}\int_0^{\omega_p}(M(\omega)-1)^2 d\omega = \frac{1}{\pi}\int_0^{\omega_p}(x^T a(\omega) - x^T a(0))(a^T(\omega)x - a^T(0)x)d\omega$$

$$= \frac{1}{\pi}x^T\left[\int_0^{\omega_p}(a(\omega)-a(0))(a(\omega)-a(0))^T d\omega\right]x = \frac{1}{\pi}x^T A_P x \quad (6.108)$$

where A_p is an $(N+1)\times(N+1)$ matrix given by:

$$A_P = \frac{1}{\pi}\int_0^{\omega_p}(a(\omega)-a(0))(a(\omega)-a(0))^T d\omega \quad (6.109)$$

The closed form solution for A_p is given by:

$$[A_P]_{ij} = \frac{1}{\pi}\int_0^{\omega_P}(\cos(i-1)\omega-1)(\cos(j-1)\omega-1)d\omega$$

$$= \frac{\sin[(i+j-2)\omega_p]}{2\pi(i+j-2)} + \frac{\sin[(i-j)\omega_p]}{2\pi(i-j)} - \frac{\sin[(i-1)\omega_p]}{\pi(i-1)} - \frac{\sin[(j-1)\omega_p]}{\pi(j-1)} + \frac{\omega_P}{\pi}$$

$$(6.110)$$

The cost function associated with the stopband is defined as:

$$E_S = \frac{1}{\pi}\int_{\omega_S}^{\pi}M^2(\omega)d\omega = \frac{1}{\pi}\int_{\omega_S}^{\pi}M(\omega)M^T(\omega)d\omega = \frac{1}{\pi}x^T\left(\int_{\omega_S}^{\pi}a(\omega)a^T(\omega)d\omega\right)x = \frac{1}{\pi}x^T A_S x$$

$$(6.111)$$

where

$$A_S = \frac{1}{\pi}\int_{\omega_S}^{\pi}a(\omega)a^T(\omega)d\omega \quad (6.112)$$

with the closed form solution given by:

$$[A_S]_{ij} = \frac{1}{\pi}\int_{\omega_S}^{\pi}\cos(i-1)\omega\cos(j-1)\omega d\omega$$

$$= \frac{\sin[(i+j-2)\pi]}{2\pi(i+j-2)} + \frac{\sin[(i-j)\pi]}{2\pi(i-j)} - \frac{\sin[(i+j-2)\omega_S]}{2\pi(i+j-2)} - \frac{\sin[(i-j)\omega_S]}{2\pi(i-j)} \quad (6.113)$$

The total cost function is given by:

$$J = \alpha E_P + (1-\alpha)E_S = x^T[\alpha A_P + (1-\alpha)A_S]x \quad (6.114)$$

Introduction to Optimization

where $0 \leq \alpha \leq 1$ is a tradeoff parameter between the passband and stopband error. Let us now minimize the cost function subject to the constraint that the DC gain of the filter is one. That is:

$$\min_{x} J = x^T[\alpha A_P + (1-\alpha)A_S]x \qquad (6.115)$$

subject to the constraint:

$$C^T x = 1 \qquad (6.116)$$

where $C = [1 \ 1 \ \ldots \ 1]^T$.

The solution is given by:

$$x = \frac{R^{-1}C}{C^T R^{-1} C} \qquad (6.117)$$

where $R = \alpha A_p + (1-\alpha)A_S$.

The MATLAB® implementation is shown in Table 6.2.

TABLE 6.2

MATLAB® Code for the Design of Low Pass FIR Filter

```
function [h] = lslpf(wp,ws,M,a)
%
% Design of a linear phase FIR filter using LS
% wp=passband frequency
% ws=stopband frequency
% M=Number of filter taps (odd)
% a=tradeoff parameter between passband and stopband
% h=Desigend filter taps.
%
N=(M-1)/2;
k=1:1:N+1;
l=k;
[y,x]=meshgrid(k,l);
AP=0.5*(sin((x-y+eps)*wp))./((x-y+eps))+0.5*(sin((x+y-2+eps)*wp))./
  ((x+y-2+eps));
AP=AP-((sin((x-1+eps)*wp))./((x-1+eps)))-((sin((y-1+eps)*wp))./
  ((y-1+eps)))+wp;
AP=AP/pi;
AS=-0.5*(sin((x-y+eps)*ws))./
  ((x-y+eps))-0.5*(sin((x+y-2+eps)*ws))./((x+y-2+eps));
AS=AS+0.5*(sin((x-y+eps)*pi))./
  ((x-y+eps))+0.5*(sin((x+y-2+eps)*pi))./((x+y-2+eps));
AS=AS/pi;
R=a*AP+(1-a)*AS;
c=ones(N+1,1);
z=inv(R)*c/(c'*inv(R)*c);
h=[0.5*z(end:-1:2);z(1)];
h=[h;h(end-1:-1:1)];
W=linspace(0,pi,4000);
[H]=freqz(h,1,W);
plot(W,20*log10(abs(H)))
grid on
xlabel('\omega')
ylabel('|H(j\omega)|')
```

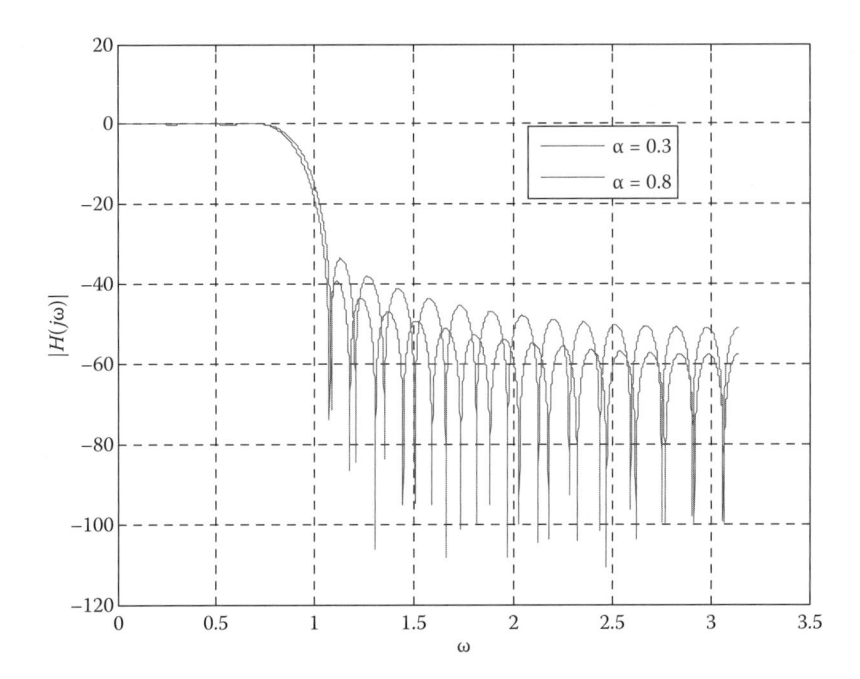

FIGURE 6.11
(See color insert following page 174.) Frequency response of the designed low pass filter.

Example 6.17

Design a linear phase low pass FIR filter of size $M=41$ to have a passband of $\omega_p = \pi/4$ and a stopband of $\pi/3$. Plot the frequency response for the following values of the tradeoff parameter α.

a. $\alpha = 0.3$
b. $\alpha = 0.5$
c. $\alpha = 0.8$

Solution: Using the MATLAB code provided in Table 6.2, the frequency response of the designed filter for different values of α is shown in Figure 6.11.

Problems

PROBLEM 6.1
Obtain the expression for the gradient vector ∇f and Hessian matrix $\nabla^2 f$ for the following functions of n variables:

a. $f(x) = 1/2 \, x^T A x + b^T x$ where A is symmetric
b. $f(x) = 1/2 \, (x^T A x)^2 + (b^T x)^2$ where A is symmetric

Introduction to Optimization

PROBLEM 6.2

Obtain an expression for the gradient vector ∇f and Hessian matrix $\nabla^2 f$ for the function of two variables given below:

$$f(x) = 100(x_2 - x_1^2)^2 + (1 - x_1)^2$$

Verify that $x^* = [1\ 1]^T$ satisfies $\nabla f = 0$ and $\nabla^2 f$ positive definite. Show that $\nabla^2 f$ is singular if and only if x satisfies the condition:

$$x_2 - x_1^2 = 0.005$$

Hence, show that $\nabla^2 f$ is positive definite for all x such that $x_2 - x_1^2 < 0.005$.

PROBLEM 6.3

Show that the function $f(x) = (x_2 - x_1^2)^2 + x_1^5$ has only one stationary point, which is neither a local maximum nor a local minimum.

PROBLEM 6.4

Find the stationary points of the function:

$$f(x) = 2x_1^3 - 3x_1^2 - 6x_1 x_2 (x_1 - x_2 - 1)$$

Which of these points are local minima, which are local maxima, and which are neither?

PROBLEM 6.5

Investigate the stationary points of the function:

$$f(x) = x_1^2 x_2^2 - 4x_1^2 x_2 + 4x_1^2 + 2x_1 x_2^2 + x_2^2 - 8x_1 x_2 + 8x_1 - 4x_2 + 6$$

PROBLEM 6.6

List all the stationary points of the function:

$$f(x) = -x_1^2 - 4x_2^2 - 16x_3^2$$

subject to the constraint $c(x) = 0$, where $c(x)$ is given by:

i. $c(x) = x_1 - 1$
ii. $c(x) = x_1 x_2 - 1$
iii. $c(x) = x_1 x_2 x_3 - 1$

PROBLEM 6.7

Solve the optimization problem below:
  Minimize $f(x,y) = x^2 + y^2 + 3xy + 6x + 19y$
  Subject to $x + 3y = 5$

PROBLEM 6.8

Let a, b, and c be positive constants. Find the least value of a sum of three positive numbers x, y, and z subject to the constraint:

$$\frac{a}{x}+\frac{b}{y}+\frac{c}{z}=1$$

by the method of Lagrange multipliers.

PROBLEM 6.9

Find the point on the ellipse defined by the intersection of the surfaces $x+y=1$ and $x^2+2y^2+z^2=1$ that is nearest to the origin.

PROBLEM 6.10

The equation $x_1+4x_2=5$ has infinite number of solutions. Find the solution that will give the minimum norm.

PROBLEM 6.11

Maximize $\dfrac{1}{3}\sum_{i=1}^{n}x_i^3$

Subject to:

$$\sum_{i=1}^{n}x_i=0, \quad \text{and} \quad \sum_{i=1}^{n}x_i^2=n$$

PROBLEM 6.12

Maximize $-\sum_{i=1}^{n}x_i\ln(x_i)$

Subject to:

$$\sum_{i=1}^{n}x_i=1.$$

PROBLEM 6.13

Maximize $2x_1x_2+x_2x_3+x_1+x_3+4$ Subject to: $x_1+x_2+x_3=3$ and determine the solution (x_1, x_2, x_3, λ). Compute the Hessian matrix and determine if the solution is a minimum, maximum, or neither.

PROBLEM 6.14

Let:

$$R=\begin{bmatrix} 4 & 4 & 6 \\ 4 & 10 & 12 \\ 6 & 12 & 20 \end{bmatrix}$$

Introduction to Optimization

a. Show that the eigenvalues of R are $\lambda_1 = \lambda_2 = 2$, and $\lambda_3 = 30$.
b. Maximize $x^T R x$, subject to: $x_1^2 + x_2^2 + x_3^2 = 1$.

PROBLEM 6.15 (MATLAB)

Consider a linear time invariant system with input $x(n)$ and output $y(n)$ given by the second order Markov model

$$z(n) = a_1 z(n-1) + a_2 z(n-2) + b x(n)$$

$$y(n) = z(n) + v(n)$$

where $v(n)$ is zero mean white Gaussian noise with variance σ_v^2. Let the system parameters be $a_1 = 0.7$, $a_2 = -0.21$, and $b = 1$.

a. Use MATLAB to generate 20 samples of $y(n)$ if $x(n) = u(n)$ in a noiseless case.
b. Use LS technique; estimate the parameters of the system. Compute MSE and compare with the true values.
c. Repeat parts (a) and (b) if SNR=80 dB and 40 dB.

PROBLEM 6.16 (MATLAB)

Design a linear phase low-pass FIR filter of size $M=31$ to have a passband of $\omega_p = \pi/2$ and a stopband of 0.67π.

PROBLEM 6.17 (MATLAB)

Design a linear phase low-pass FIR filter of size $M=41$ to have a passband of $\omega_p = \pi/3$ and a stopband of 0.4π.

PROBLEM 6.18 (MATLAB)

Consider the linear model $y = 2x + 1$. Let us assume that we have the following noisy measured samples of x and y:

$$x_n = \begin{bmatrix} 0.6598 \\ 1.1699 \\ 1.2316 \\ 1.7602 \\ 2.3432 \end{bmatrix} \quad y_n = \begin{bmatrix} 1.3620 \\ 2.3908 \\ 3.5360 \\ 5.0183 \\ 5.4835 \end{bmatrix}$$

Using the above measured noisy data, do the following:

a. Fit a straight line using LS algorithm.
b. Fit a line using TLS approach.
c. Plot the resulting fits and compare them with the true model.

PROBLEM 6.19 (MATLAB)

Consider the set of linear system of equations given by:

$$Ax = b$$

where A is an $n \times n$ symmetric positive definite matrix, b and x are $n \times 1$ vectors. We would like to solve the above equation by using the following recursive algorithm:

$$x(k+1) = x(k) - \mu A^T (Ax(k) - b)$$

a. Show that the above recursive equation minimizes the cost function $J = \|Ax - b\|^2$.
b. Find a range for the step size parameter μ that guarantees convergence of the algorithm to the true solution $x^* = A^{-1} b$ independent of the initial condition $x(0)$.
c. Implement the above algorithm in MATLAB for the following two cases

$$\text{(I)} \quad A = \begin{bmatrix} 1 & 2 \\ 2 & 5 \end{bmatrix}, b = \begin{bmatrix} 5 \\ 12 \end{bmatrix}$$

$$\text{(II)} \quad A = \begin{bmatrix} 2 & -1 & 1 \\ -1 & 3 & 1 \\ 1 & 1 & 4 \end{bmatrix}, b = \begin{bmatrix} 2 \\ 3 \\ 6 \end{bmatrix}$$

PROBLEM 6.20 (MATLAB)

The capacity of a Gaussian communication channel is given by:

$$C = \frac{1}{2} \log\left(1 + \frac{P}{N}\right) \text{ bits/channel use}$$

where P is the signal power, N is the variance of the Gaussian noise and the base of log is two. The capacity is a metric used to measure the amount of information that can be transmitted reliably over a noisy channel.

a. Suppose we have K Gaussian channels each has noise with average power (variance) of N_i, $i = 1, 2, \ldots, K$. If the total power available is P, find the optimum way of distribution the total power between the K channels to maximize the overall channel capacity given by:

$$C = \frac{1}{2} \sum_{i=1}^{K} \log\left(1 + \frac{P_i}{N_i}\right)$$

b. Write a general MATLAB code to implement the algorithm derived in part (a).
c. Find the overall channel capacity if $P = 20$ with four Gaussian channels having noise variances of $N_1 = 4$, $N_2 = 5$, $N_3 = 1$, and $N_4 = 6$ units of power.

Appendix A: The Laplace Transform

A1 Definition of the Laplace Transform

The Laplace transform of the continuous signal $x(t)$ is a mapping from the time domain to the complex s domain defined as

$$X(s) = \int_{-\infty}^{\infty} x(t)e^{-st} dt \qquad (A.1)$$

where $s = \sigma + j\omega$ with σ and ω as the real and imaginary parts, respectively. The integral in Equation A.1 may not converge for all values of s. The region in the complex plane where the complex function $X(s)$ converges is known as the region of convergence (ROC) of $X(s)$. The transform defined by Equation A.1 is referred to as the double-sided Laplace transform. In many engineering applications, the signals are defined over the time interval of 0 to ∞. Therefore, for this type of function, we are interested in the one-sided Laplace defined as:

$$X(s) = \int_{0}^{\infty} x(t)e^{-st} dt \qquad (A.2)$$

Example A.1

Find the Laplace transform and ROC of the signal $x(t) = e^{-at}u(t)$.

$$X(s) = \int_{0}^{\infty} x(t)e^{-st} dt = \int_{0}^{\infty} e^{-at}e^{-st} dt = \int_{0}^{\infty} e^{-(s+a)t} dt = \frac{-1}{s+a} e^{-(s+a)t} \bigg|_{0}^{\infty} = \frac{1}{s+a} \qquad (A.3)$$

The convergence of $X(s)$ requires that $\lim_{x \to \infty} e^{-(s+a)t} \to 0$. Thus, the ROC is the set of points in the complex s-plane for which real$(s) > -a$.

Example A.2

Find the Laplace transform and ROC of the signal $x(t) = u(t)$, where $u(t)$ is the unit step function.

321

This is a special case of Example A.1 where $a=0$. Therefore:

$$X(s) = \frac{1}{s} \quad \text{ROC: Real}(s) > 0 \tag{A.4}$$

Example A.3

Find the Laplace transform and ROC of the unit impulse function $x(t)=\delta(t)$.

$$X(s) = \int_0^\infty x(t)e^{-st}dt = \int_0^\infty \delta(t)e^{-st}dt = \delta(t)e^{-st}\Big|_{t=0} = 1 \tag{A.5}$$

Since the above integral converges for all values of s, the ROC is the entire complex s-plane.

The Laplace transforms of some elementary signals and their regions of convergence are listed in Table A.1. Properties of the Laplace transform are given in Table A.2. In Table A.2, L stands for the Laplace transform operator.

TABLE A.1

Laplace Transform of some Elementary Functions

Signal	Transform	ROC
$\delta(t)$	1	s-plane
$u(t)$	$\dfrac{1}{s}$	Real$(s) > 0$
$e^{-at}u(t)$	$\dfrac{1}{s+a}$	Real$(s) \geq$ real(a)
$te^{-at}u(t)$	$\dfrac{1}{(s+a)^2}$	Real$(s) \geq$ real(a)
$t^n e^{-at}u(t)$	$\dfrac{n!}{(s+a)^{n+1}}$	Real$(s) \geq$ real(a)
$\cos\omega_0 t u(t)$	$\dfrac{1}{s^2+\omega_0^2}$	Real$(s) > 0$
$\sin\omega_0 t u(t)$	$\dfrac{s}{s^2+\omega_0^2}$	Real$(s) > 0$
$e^{-at}\cos\omega_0 t u(t)$	$\dfrac{1}{(s+a)^2+\omega_0^2}$	Real$(s) \geq$ real(a)
$e^{-at}\cos\omega_0 t u(t)$	$\dfrac{s+a}{(s+a)^2+\omega_0^2}$	Real$(s) \geq$ real(a)

Appendix A: The Laplace Transform

TABLE A.2
Properties of the Laplace Transform

Property	
	$x(t) \xrightarrow{L} X(s)$
	$y(t) \xrightarrow{L} Y(s)$
Linearity	$ax(t) + by(t) \xrightarrow{L} aX(s) + bY(s)$
Differentiation (first derivative)	$\dfrac{dx(t)}{dt} \xrightarrow{L} sX(s) - x(0)$
Differentiation (second derivative)	$\dfrac{dx^2(t)}{dt^2} \xrightarrow{L} s^2 X(s) - sx(0) - x'(0)$
Multiplication by t	$tx(t) \xrightarrow{L} -\dfrac{dX(s)}{ds}$
Shift	$x(t - t_0) \xrightarrow{L} e^{-st_0} X(s)$
Scaling	$x(at) \xrightarrow{L} \dfrac{1}{|a|} X\left(\dfrac{s}{a}\right)$
Convolution	$x(t) * y(t) \xrightarrow{L} X(s)Y(s)$

A2 The Inverse Laplace Transform

There are different techniques, such as the inversion integral and partial fraction expansion, to find the inverse Laplace transform. Here we will only consider the partial fraction expansion approach.

A3 Partial Fraction Expansion

If $X(s)$ is a rational fraction, then:

$$X(s) = \frac{N(s)}{D(s)} = \frac{\sum_{k=0}^{M} b_k s^k}{\sum_{k=0}^{N} a_k s^k} = \frac{\sum_{k=0}^{M} b_k s^k}{\prod_{i=1}^{Q} (s - p_i)^{m_i}} \qquad (A.6)$$

where p_i, $i=1,2,\ldots,Q$ are poles of $X(s)$ or roots of polynomial $\sum_{k=1}^{N} a_k s^k$ and m_i is the multiplicity of the ith pole. Note that $m_1+m_2+\cdots+m_Q=N$. Assuming that $N>M$, the partial fraction expansion of $X(s)$ yields:

$$X(s) = \left[\frac{A_1}{s-p_1} + \frac{A_2}{(s-p_1)^2} + \cdots + \frac{A_{m_1}}{(s-p_1)^{m_1}}\right]$$
$$+ \left[\frac{B_1}{s-p_2} + \frac{B_2}{(s-p_2)^2} + \cdots + \frac{B_{m_2}}{(s-p_2)^{m_2}}\right] + \cdots \quad (A.7)$$

The residues $A_1, A_2, \ldots, A_{m_1}$ corresponding to the pole p_1 are computed using

$$A_k = \frac{d^{k-m_1}}{ds^{k-m_1}}(s-p_1)^{m_1} X(s)\bigg|_{s=p_1} \quad k=m_1, m_1-1, \ldots, 2, 1 \quad (A.8)$$

Similarly, the other set of residues corresponding to the other poles are computed. Once the partial fraction is completed, the time domain function $x(t)$ is:

$$x(t) = \left[A_1 e^{p_1 t} + A_2 \frac{t e^{p_1 t}}{1!} + A_3 \frac{t^2 e^{p_1 t}}{2!} + \cdots + A_{m_1} \frac{t^{m_1-1} e^{p_1 t}}{(m_1-1)!}\right]$$
$$+ \left[B_1 e^{p_2 t} + B_2 \frac{t e^{p_2 t}}{1!} + B_3 \frac{t^2 e^{p_2 t}}{2!} + \cdots + B_{m_2} \frac{t^{m_2-1} e^{p_2 t}}{(m_2-1)!}\right] + \cdots \quad (A.9)$$

MATLAB® can be used to perform partial fractions. The command is **[R, P, K]=residue(num,den)**, where num is the vector containing the coefficients of the numerator polynomial $N(s)$, den is the vector containing the coefficients of the denominator polynomial $D(s)$, R is the vector of residues corresponding to the poles in vector P, and the direct terms (terms like $K(1)+K(2)s+\cdots$) are in row vector K.

Example A.4

Find the inverse Laplace transform of the following rational function:

$$X(s) = \frac{3(s+3)}{(s+1)(s+2)(s+4)}$$

Using partial fraction expansion, we have:

$$X(s) = \frac{3(s+3)}{(s+1)(s+2)(s+4)} = \frac{A_1}{s+1} + \frac{A_2}{s+2} + \frac{A_3}{s+4}$$

$$A_1 = \lim_{s \to -1}(s+1)X(s) = \lim_{s \to -1}\frac{3(s+3)}{(s+2)(s+4)} = 2$$

Appendix A: The Laplace Transform

$$A_2 = \lim_{s \to -2}(s+2)X(s) = \lim_{s \to -2}\frac{3(s+3)}{(s+1)(s+4)} = -\frac{3}{2}$$

$$A_3 = \lim_{s \to -4}(s+4)X(s) = \lim_{s \to -4}\frac{3(s+3)}{(s+1)(s+2)} = -\frac{1}{2}$$

Therefore, the partial fraction expansion for $X(s)$ is:

$$X(s) = \frac{3(s+3)}{(s+1)(s+2)(s+4)} = \frac{2}{s+1} - \frac{3/2}{s+2} - \frac{1/2}{s+4}$$

Hence:

$$x(t) = \left[2e^{-t} - \frac{3}{2}e^{-2t} - \frac{1}{2}e^{-4t}\right]u(t)$$

Example A.5

Find the inverse Laplace transform of the following function:

$$X(s) = \frac{s}{(s+1)^2(s+2)}$$

Using partial fraction expansion, we have:

$$X(s) = \frac{s}{(s+1)^2(s+2)} = \frac{A_1}{s+1} + \frac{A_2}{(s+1)^2} + \frac{B_1}{s+2}$$

$$B_1 = \lim_{s \to -2}(s+2)X(s) = \lim_{s \to -2}\frac{s}{(s+1)^2} = -2$$

$$A_2 = \lim_{s \to -1}(s+1)^2 X(s) = \lim_{s \to -1}\frac{s}{s+2} = -1$$

$$A_1 = \lim_{s \to -1}\frac{d(s+1)^2 X(s)}{ds} = \lim_{s \to -1}\frac{d}{ds}\frac{s}{s+2} = \lim_{s \to -1}\frac{2}{(s+2)^2} = 2$$

Therefore:

$$X(s) = \frac{2}{s+1} - \frac{1}{(s+1)^2} - \frac{2}{s+2}$$

Hence:

$$x(t) = [2e^{-t} - te^{-t} - 2e^{-2t}]u(t)$$

Example A.6

Use MATLAB to find the inverse Laplace transform of the function:

$$X(s) = \frac{4s^5 + 57s^4 + 303s^3 + 734s^2 + 768s + 216}{s^5 + 14s^4 + 69s^3 + 144s^2 + 108s}$$

Solution: The MATLAB code is:
```
num = [4 57 303 734 768 216];
den = [1 14 69 144 108];
[R,P,K] = residue(num,den)
```
And the results are:
```
R = [2 3 -3 -1 -2];
P = [0 -2 -6 -3 -3];
K = [4];
```

Therefore, the partial fraction expansion for X(s) is

$$X(s) = 4 + \frac{2}{s} + \frac{3}{s+2} + \frac{-3}{s+6} + \frac{-1}{s+3} + \frac{-2}{(s+3)^2}$$

Hence:

$$x(t) = 4\delta(t) + 2u(t) + 3e^{-2t}u(t) - 3e^{-6t}u(t) - e^{-3t}u(t) - 2te^{-3t}u(t)$$

Example A.7

Solve the following second order differential equation using the Laplace transform:

$$\frac{dy^2(t)}{dt^2} + 3\frac{dy(t)}{dt} + 2y(t) = e^{-4t}u(t)$$

where $u(t)$ is the unit step function. The initial conditions are $y(0) = -1$ and $\left.\frac{dy(t)}{dt}\right|_{t=0} = 0$.

Solution: Taking the Laplace transform of both sides of the above differential equation yields

$$s^2Y(s) - sy(0) - y'(0) + 3(sY(s) - y(0)) + 2Y(s) = \frac{1}{s+4}$$

Using the initial conditions, we have:

$$(s^2 + 3s + 2)Y(s) + s + 3 = \frac{1}{s+4}$$

Therefore:

$$Y(s) = \frac{-s^2 - 7s - 11}{(s+4)(s^2 + 3s + 2)}$$

Appendix A: The Laplace Transform

Using partial fraction expansion, we have:

$$Y(s) = \frac{-s^2 - 7s - 11}{(s+4)(s^2 + 3s + 2)} = \frac{-s^2 - 7s - 11}{(s+4)(s+1)(s+2)}$$

$$Y(s) = \frac{A_1}{s+1} + \frac{A_2}{s+2} + \frac{A_3}{s+4}$$

The residues A_1, A_2, and A_3 are computed as:

$$A_1 = \lim_{s \to -1}(s+1)Y(s) = \frac{-s^2 - 7s - 11}{(s+2)(s+4)} = -\frac{5}{3}$$

$$A_2 = \lim_{s \to -2}(s+2)Y(s) = \frac{-s^2 - 7s - 11}{(s+1)(s+4)} = \frac{1}{2}$$

$$A_3 = \lim_{s \to -4}(s+4)Y(s) = \frac{-s^2 - 7s - 11}{(s+1)(s+2)} = \frac{1}{6}$$

Therefore:

$$y(t) = \left(-\frac{5}{3}e^{-t} + \frac{1}{2}e^{-2t} + \frac{1}{6}e^{-4t}\right)u(t)$$

Appendix B: The z-Transform

B1 Definition of the z-Transform

The z-transform of discrete-time signal x(n) is defined as

$$X(z) = \sum_{n=-\infty}^{\infty} x(n) z^{-n} \quad (B.1)$$

where z is a complex variable. The infinite sum in Equation B.1 may not converge for all values of z. The region where the complex function X(z) converges is known as the region of convergence (ROC). This is referred to as the double-sided z-transform. If the signal x(n)=0 for n<0, then we have a one-sided z-transform which is defined as:

$$X(z) = \sum_{n=0}^{\infty} x(n) z^{-n} \quad (B.2)$$

Example B.1

Find the z-transform of the one-sided signal x(n)=a^n u(n).

$$X(z) = \sum_{n=-\infty}^{\infty} x(n) z^{-n} = \sum_{n=0}^{\infty} a^n z^{-n} = \sum_{n=0}^{\infty} (az^{-1})^n = \frac{1}{1 - az^{-1}} = \frac{z}{z - a} \quad (B.3)$$

The convergence of X(z) requires that $|az^{-1}|<1$. Thus, the ROC is the set of points in the complex z-plane for which $|z|>a$. Notice that the function X(z) has a single pole located at z=a that is outside the ROC of X(z).

Example B.2

Find the z-transform of the sequence x(n)=cos(ω_0n)u(n).

$$X(z) = \sum_{n=-\infty}^{\infty} x(n) z^{-n} = \sum_{n=0}^{\infty} \cos(\omega_0 n) z^{-n} = 0.5 \sum_{n=0}^{\infty} (e^{j n \omega_0} + e^{-j n \omega_0}) z^{-n}$$

$$= 0.5 \sum_{n=0}^{\infty} (e^{j \omega_0} z^{-1})^n + 0.5 \sum_{n=0}^{\infty} (e^{-j \omega_0} z^{-1})^n$$

If $|e^{j\omega_0}z^{-1}|<1$ and $|e^{-j\omega_0}z^{-1}|<1$ or $|z|>1$, the two sums in the above equation converge and

$$X(z)=0.5\frac{1}{1-e^{j\omega_0}z^{-1}}+0.5\frac{1}{1-e^{-j\omega_0}z^{-1}}=\frac{0.5z}{z-e^{j\omega_0}}+\frac{0.5z}{z-e^{-j\omega_0}}=\frac{z(z-\cos\omega_0)}{z^2-2\cos\omega_0 z+1} \quad (B.4)$$

Example B.3

Find the z-transform of the sequence $x(n)=a^{|n|}$, where $0<a<1$.

$$X(z)=\sum_{n=-\infty}^{\infty}x(n)z^{-n}=\sum_{n=-\infty}^{-1}(a)^{-n}z^{-n}+\sum_{n=0}^{\infty}a^n z^{-n}=\sum_{n=-\infty}^{-1}(az)^{-n}+\sum_{n=0}^{\infty}(az^{-1})^n$$

$$X(z)=\sum_{n=1}^{\infty}(az)^n+\sum_{n=0}^{\infty}(az^{-1})^n=-1+\sum_{n=0}^{\infty}(az)^n+\sum_{n=0}^{\infty}(az^{-1})^n=-1+\frac{1}{1-az}+\frac{1}{1-az^{-1}}$$

Therefore:

$$X(z)=\frac{z(z+a-a^{-1})}{(z-a)(z-a^{-1})} \quad (B.5)$$

For convergence of $X(z)$, both sums in the above equation must converge. This requires that $|az|<1$ and $|az^{-1}|<1$, or equivalently, $|z|<a^{-1}$ and $|z|>a$. Therefore, the ROC of $X(z)$ is $a<|z|<a^{-1}$.

The z-transforms of some elementary functions and their ROCs are listed in Table B.1.

B2 The Inverse z-Transform

An important application of the z-transform is in the analysis of linear discrete-time systems. This analysis involves computing the response of the systems to a given input using the z-transform. Once the z-transform of the output signal is determined, the inverse z-transform is used to find the corresponding time domain sequence. There are different techniques to find the inverse z-transform from a given algebraic expression. In this appendix, we will consider the partial fraction expansion which can be used to find the inverse z-transform of $X(z)$ if it is a rational function.

B2.1 Inversion by Partial Fraction Expansion

If $X(z)$ is a rational function, then

$$X(z)=\frac{\sum_{k=0}^{M}b_k z^k}{\sum_{k=0}^{M}a_k z^k}=\frac{\sum_{k=0}^{M}b_k z^k}{\prod_{i=1}^{Q}(z-p_i)^{m_i}} \quad (B.6)$$

Appendix B: The z-Transform

TABLE B.1

z-Transform of some Elementary Sequences

$x(n)$	$X(z)$	ROC
$\delta(n)$	1	Entire z-plane
$\delta(n-k)$	z^{-k}	Entire z-plane
$u(n)$	$\dfrac{z}{z-1}$	$|z|>1$
$a^n u(n)$	$\dfrac{z}{z-a}$	$|z|>|a|$
$nu(n)$	$\dfrac{z}{(z-1)^2}$	$|z|>1$
$na^n u(n)$	$\dfrac{az}{(z-a)^2}$	$|z|>|a|$
$\dfrac{n(n-1)\cdots(n-m+1)}{m!} a^{n-m} u(n-m+1)$	$\dfrac{z}{(z-a)^{m+1}}$	$|z|>|a|$
$\cos(\omega_0 n) u(n)$	$\dfrac{z(z-\cos\omega_0)}{z^2 - 2z\cos\omega_0 + 1}$	$|z|>1$
$c \sin(\omega_0 n) u(n)$	$\dfrac{z \sin\omega_0}{z^2 - 2z\cos\omega_0 + 1}$	$|z|>1$
$a^n \cos(\omega_0 n) u(n)$	$\dfrac{z(z-a\cos\omega_0)}{z^2 - 2az\cos\omega_0 + a^2}$	$|z|>|a|$
$a^n \sin(\omega_0 n) u(n)$	$\dfrac{za \sin\omega_0}{z^2 - 2za\cos\omega_0 + a^2}$	$|z|>|a|$

where p_i, $i=1,2,\ldots,Q$ are distinct poles of $X(z)$ or roots of polynomial $\sum_{k=1}^{M} a_k z^k$ and m_i is the multiplicity of the ith pole. Note that $m_1 + m_2 + \ldots + m_Q = M$.

The partial fraction expansion of $X(z)/z$ yields:

$$\frac{X(z)}{z} = \frac{A_0}{z} + \left[\frac{A_1}{z-p_1} + \frac{A_2}{(z-p_1)^2} + \cdots + \frac{A_{m_1}}{(z-p_1)^{m_1}} \right]$$

$$+ \left[\frac{B_1}{z-p_2} + \frac{B_2}{(z-p_2)^2} + \cdots + \frac{B_{m_2}}{(z-p_2)^{m_2}} \right] + \cdots \quad (B.7)$$

or

$$X(z) = A_0 + \left[\frac{A_1 z}{z-p_1} + \frac{A_2 z}{(z-p_1)^2} + \cdots + \frac{A_{m_1} z}{(z-p_1)^{m_1}} \right]$$

$$+ \left[\frac{B_1 z}{z-p_2} + \frac{B_2 z}{(z-p_2)^2} + \cdots + \frac{B_{m_2} z}{(z-p_2)^{m_2}} \right] + \cdots \quad (B.8)$$

where

$$A_0 = z\frac{X(z)}{z}\bigg|_{z=0} = X(0) \qquad (B.9)$$

The residues $A_1, A_2, \ldots, A_{m1}$ corresponding to the pole p_1 are computed using

$$A_k = \frac{d^{k-m_1}}{dz^{k-m_1}}(z-p_1)^{m_1}\frac{X(z)}{z}\bigg|_{z=p_1} \quad k = m_1, m_1-1, \ldots, 2, 1 \qquad (B.10)$$

Similarly, the other set of residues corresponding to the other poles are computed. Once the residues are computed, the time domain function is

$$x(n) = A_0\delta(n) +$$

$$\left[A_1(p)^n u(n) + A_1 n(p_1)^{n-1} u(n) + \cdots \right.$$

$$+ A_{m_1}\frac{n(n-1)(n-2)\cdots(n-m_1+1)}{m_1!}(p_1)^{n-m_1} u(n-m_1+1)\bigg]$$

$$+ \bigg[B_1(p_2)^n u(n) + B_1 n(p_2)^{n-1} u(n) + \cdots \qquad (B.11)$$

$$+ B_{m_2}\frac{n(n-1)(n-2)\cdots(n-m_2+1)}{m_2!}(p_2)^{n-m_2} u(n-m_2+1)\bigg] + \cdots$$

MATLAB® can be used to perform partial fractions. The command is **[R, P, K]=residuez(num,den)**, where num is the vector containing the coefficients of the numerator polynomial $N(z)$, den is the vector containing the coefficients of the denominator polynomial $D(z)$, R is the vector of residues corresponding to the poles in vector P, and the direct terms (terms like $K(1) + K(2)z + \cdots$) are in row vector K.

Example B.4

Find the inverse z-transform of the following function:

$$X(z) = \frac{3z^2}{(z-0.2)(z-0.5)} \quad \text{ROC: } |z| > 0.5$$

Solution: The partial fraction expansion of $X(z)/z$ yields:

$$\frac{X(z)}{z} = \frac{-2}{z-0.2} + \frac{5}{z-0.5}$$

Therefore:

$$X(z) = \frac{-2z}{z-0.2} + \frac{5z}{z-0.5}$$

Taking the inverse z-transform yields: $x(n) = -2(0.2)^n\, u(n) + 5(0.5)^n\, u(n)$

Appendix B: The z-Transform 333

Example B.5

Find the inverse z-transform of the following function:

$$X(z) = \frac{z^2}{(z-0.2)^2(z-0.3)} \quad \text{ROC:} \quad |z| > 0.3$$

Solution: Using the partial fraction expansion, we have

$$X(z) = \frac{-30z}{z-0.2} - \frac{2z}{(z-0.2)^2} + \frac{30z}{z-0.3}$$

Taking the inverse z-transform yields:

$$x(n) = -30(0.2)^n u(n) - \frac{2n(0.2)^n}{0.2} u(n) + 30(0.3)^n u(n)$$

or

$$x(n) = [-30(0.2)^n - 10n(0.2)^n + 30(0.3)^n] u(n)$$

Example B.6

Use MATLAB to find the inverse z-transform of the following function:

$$X(z) = \frac{z^2 + 3z}{z^3 - 1.4z^2 + 0.56z - 0.064}$$

Solution: % MATLAB Code
```
num=[0  1  3  0];
den=[1 -1.4  0.56  -0.064];
[R,P,K]=residuez(num,den);
The Results:
R=[15.8333   -42.50   26.6667]
P=[0.8000   0.4000   0.2000]
K=[0]
```

Therefore, the partial fraction of $X(z)$ is:

$$X(z) = \frac{15.8333z}{z-0.8} + \frac{-42.50z}{z-0.4} + \frac{26.6667z}{z-0.2}$$

Hence:

$$x(n) = [26.667(0.2)^n - 42.5(0.4)^n + 15.883(0.8)^n] u(n)$$

Bibliography

Akaike, H. 1973. Block Toeplitz matrix inversion. *SIAM J. Appl. Math.*, 24:234–41.
Albert, A. 1972. *Regression and the Moore-Penrose Pseudoinverse*. Academic Press, New York.
Amundsen, N. R. 1996. *Mathematical Models in Chemical Engineering; Matrices and their Applications*. Prentice Hall, Englewood Cliffs, NJ.
Berry, M, T. F. Chan, J. Demmel, and J. Donato. 1987. *Templates for the Solution of Linear Systems: Building Blocks for Iterative Methods*. Society for Industrial and Applied Mathematics (SIAM), Philadelphia, PA.
Bartels, R. H., and G. W. Stewart. 1972. Solution of the equation AX+XB=C. *Comm. ACM*, 15:820–26.
Basseville, M. 1981. Edge detection using sequential methods for change in level—part 2: Sequential detection of change in mean. *IEEE Trans. Acoust., Speech, Signal Processing*, 29(1):32–50.
Bellman, R., and K. R. Cooke. 1971. *Modern Elementary Differential Equations*, 2nd edition. Addison-Wesley, Reading, MA.
Bjorck, A., R. J. Plemmons, and H. Schneider. 1981. *Large-Scale Matrix Problems*. North Holland. Amsterdam.
Blahut, R. E. 1987. *Principles and Practice of Information Theory*. Addison-Wesley, Reading, MA.
Braun, M. 1975. *Differential Equations and their Applications*. Springer Verlag. New York.
Bunch, J. R., and D. J. Rose (eds.). 1976. *Sparse Matrix Computations*. Academic Press, New York, NY.
Chan, T. F. 1982. An improved algorithm for computing the singular value decomposition. *ACM Trans. Math. Soft.*, 8:72–83.
Comon, P. 1994. Independent component analysis: A new concept? *Signal Processing*, 36:287–314.
Dantzig, G. B. 1963. *Linear Programming and Extensions*. Princeton University Press, Princeton, NJ.
Dongarra, J. J., C. B. Moler, J. R. Bunch, and G. W. Stewart. 1979. *LINPACK User's Guide*. SIAM, Philadelphia, PA.
Duff, I. S., and G. W. Stewart (eds.). 1979. *Sparse Matrix Proceedings*. SIAM, Philadelphia, PA.
Durbin, J. 1960. The fitting of time series models. *Rev. Inst. Int. Stat.*, 28:233–43.
Ferguson, T. S. 1967. *Mathematical Statistics*. Academic Press, New York, NY.
Fernando, K. V., and B. N. Parlett. 1994. Accurate singular values and differential QD algorithms. *Numerische Mathematik*, 67:191–229.
Fletcher, R. 2000. *Practical Methods of Optimization*. John Wiley & Sons, New York.
Fletcher, R., J. A. Grant, and M. D. Hebden. 1971. The calculation of linear best LP approximations. *Comput. J.*, 14:276–79.
Forsythe, G. E., and C. B. Moler. 1967. *Computer Solution of Linear Algebraic Systems*. Prentice Hall, Englewood Cliffs, NJ.
Forsythe, G. E., M. A. Malcolm, and C. B. Moler. 1977. *Computer Methods for Mathematical Computation*. Prentice Hall, Englewood Cliffs, NJ.

Frazer, R. A., W. J. Duncan, and A. R. Collar. 1938. *Elementary Matrices and Some Applicationsto Dynamics and Differential Equations.* Cambridge University Press, New York, NY.

Friedberg, S. H., A. J. Insel, and L. E. Spence. 2003. *Linear Algebra*, 4th edition. Prentice Hall, Englewood Cliffs, NJ.

Friedland, B. 1986. *Control System Design: An Introduction to State-Space Design.* McGraw-Hill, New York, NY.

Gantmacher, F. R. 1959. *The Theory of Matrices*, Vol. 1. Chelsea Publishing Company, New York, NY.

Gantmacher, F. R. 2005. *Applications of the Theory of Matrices.* Dover, New York.

Garbow, B. S., J. M. Boyle, J. J. Dongaarrag, and C. B. Moler. 1972. *Matrix Eigensystem Routines: EISPACK Guide Extension.* Springer Verlag, New York, NY.

Geroge, J. A., and J. W. Liu. 1981. *Computer Solution of Large Sparse Positive Definite Systems.* Prentice Hall, Englewood Cliffs, NJ.

Gill, P. E., W. Murray, and M. H. Wright. 1981. *Practical Optimization.* Academic Press, London, UK.

Goldberg, D. 1989. *Genetic Algorithms in Search, Optimization, and Machine Learning.* Addison-Wesley, Reading, MA.

Golub, G. H. 1973. Some modified matrix eigenvalue problems. *SIAM Rev.*, 15(2):318–34.

Golub, G. H., and C. F. Van Loan. 1980. An analysis of the total least squares problem. *SIAM J. Numer. Anal.*, 17(6):883–93.

Golub, G. H., and C. F. Van Loan. 1983. *Matrix Computations.* Johns Hopkins University Press, Baltimore, MD.

Graybill, F. A. 1961. *An Introduction to Linear Statistical Models*, Vol. I. McGraw-Hill, New York, NY.

Greville, T. N. E. 1966. On stability of linear smoothing formulas. *J. SIAM Numer. Anal.*, 3:157–70.

Haberman, R. 1977. *Mathematical Models.* Prentice Hall, Englewood Cliffs, NJ.

Hadley, G. 1962. *Linear Programming.* Addison-Wesley, New York.

Hageman, L. A., and D. M. Young, Jr. 1981. *Applied Iterative Methods.* Academic Press, New York, NY.

Halmos, P. R. 1993. *Finite Dimensional Vector Spaces.* Springer Verlag, New York, NY.

Harwit, M., and N. J. A. Sloane. 1979. *Hadamard Transform Optics.* Academic Press, New York, NY.

Heading, J. 1960. *Matrix Theory for Physicists.* John Wiley & Sons, New York.

Hildebrand, F. B. 1956. *Introduction to Numerical Analysis.* McGraw-Hill, New York, NY.

Ho, B. L., and R. E. Kalman. 1965. Effective construction of linear state-variable models from input/output data. *Proc. Third Allerton Conference*, 449–59.

Householder, A. S. 2006. *The Theory of Matrices in Numerical Analyses*, Dover, New York.

Hurty, W. C., and M. F. Rubinstein. 1964. *Dynamics of Structures.* Prentice Hall, Englewood Cliffs, NJ.

Ipsen, Ilse C. F. 1997. Computing an eigenvector with inverse iteration. *SIAM Rev.*, 39(2):254–91.

Jain, A. K. 1989. *Fundamentals of Digital Image Processing.* Prentice Hall, Englewood Cliffs, NJ.

Kahan, W. 1966. Numerical linear algebra. *Canadian Math. Bull.*, 9:757–801.

Kahng, S. W. 1972. Best LP approximation. *Math. Comput.*, 26(118):505–08.

Bibliography

Karlin, S. 1959. *Mathematical Models and Theory in Games, Programming, and Economics*, Vols. I, II. *Addison*-Wesley, New York.

Karmarkar, N. 1984. A polynomial-time algorithm for linear programming. *Combinatorica*, 4(4):373–95.

Kemeny, J. G., and J. L. Snell. 1983. *Finite Markov Chains*. Springer-Verlag, New York.

Kemeny, J. G., and J. L. Snell. 1978. *Mathematical Models in the Social Sciences*. MIT Press, Cambridge, MA.

Kemeny, J. G., J. L. Snell, and G. L. Thompson. 1957. *Introduction to Finite Mathematics*. Prentice Hall, Englewood Cliffs, NJ.

Larsen, R. and L. Marx. 2001. *An Introduction to Mathematical Statistics and Its Applications*, 3rd edition. Prentice Hall, Englewood Cliffs, NJ.

Lawrence E. S., A. J. Insel, and S. H. Friedberg. 2008. *Elementary Linear Algebra: A Matrix Approach*, 2nd edition. Prentice Hall, Englewood Cliffs, NJ.

Lawson. C. L., and R. J. Hanson. 1974. *Solving Least Squares Problems*. Prentice Hall, Englewood Cliffs, NJ.

Linnik, L. 1961. Method of Least Squares and Principles of the Theory of Observations. Pergamon Press, London, United Kingdom.

Luenberger, D. G. and Yinlu Ye 2008. *Linear and Nonlinear Programming*. Springer-Verlag, New York..

Maki, D., and M. Thompson. 1973. *Mathematical Models and Applications*. Prentice Hall, Englewood Cliffs, NJ.

Mangasarian, O. L. 1969. *Nonlinear Programming*. McGraw-Hill, New York, NY.

Marcus, M., and H. Minc. 1992. *A Survey of Matrix Theory and Matrix Inequalities*. Dover Publications, New York.

Martin, H. C. 1966. *Introduction to Matrix Methods of Structural Analysis*. McGraw-Hill. New York, NY.

Moore, B. C. 1978. Singular value analysis of linear systems. *Proc. IEEE Conf. on Decision and Control*, 66–73.

Morrison, D. F. 2005. *Multivariate Statistical Methods*, 4th edition. Brooks/Cole, Pacific Grove, CA.

Muir, T. 1960. *Determinants*. Dover Publications, New York.

Nashed, M. Z. 1976. *Generalized Inverses and Applications*. Academic Press, New York.

Naylor, A. W., and G. R. Sell. 1982. *Linear Operator Theory in Engineering and Science*. Springer-Verlag, New York, NY.

Nelder, J. A., and R. Mead. 1965. A simplex method for function minimization. *Comp. J.*, 7:308–13.

Noble, B., and J. W. Daniel. 1998. *Applied Linear Algebra*, 3rd edition. Prentice Hall, Englewood Cliffs, NJ.

Parlett, B. N. 1980. *The Symmetric Eigenvalue Problem*. Prentice Hall, Englewood Cliffs, NJ.

Pierre, D. A. 1986. *Optimization Theory with Applications*. Dover, New York, NY.

Pipes, L. A. 1963. *Matrix Methods in Engineering*. Prentice Hall, Englewood Cliffs, NJ.

Press, W. H. 1988. *Numerical Recipes in C*. Cambridge University Press, New York.

Rao, C. R. 1952. *Advanced Statistical Methods in Biometric Research*. John Wiley & Sons, New York.

Rissanen, J. 1974. Solution of linear equations with Hankel and Toeplitz matrices. *Numer. Math.*, 22:361–66.

Robinson, J. 1966. *Structural Matrix Analysis for the Engineer*. John Wiley & Sons, New York.

Rose, D., and R. Willoughby (eds.) 1972. *Sparse Matrices and Their Applications.* Springer-Verlag, New York.

Schwartz, J. T. 1961. *Lectures on the Mathematical Method in Analytical Economics.* Gordon and Breach. subsidiary of Taylor & Francis Group.

Searle, S. R. 1966. *Matrix Algebra for the Biological Sciences (Including Applications in Statistics).* John Wiley & Sons, New York.

Senturia, S., and B. Wedlock. 1975. *Electronic Circuits and Applications.* John Wiley & Sons, New York.

Shim. Y. S., and Z. H. Cho. 1981. SVD pseudoinverse image reconstruction. *IEEE Trans. Acoust., Speech, Signal Processing*, 29(4):904–09.

Simonnard, M. 1966. *Linear Programming*, translated by W. S. Jewell. Prentice Hall, Englewood Cliffs, NJ.

Smith, B. T., J. M. Boyle, J. Dongarra, B. Garbow, Y. Ikebe, V. C. Klema, and C. B. Moler. 1976. *Matrix Eigensystem Routines: EISPACK Guide*, 2nd edition. Springer Verlag, New York, NY.

Stewart, G. W. 1973. *Introduction to Matrix Computations.* Academic Press, San Diego.

Strang, G. 1986. *Introduction to Applied Mathematics.* Wellesley-Cambridge Press, Wellesley, MA.

Strang, G. 1988. *Linear Algebra and Its Applications*, 3rd edition. Harcourt Brace Jovanovich, Fort Worth, TX.

Trooper, A. M. 1962. *Matrix Theory for Electrical Engineers.* Addison-Wesley and Harrap, New York.

Van Huffel, S., and J. Vandewalle. 1985. The use of total linear least squares technique for identifi cation and parameter estimation. *Proc. IFAC/IFORS Symp. on Identifi -cation and Parameter Estimation*, 1167–72.

Varga, R. S. 1962. *Matrix Iterative Analysis.* Prentice Hall, Englewood Cliffs, NJ.

Walpole, R. E., and R. H. Myers. 1978. *Probability and Statistics for Engineers and Scientists*, 8th edition. Prentice Hall, Englewood Cliffs, NJ.

Weiss, A. V. 1964. *Matrix Analysis for Electrical Engineers.* Van Nost. Reinhold, New York, NY.

Wilkinson, J. H. 1963. *Rounding Errors in Algebraic Processes.* Prentice Hall, Englewood Cliffs, NJ.

Wilkinson, J. H. 1988. *The Algebraic Eigenvalue Problem.* Oxford University Press, New York, NY.

Young, D. 1971. *Iterative Solution of Large Linear Systems.* Academic Press, New York, NY.

Zoltowski, M. D. 1987. Signal processing applications of the method of total least squares. *Proc. XXIst Asilomar Conf. on Signals, Systems and Computers*, 290–96.

(Unauthored), *MATLAB Matrix Software.* The MathWorks, Inc., 158 Woodland St., Sherborn, MA.

Index

A

Additive zero mean white Gaussian noise channel, 125

B

Bandlimited operator, 161
Binary digital data communication system, 125; *see also* Communication channel
Block multiplication, 35–38
Bose, Chaudhuri and Hocquenghem (BCH) block codes, 91

C

Cauchy sequence, 129–131; *see also* Hilbert space
Cayley-Hamilton theorem, 232–234, 276
  matrix polynomial reduction, 234–236
  technique, 240–245
Circulant matrix, 93
Communication channel
  binary bit 0 and 1, 126
  one-bit quantizer, 125
  Schwarz inequality, 127
  SNR at input to quantizer, 127
  variance of noise, 127
  zero mean
    Gaussian random variables, 126
    white Gaussian noise, 128
Complex matrix, 2
Continuous LTI systems
  continuous-time state space equations, solution of, 256–257
  state space representation of controllable and observable canonical form, 255
    MATLAB® commands, 256
    SISO system, 254–256

Continuous time control systems; *see also* Linear time invariant (LTI) systems
  analysis and design
    matrix exponential function e^{At}, 247–248
Continuous-time dynamic system; *see also* Linear time invariant (LTI) systems
  state, 250
  equations, 251–254
Continuous-time state space equations solution of
  LTI system, 256–257
Controllability, 270–271; *see also* Linear time invariant (LTI) systems
Convolution integral, 157
  around x_1 axis
    projection on, 159
    reflection, 158
  counterclockwise and clockwise rotation by angle θ, 158
Covariance matrix, 39
Cramer's rule, 70–71

D

Discrete Fourier transform (DFT), 18–22
Discrete-time control systems
  analysis and design
    matrix function A^k, 249–250
Discrete-time systems
  definition and condition of
    controllability, 270–271
    observability, 272–275
  state equations, 265
    homogenous, solution of, 266
    solution of, 268–269
    state transition matrix, computing methods, 266–268

339

340 Index

state space representation of,
  264–265
  discrete-time LTI systems, 264–265
  state equations, 263–264

E

Eigen filters, 304
  eigenvector of signal covariance
    matrix, 306–307
  SNR, 305
Eigenvalues, 157, 165
  applications in
    image edge detection, 206–209
    signal subspace decomposition,
      217–221
    vibration systems, 214–217
  applications in, gradient based edge
    detection of
    gray scale images, 209
    RGB images, 210
  convergence of dominant
    eigenvalues, 202
  diagonalization and distinct, 173–175
  Hermitian matrices, 183–185
  Jordan canonical form, 175–176
  MATLAB® command, 166
  multiplicity, 176–180
  nonsingular transformation, 176
  numerical computation of, 199
  product and sum, 170–171
  properties of, 205–206
  Sylvester's expansion, 236–240
Elementary matrices; *see also* Linear
    time invariant (LTI) systems
  $m \times n$ matrix, 2
    diagonal, 4
    identity, 4–5
    square, 3
    upper and lower triangular, 4
  operations
    transpose and Hermitian, 5–6
Elementary row operations for
    simultaneous set of linear
    equations, 22–23; *see also*
    Simultaneous equations
    solutions
  elementary transformation matrices
    combination transformation
      matrix (E_3), 26–27

interchange transformation
  matrix (E_2), 25
scaling transformation matrix
  (E_1), 24–25
row echelon form (REF), 23–24
Equality constraints, stationary
    points with
  finite impulse response (FIR) filters,
    digital design, 312–316
  Lagrange multipliers, 307–310
  maximum entropy problem, 310–312
Euclidean norm, *see* Frobenius norm

F

Finite impulse response (FIR) filters,
    digital design, 312–316
Fourier series, 136–137
  continuous Fourier series, 137–139
  discrete Fourier transform (DFT),
    144–145
  expansion, 140–141, 143
  generalized Fourier series (GFS),
    135–137
  MATLAB® code for coefficients, 139
  periodic square wave signal, 139
  triangular periodic signal, 142
  truncated Fourier series
    expansions, 140
Frobenius norm, 195–196

G

Gantmacher's algorithm, 238–239
Gaussian elimination, 27–31
Gaussian noise, 165
Gram–Schmidt orthogonalization
    process, 131–133
  MATLAB® code implementation of
    algorithm, 134

H

Hadamard matrix, 93
  inverse Hadamard transform, 94
Hankel matrix, H = hankel(C), 91
Hermitian (symmetric) projection
    matrix, 163
Hessian matrix, 285–286
Hilbert space, 129–131

Index

Homogeneous coordinates system, 78
  in image processing
    2-d transformations, 79
    3-d transformations, 80
    MATLAB® code for, 81, 83–85
Homogenous state equations
  solution of, 257, 266

I

Ideal digital low pass filter, 312–313
Infinite matrix series, 230–231
  convergence of, 231–232
Inner product spaces, 120–121
  norm derived from, 123
  Schwarz's inequality, 121–123
Inner (scalar) product, 38–39
Iteration index, 202, 204; *see also*
    Eigenvalues

J

Jordan canonical form, 175–176
  MATLAB® commands, 178

K

Kronecker product of two matrices, 40

L

Lagrange multiplier technique, 307–310
Laplace transform
  for functions, 322
  inverse Laplace transform, 323
  partial fraction expansion, 323–327
  properties of, 323
  region of convergence (ROC), 321–322
Least-square (LS) technique, 287–288
  computation using
    QR factorization, 288–289
    singular value decomposition (SVD), 289–291
  curve fitting problem, 293–295
  image pixel values, 301
  linear filter
    continuous signal x(t), 297
    deconvolved signal in presence of noise, 297–298
    one dimensional Wiener filter, 295–298
  Q matrix and scale factor b, 298–300
  two dimensional Wiener filter, 300–301
  vector space V, 287
  weighted least square (WLS), 291–293
Legendre polynomials, 145
Linear continuous-time systems
  discrete-time system
    observability condition, 272–275
  state space modeling of
    multiple-input-multiple-output (MIMO), 250–254
    state of continuous-time dynamic system, 250
Linear equations systems solution
  Cramer's rule, 70–71
  DC and AC circuit analysis and, 75–78
  under determined systems, 32
  Gaussian elimination, 27–31
  lower-upper (LU) decomposition, 71–74
  over determined systems, 31–32
  solution of simultaneous linear equations, 67–69
  strict triangular form, 69–70
  using QR factorization, 149–150
Linear time invariant (LTI) systems
  controllability and observability, 250
  eigenfunctions for, 165
  Gaussian noise, 165
  LTI discrete-time system
    controllability of, 270–271
    observability of, 272–275
    state space and output equations, 265–269
    state space representation of, 264–265
  probability density function (PDF), 165
  state space equations, 256–257, 263
Linear transformations, 157–159
  matrices as linear operators, 160
  null space of matrix, 160–161
  projection operator, 161
Linear vector spaces, 105–107
  additional properties of, 107

set of combinations, 109–110
subspace of, 107–108
Low pass FIR filter
   code for design, using MATLAB®
      command, 215
   frequency response, 316

M

Markov matrix, 92–93
Matched filter detector, 123; *see also*
      Optimum filter
   MATLAB® code for, 130
   received signal and output of, 129
   signal-to-noise ratio (SNR), 124
MATLAB® commands
   add or subtract, 7
   characteristic polynomial of matrix
      A p = poly(A), 166
   code
      affine.m, 82–83
      to compute Fourier series
         coefficients, 139
      1-d convolution using matrices
         implementation, 11
      for design of low pass FIR filter,
         315
      edge map for gray scale image,
         creation of, 210
      edge map for RGB image,
         creation of, 211
      eigenimages, 193
      to generate circulant matrix, 19
      implementation of MUSIC
         algorithm, 222
      matched filter receiver
         implementation, 125, 130
      to perform 2-d convolution using
         matrices, 19
      power method, 202–203
   computation of
      B = f(A) B = funm(A,@name), 243
      DFT and inverse DFT, X = fft(x)
         and x = ifft(X), 144
      dot product between two vectors
         x and y in $C_n(R_n)$ dot(x,y), 121
      eigenvalues and eigenvectors of
         matrix A [U,S] = eig(A), 166
      Frobenius norm,
         norm(A,'fro'), 195

Jordan canonical form [V,J] =
      jordan(A), 178
least-square (LS) solution x = Ay,
      287–288
norm of matrix A norm(A,p), 195
observability matrix P, 273–274
to compute p norm of
   n dimensional real or complex
      vector norm(x,p), 117
   p = ∞, norm(x,inf), 117
condition number of matrix A c =
      cond(A), 197
controllability matrix Q, 271
2-d convolution of two 2-d signals, 14
to factor polynomial f(x)
   p = roots(a), 230
f = imread('cameraman.tif'), 47, 104
to fit polynomial of degree m to data
      set stored in x and y vectors
      P = polyfit(x,y,m), 294
f = round(255*round(16,16)), 47
function file
   for orthonormal set, 134
   to test controllability and
      observability of system, 275
Kronecker product, 40
multiplication C = A.*B, 8
partial fractions, [R, P, K] =
      residue(num,den), 324, 332
to perform QR factorization of
      matrix A [Q,R] = θp(A), 148
state space realization of dynamic
      system in controllable
      canonical form, 256
to transpose, 5
Matrix
   addition properties, 7
   characteristic polynomial
      Cayley–Hamilton theorem,
         232–234
      MATLAB® command, 166
      recursive algorithm, 171–172
   column space C(A), 87
   condition number, 196–198
   Cramer's rule, 70–71
   determinant of, 49
      matrix 3×3, 51–52
      matrix 1×1 and 2×2, 50
      properties of, 52–53
      row operations and, 53–55

Index

diagonalization
  distinct eigenvalues, 173–175
exponential function e^{At}, 247–248
  convergent series, 247
factorization, QR factors, 146
function A^k, 249–250
  eigenvalues and eigenvectors of, 249–250
inversion, 58–59
  formulas, 63
  Gauss-Jordan method for calculating, 60–62
  matrix inversion lemma, 63
  properties of, 60
  recursive least square (RLS) parameter estimation, 64–66
as linear transformations, 157–158
  null space, 160–161
  operators, 160
  projection operator, 161
MATLAB® command to transpose, 5
modal matrix, 173
  technique, 245–246
multiplication, 7
  discrete Fourier transform (DFT), 18–22
  in linear discrete one dimensional convolution, 9–14
  in linear discrete two dimensional convolution, 14–18
  properties of, 8
  in signal and image processing, 8–9
norm, 192–193
  and Lagrange multiplier technique, 194–195
null space N(A), 85–87
partitions, 32
  columns, 33–34
  rows, 34–35
polynomials, 229, 276
  Cayley-Hamilton theorem and, 232–236
  eigenvalues of, 238–243
  infinite series, convergence of, 230–232
  MATLAB® command, 230
rank of, 89
  properties, 90

row space R(A), 87–89
singular matrices, 55–58
singular value decomposition (SVD) of, 188–192
skew-symmetric matrix, 6
subtraction, 7
Sylvester's expansion, 236–240
symmetric matrix, 6
test for positiveness, 185–188
trace of square matrix, 6
and vector notation, 1
Maximum entropy problem, 310–312
Modal matrix technique
  Taylor series expansion of function f(A), 245–246
Multiple-input-multiple-output (MIMO) continuous-time dynamic system, 250–254
  MATLAB® command
    controllability and observability of, 274–275
Multiple signal classification (MUSIC) algorithm, 219–222

N

Negative definite and semidefinite matrices, 185
Nilpotent matrix, 94–95
Normed vector spaces, 116–117
  distance function, 117–118
  equivalence of norms, 118–119

O

Observability, 272–275; see also Linear time invariant (LTI) systems
One-bit quantizer, 125
One dimensional (1-d) discrete linear time invariant (LTI) system, 9
One dimensional Wiener filter, 295–298
Optimization
  digital finite impulse response (FIR) filters
    design of, 312–316
  eigen filters, 304–307
  equality constraints
    Lagrange multiplier technique, 307–310
  gradient vector, 284

least-square (LS) technique, 287–288
maximum entropy problem, 310–312
n variables
  continuous function of, 283–284
  Hessian matrix, 285–286
one dimensional Wiener Filter
  application of, 295–298
Q Matrix and Scale Factor b,
  operator, 298–300
total least-squares (TLS), 302–304
two dimensional Wiener Filter
  application of, 300–301
Optimum filter, 125
Orthogonality
  concept, 131
  Gram–Schmidt orthogonalization
    process, 131–134
  orthogonal matrices, 134–135
  orthogonal projection, 162
  orthonormal set, 131
    Legendre polynomials, 145
    sinc function, 146–147
Outer product between two vectors, 39

P

Parseval's identity, 182
Partial fraction expansion, 323–327
  laplace transform, 321–327
  z-transform, 330–333
Permutation matrix, 92
Positive definite and semidefinite
  matrices, 185
Probability density function (PDF), 165
Projection theorem, 163
  projection onto 2-d subspace S, 164

Q

QR factorization
  computation using, 288–289
Quadratic functions, 96–98

R

Real matrix, 2

S

Scalar function of vector
  derivative/gradient of, 95–96

Scalar infinite series, 231–232
Schwarz's inequality, 121
  application in communication
    systems for detection
      continuous signal buried in
        noise, 125
      discrete signal buried in white
        noise, 123–124
  C^n and $L(a, b)$, 123
  filter coefficients, 124
  inner product space, 122
Set of vectors
  basis vector, 113–114
  change of, 114–116
  span of set of vectors, 108–109
    linear dependence, 110–111
    spanning, 110, 113–114
    theorem, 110–113
Signal subspace decomposition
  direction of arrival estimation,
    219–221
  frequency estimation, 217–219
Simultaneous equations solutions
  gradient vector, 284–285
  n variables and stationary points,
    283
Sinc function, 146–147; see also
  Orthogonality
Singular value decomposition (SVD),
  289–291; see also Least-square
  (LS) technique
  LENA image, 191
    gradient of, 213
    reconstruction using
      eigenimages, 192
    singular values of, 189–190
      normalized with respect to
        largest singular value, 191
    and unitary matrices, 188
State space equations, 250
  continuous LTI systems,
    representation of, 254
  controllable and observable
    canonical form, 255–256
  discrete-time dynamic system, state
    of, 263
  linear continuous time invariant LTI
    system
      block diagram, 251
      continuous-time state space
        equations, solution of, 256–257

Index

MATLAB® code and plots of, 261–262
RLC circuit, series with, 252–253
spring-mass system with friction, 253
state and output equations of, 251
zero input response (ZIR), 259–260, 268–269
zero state response (ZSP), 259–261, 268–269
State transition matrix $\phi(t)$
   properties of, 258
   solution of, 257, 266
      Laplace-transform computation, 258–259
Supremum (sup) of set S (subset of real numbers), 119–120
Sylvester's expansion, 236–240
Sylvester's inequality, 90

T

Toeplitz matrix, 91
   T = toeplitz (C, R), 92
Total least-squares (TLS), 302–303
   linear fit, 303–304
Two dimensional (2-d) linear shift invariant (LSI) system, 14
Two dimensional Wiener filter (2-d Wiener filter)
   image degradation model and, 300–301

U

Unitary matrices, 180–182

V

Vandermonde matrix, 90
   V = vander(r)'; V = V(end:-1:1,:), 91

Vectors
   basis vector set, 113–114
   eigenvector, 157, 165
      gradient based edge detection of gray scale images, 209
      image edge detection, 206–209
      independence of, 168–170
      MATLAB® command, 166
      numerical computation, power method, 199–200
      properties of, 205–206
      Sylvester's expansion, 236–240
   error vector, 132
   filter coefficients and Schwarz inequality, 124
   function of
      derivative of Jacobian of f(x), 98
   Gram–Schmidt orthogonalization process for
      independent or dependent vectors, 131–134
   linearly independent, 110–113
   normed linear vector space, 116
   p-norm distance between, 117–118
   vector space R^3, 109

W

Weighted least square (WLS), 291
   covariance matrix of, 292–293

Z

Z-Transform
   elementary sequences, 331–332
   inverse Z-transform
      partial fraction expansion, inversion by, 330
   MATLAB® command, 332–333
   one-sided and double-sided, 329–330